中国东北虎种群和栖息地动态及其精准管理策略

Population and Habitat Dynamics of Amur Tigers and Their Targeted Management Strategies in China

姜广顺 华 彦 顾佳音 等 著
Authors Jiang Guangshun, Hua Yan, Gu Jiayin, et al.

马建章 主审
Chief Reviewer Ma Jianzhang

科学出版社
北 京

内 容 简 介

本书基于对中国东北虎种群和栖息地动态变化的长期研究,收集整理并分析了大量的野外和实验室数据,综合应用物种分布模型理论、食物选择理论、上行效应理论、动物行为学理论、分子生态学理论、土地分享和土地抽取理论、生态阈值理论等生态学理论,以及栖息地优先区等级和生态廊道确定技术、基于足迹影像的个体识别技术、东北虎个体花纹识别技术等探索性技术,较为全面地揭示了东北虎种群和栖息地的动态特征,将有效地促进东北虎种群和栖息地的精准管理策略的制定与实施。

本书不仅是一本面向所有野生动物爱好者的关于东北虎保护的科普读物,而且对于野生动物的科研保护也具有一定的参考意义,对于野生动物管理者对东北虎及其他濒危食肉动物种群和栖息地的保护与管理实践也具有一定的借鉴意义。

This book is based on the long-term research on the population and habitat dynamics of Amur tigers in China, collecting and analyzing lots of field and lab data. We have comprehensively applied the theories of species distribution models, food selection, bottom-up effect, animal behavior, molecular ecology, land sharing and sparing, and ecological threshold, and exploratory technologies on habitat priority zone grading and ecological corridor, digital footprint photos for individual identification and body pattern identification of Amur tigers, and so on. We have also revealed the population and habitat dynamic characteristics of Amur tigers, which will effectively prompt the establishment and application of targeted management strategies for Amur tigers population and habitat.

This book is not only a popular science book about conservation of Amur tiger to all wildlife enthusiasts but also a professional book for wildlife scientific research protection. Furthermore, this book is significant for wildlife administrators to learn about the conservation and management of population and habitat of Amur tigers and other endangered predators.

审图号:GS(2020)5799 号

图书在版编目(CIP)数据

中国东北虎种群和栖息地动态及其精准管理策略/姜广顺等著. —北京:科学出版社,2020.10
ISBN 978-7-03-063226-5

Ⅰ. ①中… Ⅱ. ①姜… Ⅲ. ①东北虎–种群–研究–中国 ②东北虎–栖息地–研究–中国 Ⅳ. ①Q959.838

中国版本图书馆 CIP 数据核字(2019)第 250720 号

责任编辑:张会格 付丽娜 / 责任校对:严 娜
责任印制:赵 博 / 封面设计:铭轩堂

科学出版社 出版
北京东黄城根北街 16 号
邮政编码:100717
http://www.sciencep.com
北京凌奇印刷有限责任公司印刷
科学出版社发行 各地新华书店经销
*
2020 年 10 月第 一 版 开本:720×1000 1/16
2025 年 1 月第三次印刷 印张:25 1/2
字数:513 000
定价:268.00 元
(如有印装质量问题,我社负责调换)

作者名单

姜广顺　华　彦　顾佳音　齐进哲　宁　瑶
周绍春　龙泽旭　王　琦　刘　丹　郭玉荣
长有德　刘培琦　张明海

Authors

Jiang Guangshun, Hua Yan, Gu Jiayin, Qi Jinzhe, Ning Yao, Zhou Shaochun, Long Zexu, Wang Qi, Liu Dan, Guo Yurong, Chang Youde, Liu Peiqi, Zhang Minghai

第一作者简介

姜广顺 男，1971年8月11日生，吉林省通榆县人。东北林业大学野生动物与自然保护地学院教授、博士生导师；中国科学院动物研究所出站博士后；国务院学位委员会和教育部"全国优秀博士学位论文"获得者、教育部"新世纪优秀人才支持计划"获得者、世界自然基金会"虎保护最佳巡护监测奖"和中国野生动物保护协会"斯巴鲁生态保护奖"获得者；兼任东北林业大学学术委员会委员、国家林业和草原局（以下简称国家林草局）猫科动物研究中心常务副主任、国际学术期刊 Journal of Zoological Research 主编（editor-in-chief）、国际动物学会全球变化的生物学效应专家工作组委员、俄罗斯滨海边疆区东北虎保护专家组成员、黑龙江省中俄虎豹国际合作指定中方负责专家、东北虎豹国家公园野生动物紧急救助专家组成员、《野生动物学报》编委、中国动物学会兽类学分会理事、中国林学会自然保护地与生物多样性分会常务委员兼秘书长等。

他主持"东北虎豹栖息地哺乳动物类群时空互作网络的研究""猎物群落与东北虎种群空间分布动态互作关系的研究""东北豹种群空间分布的生态驱动机制研究"等国家自然科学基金面上项目及其他教育部人才支持计划项目、国家林业局保护司项目、科技部重点研发项目、东北亚次区域环境合作计划"基于自动相机监测和分子遗传分析方法的东北虎豹跨境移动研究"项目、世界自然基金会虎豹保护项目等围绕东北虎豹保护研究和技术服务的国内外项目共计48项。其研究团队在大型兽类生态与保护管理方面取得了重要的研究进展，2019年获国家林草局与中国林学会评审的梁希科学技术进步奖二等奖第一名，并成为黑龙江省"头雁人才计划"成员。

他带领团队多年致力于中国虎豹精准保护与管理的研究，通过开展虎豹保护基础研究，定量分析了人为干扰对虎豹种群空间分布的影响，揭示了虎豹与猎物之间的互作关系，研发了新的适宜栖息地识别和优先保护区域确定技术等，相关成果已经在 PNAS、Biological Conservation、Ecography、Integrative Zoology、

Landscape Ecology、*Applied Animal Behaviour Science*、*Ecological Modelling*、*Oryx* 等国际主流学术期刊上发表，先后在国内外发表SCI学术论文77篇，撰写专著1部——《中国东北豹种群和栖息地研究》（科学出版社），获得野生动物监测技术发明专利6项、野生动物影像数据处理软件著作权1项。负责编制了国家林草局《中国东北虎、东北豹保护行动计划（2016—2025）》。另外，他多次赴美国、俄罗斯、印度、越南、泰国等地开展考察和技术合作，带领科研团队首次发布了清晰完整的全球濒危东北豹的自动相机影像，首次在黑龙江省完达山区拍摄到野生东北虎自动相机影像，首次在黑龙江省拍摄到东北豹自动相机影像，首次发现并报道了中国东北豹和东北虎的繁殖种群的证据，且多次被中央电视台、英国《每日电讯报》、《北京科技报》、《文汇报》等几十家国内外媒体采访报道。

Introduction of the First Author

Mr. Jiang Guangshun was born on August 11, 1971 in Tongyu County, Jilin Province. He is a professor of the College of Wildlife and Protected Area, Northeast Forestry University, doctoral tutor, and a postdoctoral fellow of the Institute of Zoology, Chinese Academy of Sciences. He is the winner of the "National Excellent Doctoral Dissertation" (NEDD) of the Academic Degrees Committee of the State Council and the Ministry of Education, the winner of the "Program for New Century Excellent Talents Support" funded by the Ministry of Education, and the winner of the "Best Patrol Monitoring Award for Tiger Conservation" granted by WWF and "SUBARU Ecological Protection Award" granted by China Wildlife Conservation Association. He is concurrently serving as a member of the Academic Membership Committee of Northeast Forestry University, executive deputy director of the Feline Research Center of the National Forestry and Grassland Administration, editor-in-chief of *Journal of Zoological Research*, a member of the Biological Consequences of Global Change (BCGC) working group of the International Society of Zoological Sciences (ISZS), a member of Primorsky Krai (Russia) Amur Tiger Conservation Expert Group, designated expert in charge of the Chinese side of the China-Russia Tiger and Leopard International Cooperation in Heilongjiang, a member of the Wildlife Emergency Rescue Expert Team of the Northeast Tiger and Leopard National Park, a member of the editorial board of the *Chinese Journal of Wildlife*, and director of the Mammalogical Society of the Chinese Zoological Society, standing committee member and secretary general of Nature Reserve and Biodiversity Branch of Chinese Forestry Society, among others.

He has led a total of 48 domestic and international projects centered on Amur tiger and leopard conservation and technical services, including National Nature Science Foundation of China (NSFC) projects (sub-class) such as "Study on the Space-Time Interaction Network of Mammal Groups in the Amur Tiger and Leopard Habitats" "Study on Ecological Driving Mechanisms for the Spatial Distribution of the Amur Leopard Population", and "Study on the Dynamic Interaction of Spatial Distributions of Prey Populations and Amur Tiger Populations". Further, key research and development projects have been supported by the Ministry of Education Talent Support Program, funded by the Department of Wildlife Conservation of the State Forestry

Administration and by the Ministry of Science and Technology, and include the North-East Asian Subregional Programme for Environmental Cooperation (NEASPEC) project, "Trans-border Movement of Amur Tigers and Amur Leopards Using Camera Trapping and Molecular Genetic Analysis", the WWF Tiger and Leopard Protection Project, and so on. His research team has made important advances in the ecology and conservation management of large animals, so that they won the second prize of the Liangxi science and technology progress award, which was appraised by the National Forestry and Grassland Administration and Chinese Society of Forestry in 2019, and became the member of "Head Goose Talent Plan" of Heilongjiang Province.

Based on basic research of tiger and leopard protection, the team led by the author has conducted quantitative research on the evaluation of the effects of anthropogenic disturbance on the spatial distribution of tiger and leopard populations and research on the interactions between tigers and leopards and their prey, and explored technologies for suitable habitat identification and priority protected area determination. He had been dedicated to the research on targeted protection management of tigers and leopards in China and had published relevant scientific results in international mainstream academic journals like *Biological Conservation*, *Ecography*, *Integrative Zoology*, *Landscape Ecology*, *Applied Animal Behaviour Science*, *Ecological Modelling*, *Oryx*, and others. He had published 77 SCI academic papers at home and abroad, one monograph entitled *Population and Habitat of Amur Leopards in China* (Science Press), obtained six invention patents relating to wildlife monitoring technologies, one wildlife image data processing software copyright. He also led the preparation of *China's Amur Tiger and Amur Leopard Conservation Action Plan (2016-2025)*. Further, He had visited the United States, Russia, India, Vietnam, Thailand and other places a number of times for inspection and technical cooperation, under his leadership, his research team had released clear and complete camera trapping-based images of globally endangered Amur leopards for the first time, captured camera trap images of wild Amur tigers in the Wandashan area, Heilongjiang Province for the first time, and discovered and reported, for the first time, evidence of breeding populations of Amur leopards and tigers, which was covered by dozens of domestic and foreign media, including CCTV, *the Daily Telegraph* (UK), *Beijing Science and Technology News*, and *Wenhui Daily*.

致　　谢

首先感谢世界自然基金会（WWF）对国家林草局猫科动物研究中心"东北虎豹种群和栖息地监测数据库"的建设，对"东北虎豹保护行动计划"等诸多虎豹保护项目的持续支持，以及对本书出版的资助；感谢国家林草局保护司为国家林草局猫科动物研究中心东北虎豹监测平台的运转提供了持续的项目资金支持，为本书的出版提供了工作基础；感谢国家自然科学基金项目"东北虎豹栖息地哺乳动物类群时空互作网络的研究"（NSFC 31872241）、"十三五"国家重点研发计划项目"珍稀动物濒危机制及保护技术研究"（2016YFC0503200）、东北虎豹种群监测关键技术研发（中央高校基本科研专项资金平台项目：2572017 PZ14），以及生态环境部"生物多样性调查、观测和评估"项目（2019HB2096001006）等的支持，为本书大型猫科动物保护前沿理论的探索提供了难得的机遇。

感谢王桂明教授、Marcel Holyoak 教授、Sky K. Alibhai 博士、Zoe C. Jewell 博士、Dale G. Miquelle 博士、Sergey Aramilev 博士、Yury Darman 博士、Aleksey V. Kostyria 博士等多名国际专家的合作、支持和帮助；感谢黑龙江和吉林两省野生动物主管部门领导的大力支持，以及珲春、汪清、天桥岭、黄泥河、绥阳、穆棱、东京城、小北湖、东方红、迎春等林业局或保护区工作一线同仁与国家林草局猫科动物研究中心野外科研团队一直给予的配合、支持；感谢魏辅文院士、张知彬研究员在科研思路方面给予的亲切指导，感谢徐艳春教授、罗述金教授、曲智林教授、严川副研究员、李欣海副研究员等在分子生物学技术和生态模型技术探索工作中给予的帮助与支持，以及窦红亮、刘辉、张雪、李治霖、杨帆、黄冲等研究团队中的博士和硕士研究生贡献的智慧与力量！

引用本书的研究成果内容，请参考如下形式：
Citing the findings from this book, it is recommended to cite as follows:
姜广顺，华彦，顾佳音，齐进哲，宁瑶，周绍春，龙泽旭，王琦，刘丹，郭玉荣，长有德，刘培琦，张明海. 2020. 中国东北虎种群和栖息地动态及其精准管理策略. 北京：科学出版社.
Jiang Guangshun, Hua Yan, Gu Jiayin, Qi Jinzhe, Ning Yao, Zhou Shaochun, Long Zexu, Wang Qi, Liu Dan, Guo Yurong, Chang Youde, Liu Peiqi, Zhang Minghai. 2020. Population and Habitat Dynamics of Amur Tigers and Their Targeted Management Strategies in China. Beijing: Science Press.

Acknowledgement

First of all, I would like to thank the World Wide Fund for Nature (WWF) for the continued support of the construction of the Amur Tiger and Leopard Population and Habitat Monitoring Database of the Feline Research Center of National Forestry and Grassland Administration, the Amur Tiger and Leopard Protection Action Plan and many other tiger and leopard conservation projects, as well as the funding of this monograph. Thanks to the protection department of the National Forestry and Grassland Administration for providing continuous project financial support for the operation of the Amur Tiger and Leopard Monitoring Platform of the Feline Research Center, which provided the work basis for the publication of this monograph; Thanks to the National Natural Science Foundation of China for projects "Study on the Space-Time Interaction Network of Mammal Groups in the Amur Tiger and Leopard Habitats"(NSFC 31872241), the 13th Five-Year National Key Research and Development Program Project "Rare Animal Endangered System and Protection Technology Research"(2016YFC0503200), the Central University Basic Research Fund Platform Project "Key Technology Research and Development of Siberian Tiger and Leopard Population Monitoring"(2572017PZ14), as well as the "Biodiversity Survey, Observation and Assessment" project (2019HB2096001006) of the Ministry of Ecology and Environment, etc., providing a rare opportunity to explore the frontier theory of big cats protection in this book.

Thanks to many international experts for their cooperation support and help with Professor Wang Guiming, Professor Marcel Holyoak, Dr. Sky K. Alibhai, Dr. Zoe C. Jewell, Dr. Dale G. Miquelle, Dr. Sergey Aramilev, Dr. Yury Darman, Dr. Aleksey V. Kostyria, etc.; thanks to the wildlife authorities in Heilongjiang and Jilin for their strong support, and thanks to colleagues from the forestry bureau or protected area Hunchun, Wangqing, Tianqiaoling, Huangnihe, Suiyang, Muling, Dongjingcheng, Xiaobeihu, Dongfanghong, Yingchun, etc. have always cooperation and support to the field research team of the Feline Research Center; thanks to Academician Wei Fuwen and Researcher Zhang Zhibin for their kind guidance in scientific research ideas, Prof. Xu Yanchun, Prof. Luo Shujin, Prof. Qu Zhilin, Associate Researcher Yan Chuan, Associate Researcher Li Xinhai, and other experts in the molecular biology and ecological model technology exploration, as well as Dou Hongliang, Liu Hui, Zhang Xue, Li Zhilin, Yang Fan, Huang Chong, etc. these doctors and masters' wisdom and strength for the research teams.

序 一

　　自古以来，虎对中国的影响是世界上其他任何动物都无法比拟的。人们既喜爱它、崇拜它，又敬畏它，不仅因为其具有美丽的斑纹，更因为它一直是人们心目中神圣的图腾。然而，随着社会的发展，人类已侵占大部分本应由虎主宰的森林，与曾经的"森林之王"渐行渐远。20世纪中叶，我国东北地区还"诸山皆有虎"，而如今东北虎仅残存于中俄边境的狭小区域。东北林区为新中国最困难时期的经济建设做出了不可磨灭的贡献，这是东北林区特定历史年代的责任使然。但是，由于森林资源过度开发，东北虎栖息地丧失殆尽，东北虎的生存岌岌可危。

　　值得庆幸的是，党和政府审时度势，开始实施生态保护工程，特别是天然林保护工程的稳步推进、东北林区全面禁伐、成立东北虎豹国家公园等一系列重大决策使东北林区的森林得到休养生息，以东北虎为旗舰物种的生物多样性恢复成效日益凸显。在此过程中，一批国内外保护研究机构及科技工作者发挥了重要的技术支撑作用，成为国家重大生态保护决策制定的"智囊团"。国家林业和草原局猫科动物研究中心已批建成立7年，一直围绕中国东北虎保护管理面临的实际问题开展工作，开展了东北虎监测技术的研发、有效栖息地评估技术的探索、东北虎种群资源本底调查、保护优先区域斑块和生态廊道的识别、虎友好型森林经营管理，以及东北虎保护行动计划的制定等一系列卓有成效的科研与技术服务工作。本书就是这些具体研究工作成果的整理和浓缩，与《中国东北豹种群和栖息地研究》形成姐妹篇，内容涵盖了从栖息地评估到土地利用策略，从非损伤监测技术到东北虎的个体行为筛选，从保护优先区域的确定到生态廊道的设计，从生态系统保护的生态阈值到国家战略的科学建议等，比较详尽地介绍了国家林业和草原局猫科动物研究中心科研团队东北虎保护研究最新的研究成果与重要进展。

　　作为一名耄耋之年的科技工作者，我有幸见证了中国野生动物管理从无到有、从粗放经营逐渐走向科学与精准的发展。在这一过程中，我们收获了很多经验与教训。我们讲的管理不是死看死守，更不是极端保护，而是要通过科学探索逐步实现人与自然的和谐统一、人与野生动物的长期共存。要做到这一点，就要落实到"精准"的管理与保护上。尽管我国通过一系列的政策和措施逐步恢复了野生动物的栖息地，但历史上长达半个世纪的人工景观的建立和人为干扰的存在对野生动物的影响很难根本消除。以东北虎保护为例，由于老虎的家域广，对生存空间的需求大，要完全实现栖息地全区域的恢复极其困难，因为这不仅涉及生态移民与生态补偿，还涉及边境地区的宗教、民族、文化、国防等诸多因素。那么，如何实现人与虎的和谐共存呢？精准保护就是有效手段。"精"就是我们能探测到保护工作中的细微之处，"准"就是保护措施能注重实际、做到有的放矢，保护不仅应实现野生动物生态保护的需求，还应兼顾地方经济、社会、文化客观发展的需求。

东北虎的保护管理不仅涉及国际合作，还涉及社会、经济、政治和文化的方方面面，问题错综复杂，不是一个国家或某个机构可以独立完成的。该书更不能涵盖其精准保护的全部理论和技术，仅为抛砖引玉，逐步促进野生动物保护管理的科学发展。野生动物保护管理是国家生态建设的重要组成部分，我们应以习近平生态文明思想为指导，紧紧抓住"生态文明建设"和"美丽中国建设"国家战略实施的大好机遇，将东北虎保护工作推向一个新阶段。虽然东北虎保护任重而道远，但希望全社会广泛关注、业界同行共同参与、国际通力合作，让东北虎保护工作率先垂范，并引领我国野生动物保护与管理水平走向世界前列！

期待该书能成为管理部门的参考书、业内同行的工具书、青年学生的教科书。"路漫漫其修远兮，吾将上下而求索"，谨与青年学者共勉，欣然为之作序！

马建章　教授

中国工程院　院士

国家林业和草原局猫科动物研究中心　主任

东北林业大学野生动物与自然保护地学院　名誉院长

二〇一九年三月　哈尔滨

Foreword I

Since ancient times, the impact of tigers on China has been unmatched by any other animals in the world. People love them, worship them and fear them, not only because they have beautiful stripes but also because they have always been a sacred totem in people's minds. However, with the development of society, mankind has invaded most of the forests that should have been dominated by tigers. The former "King of the Forest" has gradually drifted away. In the middle of the last century, "tigers were found in every mountain" in Northeast China, but now, Amur tigers are only found in a narrow region around the Sino-Russian border. The northeast forest area in china made an indelible contribution to economic construction in the most difficult period of the new China, which was the responsibility of the northeast forest area in china in the specific historical period. However, due to the overexploitation of forest resources, the habitat of Amur tigers has been lost, reflecting a very serious ecological problem caused by "excessive blood transfusion" to the national economy, putting the survival of Amur tigers in jeopardy.

Fortunately, the Communist Party of China (CPC) and the government have taken stock of the situation and started to implement ecological protection projects, in particular, a series of major decisions like the steady advancement of the natural forest protection project, comprehensive ban on logging in the northeast forest area in china and the establishment of national parks for Amur tigers and leopards have contributed to the restoration of the forests in the northeast forest area in china, and biodiversity restoration with Amur tigers as the flagship species is increasingly prominent. In this process, a number of domestic and foreign protection research institutions and scientific and technological colleagues have played an important technical support role and become a "think tank" for nationally important ecological protection decision-making. Since the Feline Research Center of the National Forestry and Grassland Administration was established seven years ago, it has been working on the practical problems faced by the protection and management of Amur tigers in China and has conducted a series of fruitful research and technical services. This includes the research and development of Amur tiger monitoring technologies, exploration of effective habitat assessment techniques, background investigation of Amur tiger resources, identification of conservation priority area patches and ecological corridors, tiger-friendly forest management and the development of Amur tiger protection action plans. This book is the collection of the achievements of the above specific research work, forming a companion piece to the *Population and Habitat of Amur Leopards in China*, with its contents ranging from habitat assessment to land use strategies, from non-destructive monitoring techniques to the screening of individual behavior of Amur tigers, from the determination of conservation priority areas to the design of ecological corridors, and from the ecological threshold of ecosystem protection to the scientific recommendations for national strategies, giving a detailed description of the latest research findings of and important progress in the research on Amur tiger protection made by the scientific research team of the Feline Research Center of the National Forestry and Grassland Administration.

As an elderly scientist, I have the privilege to witness the development of wildlife management in China from scratch, from extensive management to science and precision. In the process, we have gained a lot of experience and lessons. The management referred to here is neither keeping an eye on wildlife all the time nor extreme protection but gradual realization of harmonious unity of man and nature through scientific exploration, and the long-term coexistence of humans and wildlife. To achieve this purpose, it is required to implement "targeted" management and protection. Though China has gradually restored the habitats of wildlife through a series of policies and measures, the effect of establishment of artificial landscapes and the existence of anthropogenic disturbance over half a century on wildlife is hard to eliminate. Taking the protection of Amur tigers as an example, due to the wide home range of tigers, they require a large space for survival, and it is extremely difficult to fully realize the restoration of the entire habitat because it involves not only ecological migration and ecological compensation, but also many other factors like religion, ethnic groups, culture and national defense in the border area. So how to achieve the harmonious coexistence between human beings and tigers? Targeted protection is an effective means. "Targeted" means that we can detect the subtleties in the protection work and that our protective measures can focus on the actual situation and protection in a targeted manner, with protection not only to meet the needs of wildlife ecological protection, but also take into account the objective development needs of local economy, society and culture.

The protection of the Amur tigers involves not only international cooperation, but also all aspects of society, economy, politics and culture, problems faced complicated. The protection of Amur tigers cannot be independently completed by a single country or a certain institution, so this book cannot cover all targeted protection-related theories and technologies. It is only intended to bring up the problems and to gradually promote the scientific progress of wildlife conservation management. Wildlife conservation management is an important part of the national ecological construction. We should take the thought of ecological civilization of General Secretary Xi Jinping as a guide and firmly grasp the great opportunity of the implementation of the national strategies, including the "ecological civilization construction" and "beautiful China construction" to elevate the Amur tiger protection work to a new level. Although the Amur tiger protection work has a long way to go, hopefully, the Amur tiger protection work can lead the way to arouse the widespread interest in the entire society and attract the participation of industry peers and close international cooperation to make China's wildlife conservation and management level at the forefront of the world!

The publication of this book is intended to be used as a reference book for management departments, a tool book for colleagues and a text book for young students. "The road ahead is long and endless in sight, yet I will search high and low with my will unbending", I am glad to write a preface to this book to share it with young scholars!

Professor Ma Jianzhang
Academician of the Chinese Academy of Engineering
Director of Feline Research Center of the National Forestry and Grassland Administration
Honorary Dean of College of Wildlife and Protected Area, Northeast Forestry University
Harbin, March 2019

序　二

东北虎是全球体型最大的猫科动物亚种，是被列入《濒危野生动植物种国际贸易公约》附录Ⅰ、《中国濒危动物红皮书》的濒危亚种和我国国家Ⅰ级保护野生动物。历史上，东北虎曾广泛分布于东北亚地区，但受各种因素影响，当前仅分布在我国黑龙江和吉林两省东部与俄罗斯、朝鲜接壤的边疆地区。东北虎在其分布国的传统文化中所具有的特殊地位及其对分布区生态系统健康的标志作用，使其成为全球生物多样性保护的旗舰物种之一，受到国内外保护界的高度关注。加强对东北虎及其栖息地的保护，不仅对我国开展生态文明建设和美丽中国建设具有重要意义，还对我国在国际上树立野生动物保护大国形象具有积极影响。

我国政府一直高度重视东北虎的保护工作，制定了《中国野生虎保护行动计划》，在东北虎栖息地实施了天然林保护、退耕还林等生态保护工程，并在东北虎栖息地内建立了20多处自然保护区，逐步恢复了森林生态系统的物种多样性，为东北虎种群恢复奠定了坚实的生态基础。特别是党的十八大以来，习近平总书记等中央领导同志多次做出重要批示、指示，要求遵循自然规律加强对东北虎及其栖息地的保护，中央全面深化改革领导小组专门审议批准了建立东北虎豹国家公园的方案，为全面实施东北虎及其栖息地保护政策指明了方向。当前国家林业和草原局已经按照国家机构改革的要求，承接了各类自然保护地的统一管理工作，正在整合优化包括东北虎栖息地在内的自然保护地，统筹推进东北虎豹国家公园建设，保护和修复东北虎赖以生存的森林生态环境，加快构建生态廊道和生物多样性保护网络，为东北虎等濒危野生动物的生存和发展营造良好的栖息地环境。

加强对东北虎栖息地的科学保护和精准管理是保护东北虎的最核心的手段与最基础的工作。东北林业大学是我国最早设立野生动物相关学科的高等院校。长期以来，一代一代东林专家不畏艰险、甘于奉献、攻坚克难、坚持不懈地开展东北虎保护研究，不仅取得了丰硕成果并积累了大量技术资料，还为全国培养了大批野生动物保护的专业人才。特别是原国家林业局在该校设立的"国家林业局猫科动物研究中心"成为东北虎保护研究领域的领军团队，先后承担了原国家林业局"国家虎豹监测平台建设""中国大型猫科动物保护行动计划编制""全球环境基金（GEF）东北虎保护"和"基于自动相机监测和分子遗传分析方法的东北虎豹跨境移动研究"等虎保护科学研究和技术服务项目，为推进东北虎等野生动物的保护做出了突出贡献。该中心常务副主任姜广顺教授是东北林业大学培养的年轻一代专家，他从我国生态保护与修复的重大国家战略需求出发，在充分总结其带领的科研团队研究成果的基础上于2016年撰写了专著《中国东北豹种群和栖息地研究》，填补了我国东北豹野外保护研究专著的空缺。今天，我们又欣然看到姜广顺教授及其科研团队撰写的《中国东北虎种群和栖息地动态及其精准管理策略》即将出版。该书的内容紧密结合东北虎保护管理的实际问题，对东北虎种群分布规律、行为特征、繁殖特性、种群监测技术、栖息地预测和评估技术、优先保护区域的确定、人为干扰管理的生态阈值、生态廊道的构建等保护管理上迫切需要解决的问题进行了深入探索，获得了新的进展和成果，为推动我国东北虎国家保护战略的有效实施提供了重要的理论基础和技术保障。该书理论水平高，信息量大，文字通俗易懂，可供从事野生动物保护和自然保护地管理的科研人员、管理人员与野生动物保护爱好者学习借鉴，特此推荐，以飨读者。

杨　超　司长
国家林业和草原局自然保护地管理司
二〇一九年三月二十八日

Foreword II

Amur tigers are the largest feline in the world and have been listed as a subspecies in Appendix I of the *Convention on International Trade in Endangered Species of Wild Fauna and Flora* and as endangered subspecies in the *China Red Data Book of Endangered Animals* and China's Grade I protected wild animals. Historically, Amur tigers were widely distributed in Northeast Asia. Affected by various factors, now they are only distributed in the areas in the east of Heilongjiang and Jilin provinces where they border with Russia and North Korea. Amur tigers have their special status in the traditional cultures in the countries where they are distributed, and have a symbolic role in the health of the forest ecosystem, making them one of the flagship species of global biodiversity conservation, as a species of high concern at home and abroad. Strengthening the protection of Amur tigers and their habitat is the concrete manifestation of China conducting ecological civilization construction and practicing ecological civilization thought, which is not only of great significance for creating a beautiful China, but also of positive significance for establishing China's image as a large wildlife protection country.

The Chinese government has always attached great importance to the protection of Amur tigers, developed *China's Wild Tiger Conservation Action Plan*, implemented ecological protection projects in the habitat of Amur tigers, including the natural forest protection project and the returning of farmland to forests project, and successively established more than 20 nature reserves, gradually restoring the biodiversity of forests in Northeast China and laying a solid ecological foundation for the restoration of the Amur tiger population. In particular, since the 18th National Congress of the CPC, leading comrades in the Central Committee, including Xi Jinping, have issued important instructions to require strengthened protection of Amur tigers and their habitats based on the laws of nature, The Central Leading Group for Comprehensive and Deepening Reform has specifically considered and approved the plan for the establishment of the National Park for Amur Tiger and Amur Leopard, which pointed out the direction for the full implementation of the protection of Amur tigers and their habitats. Currently, the National Forestry and Grassland Administration have undertaken the unified management of various types of nature reserves in accordance with the requirements of the China Institutional Reform. It is integrating and optimizing nature reserves which include the habitat of Amur tigers, conducting overall advancement of national parks of Amur tigers and leopards, protecting and restoring the forest ecological environment on which Amur tigers depend, and constructing an ecological corridor and a biodiversity conservation network, as well as creating a good habitat for the survival and development of endangered wild animals like Amur tigers.

Strengthening the scientific protection and targeted management of the habitat of Amur tigers is the basic work to protect Amur tigers and its premise lies in the basic research on Amur tigers and their habitat. Northeast Forestry University is the earliest university to establish wildlife disciplines in China. For a long time, from generation to generation, experts in Northeast Forestry University (NEFU) have been sticking to conduct research on protection of Amur tigers by bravely facing hardships, making contributions willingly and overcoming various difficulties, and have achieved fruitful results and accumulated a large quantity of technical data in the research on protection of Amur tigers and cultivated a large number of wildlife protection talents for China. In particular, the former State Forestry Administration established the "Feline Research Center of the State Forestry Administration" in this university, which has established a capable research team which has successively undertaken tiger protection-related scientific research and technical service projects of the former State Forestry Administration, including the "National Tiger and Leopard Monitoring Platform Construction", "China's Large Feline Protection Action Plan Preparation", "Global Environment Fund-based Amur Tiger Conservation" and "Study of Transboundary Movements of the Amur Tiger and Amur Leopard Using Camera Traps and Molecular Genetic Analysis", making outstanding contributions to advancing the protection of wildlife like Amur tigers. Professor Jiang Guangshun, executive deputy director of the center, is a young expert in feline conservation cultivated by NEFU, and based on China's major strategy needs of ecological protection and restoration, he had written a monograph titled *Population and Habitat of Amur Leopard in China* in 2016 on the basis of summarizing the research achievements of the scientific research team led by him, filling in a gap in the research on protection of Amur leopards in the wild in China. Today, I am relieved to learn that the book *Population and Habitat Dynamics of Amur Tigers and Their Targeted Management Strategies in China* written by Professor Jiang Guangshun and his research team is to be published. This book is closely bound up with the actual problems existing in the management of Amur tiger protection, conducting in-depth exploration of protection management problems to be solved urgently in terms of the Amur tiger population's distribution thresholds, behavioral characteristics, reproductive characteristics, population monitoring techniques, habitat prediction and assessment techniques, priority protection area determination, ecological thresholds of anthropogenic disturbance management and ecological corridor construction, obtaining new progress and achievements, and providing an important theoretical basis and technical guarantee for advancing the effective implementation of China's national Amur tiger protection strategies. This book has a high theoretical level, makes a large amount of information available, and its content is easy to be understanded. It can be used for reference by scientific researchers and management personnel engaged in wildlife protection and nature protected area management, and by wildlife protection enthusiasts. Hereby, it is recommended to readers.

Director Yang Chao
Nature Protected Area Management Division of
the National Forestry and Grassland Administration
March 28, 2019

序 三

野生虎的保护是世界自然基金会在全球开展的重点项目。世界自然基金会一直关注和支持全球13个老虎分布国的野生虎保护工作。虎是一个神奇并令人敬畏的物种，它位于食物链的顶端，它的存在与否，标志着人类与这一物种共同赖以生存的生态系统是否健康。虎体型硕大，威猛无敌，成员曾遍布于亚洲大陆的广袤大地。然而，即使是这样一个强大的物种，面对人类社会经济和生态的不平衡发展，也正一步步地走向灭绝的边缘。从这个角度来看，人类完胜老虎，人类的力量远强于老虎。然而，当人类沉湎于自身的强大并为此沾沾自喜的同时，也逐渐认识到为此所付出的巨大代价，即随着虎一同消失的还有人类赖以生存的生态环境。

近几十年来，虎的存亡牵动着全球亿万人民的心。人们为保护老虎、恢复其野生种群做着长期不懈的努力，期望通过对这一具有"神灵"象征的物种的保护，来修复岌岌可危的人类生存环境。野生虎保护工作的意义重大，任务艰巨，道路曲折漫长。我们不但要不断提高全民生态保护意识，改变普遍存在的非生态友好型社会行为，还要积极探索虎及其栖息地保护的技术和理论，切实提高保护成效。

在世界自然基金会的支持与主导下，10年前国内外科学家共同努力，联合对长白山区东北虎栖息地进行了研究与评估，开启了我国对虎栖息地研究与保护的先河。10年后的今天，世界自然基金会又支持并参与了这部《中国东北虎种群和栖息地动态及其精准管理策略》专著的编撰工作。这部专著基于我国东北虎保护工作者多年的工作实践和理论探索，凝聚了我国东北虎保护工作者的智慧和心血，较全面地阐述了目前我国在东北虎种群及栖息地、社区、猎物等方面的一些管理措施和方法论，并首次提出了东北虎种群和栖息地精准管理的概念，对于我国东北虎保护管理具有重要的理论和实践指导意义，标志着我国对东北虎及其栖息地的保护管理正朝着适应性管理的科学方向迈进，必将提高我国东北虎保护与管理的水平。希望这部专著的出版，能够起到抛砖引玉的作用，带动更多的保护工作者在老虎栖息地有效管理方面积极探索，勇于克难攻坚，取得更大的成绩，为虎保护做出贡献。

周 非

世界自然基金会（瑞士）北京代表处副总干事

二〇一九年二月二十七日

Foreword III

The protection of the global wild tigers is a key project implemented by the World Wide Fund for Nature (WWF), which has been focusing on and supporting the conservation of wild tigers in the 13 range countries. Tigers are a magical and awesome species, located at the top of the food chain, and whether they are present or not indicates whether the ecosystem that humans and this species depend on for survival is sound. Tigers are huge, powerful and invincible, with members distributed on the vast land of the Asian continent. However, even such a powerful species is gradually walking towards the verge of extinction due to the unbalanced socio-economic and ecological development of human beings. Seen from this angle, human beings defeat tigers and human beings are far more powerful than tigers. However, when human beings indulge in their own power complacently, they have gradually realized the enormous cost they have paid for this, that is when tigers disappear, the ecological environment on which humans depend will disappear as well.

In recent decades, the survival of tigers has aroused the concern and affected the hearts of hundreds of millions of people around the world. In order to protect tigers and restore their wild populations, people have made long-term and unremitting efforts to restore the human living environment by protecting the species symbolizing 'God'. The protection of wide tigers is of great significance and is an arduous task, with a long and winding road to go. We must not only constantly increase all people's awareness of ecological protection but also change the non-eco-friendly social behaviors that prevail and actively explore the technologies and theories relating to the protection of tigers and their habitats to effectively improve the effectiveness of protection.

Under the support and guidance of WWF, some Chinese and foreign scientists worked together to study and evaluate the habitat of Amur tigers in the Changbai Mountain area 10 years ago, creating a precedent for tiger habitat research and protection in China. Today, 10 years later, WWF has once again supported and participated in the compilation of this monograph *Population and Habitat Dynamics of Amur Tigers and Their Targeted Management Strategies in China*. This book is based on many years of work practice and theoretical exploration of the protection of Amur tigers in China, condensing the wisdom and efforts of protection workers of Amur tigers in China, comprehensively expounding some management measures and methodologies in terms of Amur tigers population and their habitats, communities, and prey species. It puts forward the concept of targeted management of Amur tigers population and habitat for the first time, having important theoretical and practical guiding significance for Amur tiger protection management in China. It indicates that China's protection management of Amur tigers and their habitat is moving towards the scientific direction of adaptive management. Hopefully, the publication of this monograph can serve as a catalyst to drive more protection workers to actively explore the effective management of tiger habitats, overcome difficulties bravely and achieve greater results to make contributions for tiger protection.

Zhou Fei
Deputy Director-General, Beijing Representative Office,
WWF (Switzerland)
February 27, 2019

聚焦管理实践 走向科学保护
——写在前面的话

本书作为科学出版社第一部关于中国东北虎生态与管理研究的学术著作，以期为管理部门和同行提供些许启发与参考。但我心中忐忑，唯恐大家读后收获寥寥。因为老虎不仅是全球生物多样性保护的旗舰物种和保护伞物种，还是国际关注全球濒危物种的焦点，而且对其进行保护已上升为我国的国家战略，成为国家生态建设与国家形象的重要标志和政治、经济、生态、社会发展的缩影！

◆ **东北虎保护的国家进程**

2010 年，温家宝总理与 13 个虎分布国首脑共同签署了《圣彼得堡保护老虎宣言》，并承诺力争在下一个虎年让老虎数量翻倍，保护虎的重要性和紧迫性被提升到前所未有的高度，为国际社会所瞩目。同年，中国工程院院士联名向国务院递交了《关于将东北虎保护纳入国家可持续发展战略优先领域的建议》；2011 年，国家林业局发布《中国野生虎恢复计划》；2013 年，国家林业局启动《中国大型猫科动物保护行动计划》的编制工作；2015 年，中国工程院院士联名向国务院递交了《关于推进我国老虎及其栖息地保护的建议》；2016 年，《东北虎豹国家公园体制试点方案》获中央全面深化改革领导小组审议通过，建立了面积达 1.46 万 km^2 的东北虎豹国家公园，并要求应用现代高新技术进行保护与管理，快速促进中国东北虎种群和栖息地的恢复。

◆ **东北虎保护管理面临的困境**

东北虎种群数量和分布区的动态变化监测面临困难。东北虎等大型猫科动物活动范围大、活动隐蔽，对其种群和栖息生境的有效调查监测技术缺乏。在当前10 万多平方千米东北虎分布区的大空间尺度下，掌握其种群和栖息地资源动态相当困难，给保护措施的科学制定和保护成效的科学评估带来障碍。

东北虎适宜栖息地面临着面积不断缩小和退化、扩散通道受阻等问题。尽管我国大面积的天然林在天然林保护工程的实施下正在恢复，但由于东北虎家域较大（雌性东北虎 450～500km^2、雄性可达 1000km^2 以上），而目前确定的保护地面积小、分布区隔离严重，与俄罗斯种群也存在一定程度的人为阻隔，种群基因交流困难，小的隔离种群可能面临近亲繁殖、免疫力低下而导致疾病暴发乃至种群崩溃的风险。

东北虎栖息地内的人为干扰活动缺乏科学管控。东北虎栖息地中林下经济开发、围栏、道路、村屯、矿业、农业、牧业等方面的持续人为干扰活动使栖息地斑块质量不断下降、面积不断丧失，猎物种类结构和丰富度无法满足老虎生态需求等问题突出。如何定量化地评估和管理这些人为干扰活动，如何找到野生动物和人能够相互容忍的生态阈值，促进一定区域内的人和野生动物共存，是国家野生动物栖息地管理中普遍制约管理成效的关键问题。

保护地资源的空间配置技术缺乏。人虎争夺空间资源的冲突严重，特别是保护地与当地社区经济发展间的空间资源配置矛盾突出。大尺度栖息地景观缺乏科学规划，保护地景观保证东北虎可自由迁移的渗透性不强，阻碍大型猫科动物的自由迁移和扩散，无法保障东北虎种群长期生存繁衍对大尺度生态景观资源的需求。因此，如何规划保护地体系的空间格局，如何满足种群持续生存的生态需求等精准保护管理技术的缺乏，将严重制约保护管理成效。

◆ 东北虎精准管理的理论和技术需求

作为世界虎种的起源地、虎亚种分布最多的国家，我国肩负着恢复全球老虎种群的重要使命与责任。2016年年底，中央全面深化改革领导小组审议通过了《关于健全国家自然资源资产管理体制试点方案》和《东北虎豹国家公园体制试点方案》，这两项试点工作的正式启动，是中央通过保护老虎推进生态文明建设的重大举措。建立以国家公园体制为主体的新型自然保护地管理体系，必将对加强自然资源保护管理、维护国家生态安全产生重大而深远的影响。

新型的自然保护地体系已经由以前的区域性或单一物种的保护走向大空间尺度的生态系统视角下的保护。零星隔离的保护区逐步相互整合、连通，有利于物种的迁移和交流，我国科学家也更多地关注与国际种群的交流、生态廊道科学规划与设计、大空间尺度资源的科学监测技术需求等重大保护管理问题，而这些问题的解决迫切需要加速实际精准管理理论和技术的研发日程，也涉及国家公园内外如何将人为干扰活动控制在生态系统承载能力之内，如何缓解人与动物的资源利用冲突等一系列精准管理技术的探索。

◆ 东北虎精准管理问题的探索

东北虎种群数量少、活动范围广，很难掌握其生态规律，想要确定科学、可靠的保护管理措施更是不易，需要针对实际保护问题开展长期基础研究工作。例如，大家普遍认同道路、农田、村庄等都会干扰东北虎的生存和分布，但有多大程度的影响？如何识别和管控适宜的栖息地？天然林保护工程促进了东北虎的恢复，是如何促进的？哪些因素在发挥关键作用？为什么东北虎栖息地内有蹄类猎物种群密度低？人为干扰在东北虎生存的生态系统中是如何直接或间接影响老虎分布的？中国和俄罗斯的东北虎猎物的组成与取食习性相似吗？东北虎的繁殖习性如何？东北虎有哪些固有的天然本能和个性，如何进行保育？可以通过最易获取的雪地足迹影像识别东北虎个体吗？如何在当前我国 10 万多平方千米野生东北虎分布区的大空间尺度上，平衡当地经济发展和物种生存需求，有效确定优先保护区域与生态廊道的位置和面积？如何在国家层面制定保护策略？这一系列基础科学问题都是迫切需要攻克的。

本书作者 10 多年来一直致力于解决上述东北虎保护过程中面临的基础保护问题，与世界自然基金会（WWF）、俄罗斯科学院、美国杜克大学、美国密西西比州立大学、美国加利福尼亚大学、国际野生生物保护学会（WCS）、中国科学院动物研究所、黑龙江野生动物研究所、黑龙江东北虎林园、吉林珲春自然保护区等单位的国内外专家联合攻关，应用动物行为学、种群生态学、景观生态学、保护生物学、物种分布模型技术、物种识别技术等，获得了些许理论与技术的成果和突破，可为东北虎种群和栖息地的实际管理与同行的研究提供参考及启示。

本书共分为 12 章，各章撰写具体分工如下。

第 1 章东北虎栖息地分布的环境因子驱动由姜广顺、齐进哲、长有德完成；第 2 章东北虎保护中的生态与社会的协同效应由姜广顺、齐进哲完成；第 3 章有蹄类动物死亡原因和生境质量评价综合分析由周绍春、张明海完成；第 4 章人为干扰对森林植被、东北虎及其猎物的影响由顾佳音、刘培琦、姜广顺完成；第 5 章中国珲春地区与俄罗斯 Primorskii Krai 西南部东北虎食性分析的比较由华彦、顾佳音、姜广顺完成；第 6 章野生和圈养雌性东北虎繁殖参数的比较由顾佳音、郭玉荣、姜广顺完成；第 7 章东北虎对自然猎物的偏好本能及其与个性关系的探测研究由王琦、刘丹、姜广顺完成；第 8 章利用东北虎粪便 DNA 进行基因分型的风险评估由宁瑶、姜广顺完成；第 9 章应用雪地足迹影像进行东北虎的性别鉴定由顾佳音、姜广顺完成；第 10 章中国东北虎栖息地等级保护优先区域和廊道确定的适宜性模型技术由姜广顺、龙泽旭、顾佳音完成；第 11 章东北虎种群和栖息地精准管理的生态阈值由齐进哲、姜广顺完成；第 12 章东北虎精准保护管理工作中的挑战、机遇与举措由姜广顺、顾佳音、华彦完成。全书由姜广顺和华彦负责统稿，马建章对本书进行了审定。

然而，国际野生动物保护理论和技术发展迅猛，保护生物学新理论和新技术不断涌现，特别是交叉学科的发展，如人工智能技术等有可能为野生动物现代精准管理技术的发展插上腾飞的翅膀。基于此，限于作者水平，本书的出版仅为抛砖引玉，期望与更多的同行一道探索，为解决中国东北虎保护面临的实际问题继续扎实开展工作，以贡献微薄之力。

姜广顺

东北林业大学野生动物与自然保护地学院　教授

国家林业和草原局猫科动物研究中心　常务副主任

2019 年 2 月 26 日

Focusing on Management Practice and Moving Towards Scientific Protection
—Preface

As the first academic book of the Science Press on the study of the ecology and management of Amur tigers in China, it is intended to provide management departments and peers with some enlightenment and reference. However, I worried that readers achieved little after reading. Tigers are not only the flagship species and umbrella species under global biodiversity protection but also the international focus of global endangered species. Further, the protection of tigers has been upgraded to China's national strategy and has become an important symbol of national ecological construction and national image, and is the epitome of political, economic, ecological and social development!

♦ **National Process of Amur Tiger Protection**

In 2010, Premier Wen Jiabao and the heads of 13 tiger range countries jointly signed the *St. Petersburg Declaration on the Protection of Tigers* and promised to strive to double the number of tigers before the next year of the tiger, elevating the importance and urgency of protecting tigers to an unprecedented height, attracting the attention of the international community. In the same year, the academicians of the Chinese Academy of Engineering jointly submitted the *Proposal on Integrating Amur Tiger Protection into the Priority Areas of the National Sustainable Development Strategy* to the State Council. In 2011, the State Forestry Administration issued *China's Wild Tiger Recovery Plan*. In 2013, the State Forestry Administration launched the preparation of the *China's Large Feline Protection Action Plan*. In 2015, academicians of the Chinese Academy of Engineering jointly submitted the *Proposal on Advancing the Protection of Tigers and Their Habitats in China*. In 2016, the *Pilot Program for the Amur Tiger and Leopard National Park System* was reviewed and approved by the Reform Deepening Team of the Central Government, with an Amur tiger and leopard park covering an area of 14 600km^2 established, modern advanced technology is required to be used for the protection and management, and to rapidly promote the restoration of the population and habitat of Amur tigers in China.

◆ **Dilemma Faced by the Protection Management of Amur Tigers**

It is difficult to monitor the dynamic changes of the number and distribution area of the Amur tiger population. Large feline like Amur tigers have a wide range and move stealthily and there is a lack of techniques for the effective investigation and monitoring of their population and habitat. Due to the large spatial scale of the currently more than 100 000km^2 distribution area of Amur tigers, it is quitely difficult to grasp the dynamics of their population and habitat resources, thereby bringing obstacles to the scientific formulation of protection measures and the scientific evaluation of protection effectiveness.

The habitat suitable for Amur tigers is facing increasing spatial reduction, degradation and hindrance to diffusion channels. In spite of ongoing restoration of large-scale natural forests in China with the implementation of the Natural Forest Protection Project, as Amur tigers have a large home range (450–500km^2 for a female tiger and over 1000km^2 for a male tiger), the current protected area is small, their distribution ranges are isolated and there exists a certain degree of human-related separation from the international Russian population. Thus, the genetic exchange between populations is challenged, resulting in the fact that small isolated populations may face the risk of inbreeding and low immunity, thereby posing a risk of disease outbreak and even population collapse.

There is a lack of scientific management and control techniques for anthropogenic disturbance activities in the habitat of Amur tigers. The continuous anthropogenic disturbance activities like economy development, fencing, roads, villages, mining, agriculture, animal husbandry, etc., have caused the continuous loss in habitat patch area and decline in quality, resulting in some prominent problems like the structure and richness of prey species, thus failing to meet the ecological needs of tigers. How to conduct quantitative evaluation and management of these human disturbances and how to find out a threshold that can make wild animals and human beings tolerate each other to realize the coexistence of humans and wildlife in a certain area are the common problems restricting management effectiveness in the management of national wildlife habitats.

There is a lack of techniques for spatial allocation of protected land resources. The conflicts between humans and tigers arising out of the fight for space are serious, especially the prominent contradiction between the protected area and the spatial resource allocation of the local community economic development. There is a lack of scientific planning for large-scale habitat landscapes and the landscapes in the protected areas are weak in their permeability to ensure free migration of Amur tigers, thereby hindering movements and the spread of these large feline, and render landscapes unable to satisfy the needs for large-scale and long-term survival and reproduction of the Amur tiger population. Therefore, there is a lack of precise protection management techniques for planning the spatial pattern of the protected land system and satisfying the ecological needs of the sustainable survival of the population, which will seriously restrict the effectiveness of protection management.

- **Theoretical and Technical Requirements for Precise Management of Amur Tigers**

As the country where the tiger species originates and tiger subspecies have the largest distribution in the world, China shoulders the important mission and responsibility of restoring the global tiger population. At the end of 2016, the Reform Deepening Team of the Central Government reviewed and approved the *Pilot Program for Improving the National Natural Resources Asset Management System* and the *Pilot Program for the Amur Tiger and Leopard National Park System*, and the two pieces of pilot work officially started, which is an important measure of the Central Government to advance the construction of ecological civilization through tiger protection. The establishment of a new nature protected area management system with the national park system as the main body will definitely produce a significant and far-reaching impact on strengthening natural resource protection management and safeguarding national ecological security.

The new nature protected area system has shifted from the previous regional or single species protection to that of a large spatial scale ecosystem perspective. Sporadically isolated protected areas are going through the process of gradual integration and connectivity, conducive to the migration and exchange of species, with more focus placed on major protection management issues concerning the exchanges with international populations. It also pays greater attention to scientific planning, the design of ecological corridors and the demand for technologies for scientific monitoring of resources at large spatial scales. The solutions to these issues urgently call for the acceleration of the research and development agenda of actual precision management theory and technology. This involves the exploration of a series of precise technologies for controlling human disturbance activities inside and outside of the national parks within the bearing capacity of the ecological system, and easing the conflict between humans and wildlife in resource utilization, etc.

- **Exploration of the Precious Management of Amur Tigers**

The population of Amur tigers is small, and it is hard to grasp their ecological laws due to their large range of activities. It is more difficult to determine scientific and reliable protection management measures, which requires long-term basic research work targeted at actual protection issues. For example, it is generally agreed that roads, farmlands, villages, etc. interfere with the survival and distribution of Amur tigers, but to what degree do they affect their survival and distribution? How to control and identify suitable habitats? How does the natural forest protection project promote the recovery of Amur tigers? What factors are playing a key role? Why is the population density of hoofed prey low in the habitat of Amur tigers? How does anthropogenic disturbance directly or indirectly affect the distribution of Amur tigers in the ecosystem they live in? Are the composition and feeding habits of Amur tigers in China and Russia similar? What is the breeding habit of Amur tigers? What are the inherent natural instincts and personalities of Amur tigers? How to conserve them? Can an individual Amur tiger be identified through the most readily available snow footprint images? How to balance the local economic development and the survival needs of species at the large spatial scale at which the Amur tiger occurs in China (more than 100 000km^2) to effectively determine the location, size and shape of priority protection areas and ecological corridors? How to formulate protection strategies at the national level? This series of basic scientific issues urgently require solutions.

The authors of this book have been working to solve the above basic protection issues faced in the Amur tiger protection process for more than a decade. They have cooperated with domestic and foreign experts from WWF, Russian Academy of Sciences, Duke University, Mississippi State University, University of California, WCS, Institute of Zoology, Chinese Academy of Sciences, Heilongjiang Wildlife Research Institute, Heilongjiang Siberian Tiger Park, Jilin Hunchun Nature Reserve and others to jointly overcome the above issues and have obtained some theoretical and technical achievements and breakthroughs applying animal behavior, population ecology, landscape ecology, conservation biology, species distribution model technology, species identification technology, etc. This can provide reference and enlightenment for the actual management of the Amur tigers population and habitat, and for research conducted by peers.

This book has 12 chapters, with the specific division of each chapter as follows.

Chapter 1: Environmental factors drive the habitat distribution of Amur tiger by Jiang Guangshun, Qi Jinzhe, Chang Youde; Chapter 2: The synergy between ecology and society in Amur tiger conservation by Jiang Guangshun, Qi Jinzhe; Chapter 3: An integrated analysis of the causes of ungulate mortality in the Wanda Mountains (Heilongjiang Province, China) and an evaluation of habitat quality by Zhou Shaochun, Zhang Minghai; Chapter 4: Effects of human disturbance on vegetation, prey and Amur tigers in Hunchun Nature Reserve, China by Gu Jiayin, Liu Peiqi, Jiang Guangshun; Chapter 5: A comparison of food habits and prey preferences of Amur tiger at the south-west Primorskii Krai in Russia and Hunchun in China by Hua Yan, Gu Jiayin, Jiang Guangshun; Chapter 6: A comparison of reproductive parameters of female Amur tigers in the wild and captivity by Gu Jiayin, Guo Yurong, Jiang Guangshun; Chapter 7: Innate preference for native prey and personality implications in captive Amur tigers by Wang Qi, Liu Dan, Jiang Guangshun; Chapter 8: Risks involved in fecal DNA-based genotyping of microsatellite loci in the Amur tiger: a pilot study by Ning

Yao, Jiang Guangshun; Chapter 9: Sex determination of Amur tiger from footprints in snow by Gu Jiayin and Jiang Guangshun; Chapter 10: An adaptive model for determining hierarchical priority conservation areas by Jiang Guangshun, Long Zexu, Gu Jiayin; Chapter 11: Ecological threshold on precisional management of Amur tiger population and habitat by Qi Jinzhe, Jiang Guangshun; Chapter 12: Challenges and measures for effective protection and management of the Amur tiger by Jiang Guangshun, Gu Jiayin and Hua Yan. This book was complied by Jiang Guangshun and Hua Yan, and reviewed by Prof. Ma Jianzhang.

However, the international wildlife protection theories and technologies are developing rapidly, and new theories and new technologies of conservation biology are constantly emerging, especially the development of interdisciplinary fields, such as artificial intelligence technology, which may boost the development of modern targeted management of wildlife. Based on this, restricted by the authors' level, the publication of this book is just intended to present some ideas. Hopefully, we can contribute our meager strength to solving the actual problems faced by the Amur tiger protection in China by jointly exploring with more peers and continuing to work hard.

Prof. Jiang Guangshun

College of Wildlife and Protected Area, Northeast Forestry University

Executive Deputy Director of Feline Research Center of the National Forestry and Grassland Administration

February 26, 2019

目 录

第1章 东北虎栖息地分布的环境因子驱动···1
1.1 引言···1
1.2 材料与方法···2
1.2.1 研究区域···2
1.2.2 东北虎种群数据···3
1.2.3 环境变量和人为干扰变量··4
1.2.4 空间自相关检测···5
1.2.5 统计分析···6
1.3 结果···7
1.3.1 长白山模型···7
1.3.2 完达山模型··10
1.3.3 分区比较和定量化响应··12
1.4 讨论··13
1.4.1 环境因素的影响··13
1.4.2 人为干扰的影响··14
1.4.3 空间自相关效应··15
1.4.4 保护意义··16
1.5 本章小结···17
参考文献···17

第2章 东北虎保护中的生态与社会的协同效应··22
2.1 引言··22
2.2 研究方法···24
2.2.1 森林管理数据的收集与分析··24
2.2.2 东北虎和东北豹种群大小与生境分布···27
2.3 结果··28
2.4 讨论··33

2.5	本章小结	37
参考文献		38

第3章 有蹄类动物死亡原因和生境质量评价综合分析 ... 40

3.1	引言	40
3.2	研究地区概况	41
3.3	数据收集和分析方法	42
	3.3.1 有蹄类动物死亡点数据来源	42
	3.3.2 死亡原因分析	42
	3.3.3 有蹄类动物出现点数据来源	42
	3.3.4 生境选择分析	42
3.4	研究结果	45
	3.4.1 有蹄类动物死亡原因	45
	3.4.2 生境选择分析	45
	3.4.3 生境评价	45
3.5	讨论	48
	3.5.1 有蹄类动物死亡原因及比例	48
	3.5.2 生境选择和质量评价	48
3.6	本章小结	50
参考文献		50

第4章 人为干扰对森林植被、东北虎及其猎物的影响 ... 54

4.1	引言	54
4.2	研究地区与方法	57
	4.2.1 研究区域	57
	4.2.2 数据收集	57
	4.2.3 数据分析	58
4.3	结果	59
	4.3.1 人为干扰对植被的影响	59
	4.3.2 人为干扰对猎物的影响	63
	4.3.3 人为干扰对东北虎的影响	63
4.4	讨论	63
	4.4.1 放牧的影响	63
	4.4.2 道路的影响	64

| 4.4.3   人参种植的影响 ····································· 65
| 4.4.4   GAM、GLM、SEM 的模型比较 ····················· 66
| 4.5   结论 ··· 66
| 4.6   本章小结 ··· 66
| 参考文献 ··· 67

第 5 章   中国珲春地区与俄罗斯 Primorskii Krai 西南部东北虎食性分析的比较 ··· 70
| 5.1   引言 ··· 70
| 5.2   材料与方法 ··· 71
| 5.2.1   研究区域 ··· 71
| 5.2.2   野外研究方法 ······································ 72
| 5.2.3   粪便分析 ··· 72
| 5.2.4   数据分析 ··· 73
| 5.3   结果 ··· 74
| 5.4   讨论 ··· 77
| 5.5   本章小结 ··· 79
| 参考文献 ··· 79

第 6 章   野生和圈养雌性东北虎繁殖参数的比较 ················ 82
| 6.1   引言 ··· 82
| 6.2   材料与方法 ··· 83
| 6.2.1   研究种群 ··· 83
| 6.2.2   数据分析 ··· 84
| 6.3   结果 ··· 84
| 6.3.1   圈养东北虎和野生东北虎种群的繁殖参数 ··· 84
| 6.3.2   幼崽死亡率 ··· 85
| 6.3.3   繁殖的季节性 ······································ 85
| 6.3.4   幼崽死亡后再次受孕雌虎的繁殖特征 ········ 87
| 6.4   讨论 ··· 87
| 6.5   本章小结 ··· 88
| 参考文献 ··· 89

第 7 章   东北虎对自然猎物的偏好本能及其与个性关系的探测研究 ··· 91
| 7.1   引言 ··· 91
| 7.2   材料与方法 ··· 92

7.2.1	实验动物	92
7.2.2	实验设备	92
7.2.3	刺激物	92
7.2.4	猎物识别测试	93
7.2.5	个性评估	93
7.2.6	数据分析	95
7.3	结果	97
7.3.1	对自然猎物视觉刺激的偏好反应	97
7.3.2	对自然猎物听觉刺激的偏好反应	98
7.3.3	对自然猎物嗅觉刺激的偏好反应	99
7.3.4	对自然猎物的本能偏好及其与个性特征的关系	100
7.4	讨论	101
7.5	本章小结	103
	参考文献	103

第 8 章　利用东北虎粪便 DNA 进行基因分型的风险评估　106

8.1	引言	106
8.2	材料与方法	107
8.2.1	取样	107
8.2.2	DNA 提取	107
8.2.3	微卫星数据分析	108
8.2.4	基因分型的正确性评估	108
8.2.5	使用粪便样本评估种群遗传学参数	109
8.3	结果	109
8.3.1	微卫星位点分型正确性评估	109
8.3.2	基因分型误差对估算种群遗传学参数的影响	109
8.4	讨论	113
8.5	本章小结	114
	参考文献	115

第 9 章　应用雪地足迹影像进行东北虎的性别鉴定　117

9.1	引言	117
9.2	材料与方法	118
9.2.1	名词定义	118

  9.2.2 足迹照片数据采集和分析方法 ······ 119
  9.2.3 圈养东北虎的足迹影像采集 ······ 119
  9.2.4 野生东北虎的足迹影像采集 ······ 121
  9.2.5 足迹几何轮廓的提取 ······ 122
  9.2.6 数据分析 ······ 122
 9.3 结果 ······ 123
  9.3.1 圈养东北虎的性别判别模型 ······ 123
  9.3.2 基于足迹识别技术对野生东北虎进行性别鉴定 ······ 124
 9.4 讨论 ······ 125
  9.4.1 采集和测量足迹 ······ 125
  9.4.2 雪地足迹数据的客观性 ······ 125
  9.4.3 样本量和建模方法对性别判别的影响 ······ 125
  9.4.4 管理启示 ······ 126
 9.5 本章小结 ······ 126
 参考文献 ······ 127

第10章 中国东北虎栖息地等级保护优先区域和廊道确定的适宜性模型技术 ······ 129
 10.1 引言 ······ 129
 10.2 研究方法 ······ 130
  10.2.1 数据收集与栖息地斑块确定 ······ 130
  10.2.2 栖息地廊道识别 ······ 134
  10.2.3 优化算法模型、适宜栖息地斑块与廊道内优先核心栖息地识别 ······ 134
 10.3 结果 ······ 135
  10.3.1 中国东北虎种群和栖息地分布 ······ 135
  10.3.2 东北虎栖息地模拟及最小需求面积 ······ 137
  10.3.3 等级优先保护区域确定 ······ 139
  10.3.4 潜在保护功能的比较 ······ 142
 10.4 讨论 ······ 144
 10.5 本章小结 ······ 145
 参考文献 ······ 146

第11章 东北虎种群和栖息地精准管理的生态阈值 ······ 148
 11.1 引言 ······ 148
 11.2 基于上行效应确定东北虎保护中的生态阈值 ······ 150

  11.2.1 气候和极端天气引起的阈值效应 ·· 150
  11.2.2 栖息地丧失和人为干扰引起的阈值效应 ······························· 151
  11.2.3 植被的生态阈值 ·· 153
  11.2.4 猎物或食草动物的生态阈值 ·· 153
  11.2.5 东北虎和东北豹的共存阈值 ·· 155
 11.3 未来生态阈值研究应用和面临的挑战 ··· 155
 11.4 本章小结 ·· 157
 参考文献 ·· 158

第 12 章　东北虎精准保护管理工作中的挑战、机遇与举措 ·················· 163
 12.1 引言 ·· 163
 12.2 顶层设计、统一规划 ··· 164
 12.3 专家咨询、科学规范 ··· 166
 12.4 创新举措、人虎和谐 ··· 168
 12.5 本章小结 ·· 169
 参考文献 ·· 169

Contents

Chapter 1 Environmental factors drive the habitat distribution of Amur tiger ·········· 171
1.1 Introduction ·········· 171
1.2 Materials and methods ·········· 172
   1.2.1 Study area ·········· 172
   1.2.2 Amur tiger population data ·········· 173
   1.2.3 Environmental and anthropogenic variables ·········· 174
   1.2.4 Spatial autocorrelation examination ·········· 176
   1.2.5 Statistical analysis ·········· 177
1.3 Results ·········· 178
   1.3.1 Changbaishan model ·········· 178
   1.3.2 Wandashan model ·········· 182
   1.3.3 Subregional comparison and quantitative responses ·········· 184
1.4 Discussion ·········· 186
   1.4.1 Effects of environmental factors ·········· 186
   1.4.2 Effects of anthropogenic factors ·········· 187
   1.4.3 Effects of spatial autocorrelation ·········· 188
   1.4.4 Conservation implications ·········· 189
1.5 Summary ·········· 190
References ·········· 190

Chapter 2 The synergy between ecology and society in Amur tiger conservation ·········· 195
2.1 Introduction ·········· 195
2.2 Methods ·········· 197
   2.2.1 Collection and analysis of forest management data ·········· 197
   2.2.2 Population size and habitat distribution of Amur tiger and Amur leopard ·········· 199
2.3 Results ·········· 202
2.4 Discussion ·········· 207
2.5 Summary ·········· 211
References ·········· 212

Chapter 3 An integrated analysis of the causes of ungulate mortality in the Wandashan Mountains (Heilongjiang Province, China) and an evaluation of habitat quality ···········215
- 3.1 Introduction ···········215
- 3.2 Study area and background ···········216
- 3.3 Materials and methods ···········217
  - 3.3.1 Mortality data ···········217
  - 3.3.2 Cause of mortality in the three ungulate species ···········217
  - 3.3.3 Survey design and data collection ···········218
  - 3.3.4 Analysis of habitat selection ···········218
- 3.4 Results ···········221
  - 3.4.1 Causes of mortality ···········221
  - 3.4.2 Habitat selection based on the occurrence and mortality locations of ungulate ···········222
  - 3.4.3 Habitat evaluation ···········222
- 3.5 Discussion ···········224
  - 3.5.1 Causes and percentage of mortality ···········224
  - 3.5.2 Habitat selection and quality evaluation ···········225
- 3.6 Summary ···········227
- References ···········227

Chapter 4 Effects of human disturbance on vegetation, prey and Amur tigers in Hunchun Nature Reserve, China ···········231
- 4.1 Introduction ···········231
- 4.2 Methods and techniques ···········234
  - 4.2.1 Study area ···········234
  - 4.2.2 Data collection ···········234
  - 4.2.3 Data analysis ···········236
- 4.3 Results ···········236
  - 4.3.1 Human disturbance on vegetation ···········238
  - 4.3.2 Human disturbance on prey ···········240
  - 4.3.3 Human disturbance on Amur tigers ···········241
- 4.4 Discussion ···········241
  - 4.4.1 Effects of grazing ···········241
  - 4.4.2 Effects of roads ···········243
  - 4.4.3 Effects of ginseng planting ···········243
  - 4.4.4 Results discrimination of GAMs, GLMs and SEMs ···········244
- 4.5 Conclusions ···········244
- 4.6 Summary ···········244

References ··· 245
Chapter 5 A comparison of food habits and prey preferences of Amur tiger at the southwest Primorskii Krai in Russia and Hunchun in China ········ 248
5.1 Introduction ·· 248
5.2 Materials and methods ··· 249
　5.2.1 Study area ·· 249
　5.2.2 Field research methods ·· 250
　5.2.3 Scat analysis ·· 251
　5.2.4 Data analysis ··· 251
5.3 Results ··· 252
5.4 Discussion ··· 256
5.5 Summary ··· 258
References ··· 258

Chapter 6 A comparison of reproductive parameters of female Amur tiger in the wild and captivity ··· 261
6.1 Introduction ·· 261
6.2 Materials and methods ··· 262
　6.2.1 Study population ·· 262
　6.2.2 Data analysis ··· 263
6.3 Results ··· 263
　6.3.1 Reproductive parameters of captive and wild Amur tigers population ··· 263
　6.3.2 Cub mortality ·· 264
　6.3.3 Seasonality of reproduction ·· 264
　6.3.4 Reproductive characteristics of females after losing cubs ······················· 267
6.4 Discussion ··· 267
6.5 Summary ··· 268
References ··· 269

Chapter 7 Innate preference for native prey and personality implications in captive Amur tigers ·· 270
7.1 Introduction ·· 270
7.2 Materials and methods ··· 271
　7.2.1 Experimental animals ·· 271
　7.2.2 Experimental apparatus ··· 271
　7.2.3 Stimuli ·· 272
　7.2.4 Prey recognition test ·· 272
　7.2.5 Personality assessment ·· 273
　7.2.6 Data analysis ··· 274
7.3 Results ··· 277

7.3.1 Preferential response to native prey visual stimuli ·············· 277
7.3.2 Preferential response to native prey auditory stimuli ·············· 279
7.3.3 Preferential response to native prey olfactory stimuli ·············· 280
7.3.4 Correlation between innate preference for native prey and personality ·· 281
7.4 Discussion ·············· 282
7.5 Summary ·············· 284
References ·············· 284

Chapter 8 Risks involved in fecal DNA-based genotyping of microsatellite loci in the Amur tiger a pilot study ·············· 287
8.1 Introduction ·············· 287
8.2 Materials and methods ·············· 288
  8.2.1 Samples ·············· 288
  8.2.2 DNA extraction ·············· 288
  8.2.3 Microsatellite data analysis ·············· 289
  8.2.4 Evaluation of genotyping correctness at each locus ·············· 290
  8.2.5 Evaluation of population genetic parameters using fecal samples ·············· 290
8.3 Results ·············· 291
  8.3.1 Evaluation of genotyping correctness at each locus ·············· 291
  8.3.2 Effects of genotyping error on estimation of population genetic parameters ·············· 291
8.4 Discussion ·············· 295
8.5 Summary ·············· 296
References ·············· 297

Chapter 9 Sex determination of Amur tiger from footprints in snow ·············· 299
9.1 Introduction ·············· 299
9.2 Materials and methods ·············· 300
  9.2.1 Definition of terms ·············· 300
  9.2.2 Footprint image data collection and analysis methods ·············· 301
  9.2.3 Footprint images collection from captive Amur tigers ·············· 302
  9.2.4 Collection of footprints images from free-ranging Amur tigers ·············· 303
  9.2.5 Extracting a geometric profile from digital images ·············· 304
  9.2.6 Data analysis ·············· 305
9.3 Results ·············· 306
  9.3.1 Sex discrimination model for captive Amur tiger ·············· 306
  9.3.2 Application in wild Amur tigers ·············· 307
9.4 Discussion ·············· 308
  9.4.1 Collecting and measuring footprints ·············· 308
  9.4.2 Data objectivity of footprint in snow ·············· 308

9.4.3 The effects of sample size and modelling methodology for sex discrimination ·········309
9.4.4 Management implications ·········309
9.5 Summary ·········310
References ·········311

Chapter 10 An adaptive model for determining hierarchical priority conservation areas ·········313
10.1 Introduction ·········313
10.2 Research methods ·········314
  10.2.1 Data collection and habitat patch determination ·········314
  10.2.2 Habitat corridor identification ·········318
  10.2.3 Optimization algorithm model, and identification of prioritized core habitat of suitable patches and corridors ·········319
10.3 Results ·········320
  10.3.1 Amur tigers population and habitat distribution in China ·········320
  10.3.2 Habitat modeling and MAR of Amur tiger ·········322
  10.3.3 Hierarchical priority conservation areas determination ·········324
  10.3.4 Potential conservation function comparison ·········328
10.4 Discussion ·········330
10.5 Summary ·········331
References ·········332

Chapter 11 Ecological threshold on precise management of Amur tiger population and habitat ·········334
11.1 Introduction ·········334
11.2 Possible ecological thresholds in Amur tiger conservation based on bottom-up effects ·········336
  11.2.1 Thresholds caused by climate and weather events ·········336
  11.2.2 Thresholds due to habitat loss and anthropogenic disturbance ·········338
  11.2.3 Thresholds of vegetation ·········340
  11.2.4 Thresholds of prey or herbivores ·········341
  11.2.5 Thresholds in coexistence of the Amur tiger and Amur leopard ·········342
11.3 Future challenges of threshold research and applications ·········343
11.4 Summary ·········346
References ·········346

Chapter 12 Challenges, opportunities, and measures for precise protection and management of the Amur tiger ·········351
12.1 Introduction ·········351
12.2 Top-level design and unified planning ·········352

12.3	Expert consultation and scientific norms	356
12.4	Innovative measures and human-tiger harmony	357
12.5	Summary	359
	References	359

第1章 东北虎栖息地分布的环境因子驱动

1.1 引　　言

　　人为干扰和环境因素是导致哺乳动物生存环境丧失的主要原因（Shenko et al., 2012），人类干扰压力指数可以很好地反映人为干扰对大型哺乳动物的威胁程度（Yackulic et al., 2011）。环境条件的空间变化会引起局部种群大小和其他种群参数的变化，影响其对人为干扰的承受能力（Lawton, 1993; Channell and Lomolino, 2000）。由于边缘种群比核心种群更容易受到影响（Yackulic et al., 2011），物种的分布往往是从分布的边缘开始缩减。东北虎（$Panthera\ tigris\ altaica$）是世界上最大的猫科动物（Kitchener and Yamaguchi, 2010），它们具有较强的适应性，在生态系统中扮演着重要的角色（Seidensticker, 1996）。在中国，东北虎的历史分布范围很大，包括从大兴安岭北部到燕山山脉南部，从乌苏里江东部到额尔古纳河西部的所有森林和草原地带（Ma, 2005；田瑜等, 2009）。长白山、完达山及大兴安岭和小兴安岭的森林区域曾是东北虎种群分布的核心区域（Kang et al., 2010）。在20世纪50年代中期，中国境内大约有200只东北虎，而俄罗斯的个体数量还不到50只（田瑜等, 2009）。在过去的60年里，东北虎在中国东北地区的分布区域急剧减少，到20世纪末仅有20只左右的个体生存在两个核心栖息地斑块，即老爷岭南部和完达山东部靠近中俄边境地区（Li et al., 2008; Zhou et al., 2008）。值得庆幸的是，过去10年中的多次调查数据显示，中国东北虎种群数量在经历了急剧减少后，已基本趋于稳中有升的态势（Li et al., 2008; Zhou et al., 2008）。更重要的是，在这两个栖息地斑块中仍然存在大片的森林，可以满足东北虎种群及其猎物的持续生存。与此同时，与我国毗邻的俄罗斯东北虎数量有所增加，现在已达500只左右（Carroll and Miquelle, 2006）。

　　Hemmer（1976）认为，中国是世界虎种群的发源地，遗传学研究也部分证实了这一假设（Luo et al., 2004）。如果现有的林区能够被很好地保持和管理，在中国当前这些分布区老虎种群的复壮将存在很大的可能性，从而在整个东北虎种群的恢复中发挥关键作用（Kang et al., 2010）。因此，迫切需要了解东北虎潜在适宜栖息地的分布，以及影响栖息地适宜性和老虎出现概率的各种因素，以保证野外保护实践工作的科学实施。

　　Li等（2010）和Hebblewhite等（2012）结合基于专家评分的老虎栖息地专

家评价模型（Luan et al.，2011）、基于邻近俄罗斯远东地区老虎个体出现和不出现（区分不出现和未调查到）数据的环境生态位因子分析模型（ENFA），以及资源选择函数模型（RSF）三种模型，通过这三种模型的区间预测，对中国长白山区域东北虎的重要潜在栖息地斑块进行了评估。我们认为，该研究基于邻近俄罗斯远东地区稳定的抽样调查数据来预测中国栖息地的适宜性，并没有利用中国境内东北虎的实际出现信息来验证预测结果的真实性，而且仅预测了中国境内两大主要东北虎栖息地斑块的其中之一。此外，前面提到的研究虽然确定了某些栖息地因素积极或消极的影响作用，但并没有对这些因素尤其是人为干扰的影响作用进行定量化研究，以致保护管理人员无法理解并有效地控制、减少和管理这些干扰因素。因此，迫切需要对我国东北虎的生境适宜性进行深入评价，并根据该地区老虎的实际出现情况，对威胁因素的影响程度进行定量化分析。此外，野生动物种群分布往往呈现出不同的群聚模式，对于像东北虎这样家域较大的物种，其栖息地斑块的分布情况可能造成在栖息地适宜性预测时采样的空间自相关导致的假重复问题（José and Mariana，2002），也是科学识别物种适宜栖息地急需解决的技术问题。

基于中国东北虎分布的实际情况，本研究假设中国东北虎的出现受邻近俄罗斯源种群的扩散、气候、地形、植被和人为干扰的影响，并假设应用空间模型技术进行东北虎种群栖息地适宜性的预测可能获得更符合生物学解释的结果。此外，本研究通过建立中国境内东北虎的现有信息分布与中俄边境距离的关系函数，量化老虎出现情况与环境或人为干扰因素之间的关系，以期得到一个可靠的栖息地适宜性等级分布图。

1.2 材料与方法

1.2.1 研究区域

根据 2000~2012 年中国境内的东北虎出现信息，把研究限定在 42.358°N~47.605°N、126.631°E~134.290°E 的 158 564km^2 范围内。其中包括长白山地区 107 604km^2（42.358°N~46.655°N、126.659°E~131.708°E），以及完达山地区 50 764km^2（45.027°N~47.605°N、130.257°E~134.277°E）（图 1.1）。长白山老虎种群分布区与俄罗斯滨海边疆区西南部的老虎种群分布区相连，完达山老虎种群分布区与俄罗斯锡霍特—阿林山脉地区相连。

研究地属温带大陆性季风气候。植被类型包括以红松、白桦、蒙古栎为优势种的针阔混交林、草地、灌木、人工针叶林及农田。大多数森林已被采伐过，许多低海拔森林已经转变为次生阔叶林（Hebblewhite et al.，2012）。研究区域的主要动物种类有：东北虎（*Panthera tigris altaica*）、东北豹（*Panthera pardus orientalis*）、

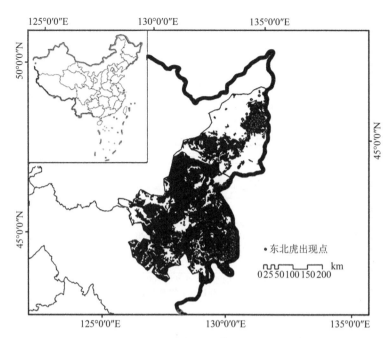

图 1.1 位于中国东北的研究区域示意图（黑色区域代表天然林覆盖区域）（彩图请扫封底二维码）

狼（*Canis lupus*）、欧亚猞猁（*Lynx lynx*）、马鹿（*Cervus elaphus*）、野猪（*Sus scrofa*）、梅花鹿（*Cervus nippon*）、狍（*Capreolus pygargus*）等。

1.2.2 东北虎种群数据

自 2000 年以来，森林管理部门和自然保护区的专家与工作人员通过实地调查 (Li et al., 2001; Zhou et al., 2008)、牲畜捕食情况调查 (Li et al., 2009) 及线人网络信息收集 (Zhang et al., 2012) 等渠道全面系统地收集了东北虎活动信息，另外，还记录了雪地、泥地东北虎个体前足掌垫宽度、出现点 GPS 坐标及老虎捕食有蹄类动物或家畜的时间等其他证明东北虎出现的信息。

为了揭示威胁东北虎的主要因素，并检验不同区域的生境差异，我们使用雌虎的平均日活动距离（即 7km）作为各区域东北虎占有模型中确定空间尺度的栅格宽度的一半 (Goodrich et al., 2005; Carroll and Miquelle, 2006)。由于数据大部分来源于森林采伐活动中当地职工的偶然发现，我们将每个栅格（14km×14km= 196km^2）2 周内多次的老虎出现信息作为 1 次记录，以将伪重复取样的风险最小化 (Hurlbert, 1984)。通过长期的数据收集，2000～2012 年共记录老虎出现信息 600 条，最终有 584 条有效信息用于本研究，然后统计出现频率，以及确定每个栅格内是否有东北虎出现，并将取样单元分成长白山和完达山两个区域分别进行分析。

1.2.3 环境变量和人为干扰变量

空间分布模型必须考虑到与物种分布相关的栖息地因子的不同特征尺度（Latimer *et al.*，2006）。本研究考虑了 5 类景观尺度的栖息地变量：气候、俄罗斯虎种群、植被、人为干扰、地形和河流（表 1.1）。气候变量[即 2000～2010 年平

表 1.1 用于分析中国东北地区东北虎出现状况的相关栖息地变量

生境因子	生境因子描述	数据类型	单位
气候			
雪深	2000～2010 年平均年最大雪深（源数据分辨率为 25km）	连续	cm
俄罗斯虎种群			
到中俄边境的距离	每个栅格中心点（14km×14km）到通过最近 10 年的调查而确定的东北虎跨境移动经常利用区域的中俄边境线的距离。一段位于完达山，另一段位于长白山	连续	m
植被			
林型	每个栅格内各种森林类型的面积。通过将 2000 年中国植被矢量图（1∶1 000 000）转换为 1km 分辨率的栅格数据进行计算	连续	km^2
针叶林	每个栅格内的针叶林面积。数据来源于转换的分辨率为 1km 的栅格图	连续	km^2
阔叶林	每个栅格内的阔叶林面积。数据来源于转换的分辨率为 1km 的栅格图	连续	km^2
人为干扰			
农田	农田在每个栅格中所占的面积比例	连续	%
道路密度	每个栅格内的道路总长度。道路分为一级公路、二级公路、改进型轻质铺路、改进型轻质碎石路、改进型轻质土路	连续	km/km^2
到铁路的距离	各栅格中心点到铁路的距离	连续	m
到村庄的距离	各栅格中心点到村庄的距离	连续	m
到城市的距离	各栅格中心点到城市（包括小城镇）的距离	连续	m
地形和河流			
海拔	分辨率为 1km 的海拔栅格数据	连续	m
海拔标准差	栅格内海拔的标准差。数据来源于分辨率为 1km 的海拔栅格数据	连续	m
坡度	从上面的数字高程图获得的 1km 分辨率的坡度栅格图	连续	(°)
坡向	从上面的数字高程图获得的 1km 分辨率的坡向栅格图	连续	
隐蔽级	隐蔽级不仅表示一个或多个观察者可以看到一个表面的哪些区域，还表示对于任何可见的位置有多少观察者可以看到这个位置。这个位置即每个栅格的中心	连续	%
河流密度	每个栅格内的河流总长度，包括主要河流及其支流	连续	km
空间自相关	空间自相关是通过"空间"模型检测的。每个样本单位观测值（0，1）在 342km（在长白山+完达山）、162km（在长白山）和 94km（在完达山）的参考样本单位，通过其与样本单位的反平方距离加权，分别以 342km、162km、94km 区域内所有样本单元的权重之和归一化	连续	%

注：数据处理采用 ArcGIS 9.0 栅格格式（14km×14km 分辨率），基于重采样或插值测量。数据来源于中国西部环境与生态科学数据中心（EESDCWC）（网址：www.resdc.cn）、国家自然科学基金（NSFC）和黑龙江流域信息中心（ICAHRB）（网址：http://amur-heilong.net/）数据库

均年最大雪深（积雪深度）]来源于中国历史雪深数据集（1978～2010 年），通过重采样将其分辨率由 25km 调整为 14km（http://www.datatang.com/dataset）。将每个栅格中心到中俄边境的距离作为源种群参数，表示该东北虎种群与俄罗斯虎种群通过迁入或迁出联系起来的位置。由于植被类型的差异会影响东北虎的出现（Carroll and Miquelle，2006），特根据中国东北地区典型的植被类型，将植被划分成 3 种：阔叶林、针叶林、针阔混交林；由于针阔混交林所占比例较小，本研究未考虑。与农田或其他土地类型相比，森林是东北虎最为偏好的生境，因此将是否是森林也视为一个变量。所有森林类型的矢量数据都被转换为 1km 分辨率的栅格数据。东北虎栖息地内有蹄类猎物的数据是非常重要的（Karanth et al.，2004）。因此，考虑到在研究区域内获得可靠的猎物密度数据的难度，我们将每个栅格内森林面积比例作为表示森林有蹄类状况的参数。人为干扰因素包括：以每个栅格内农田面积比例表示的农田侵占森林的压力指数变量；每个栅格的中心点离最近的铁路、村庄和城市的距离。偷猎可能是影响东北虎分布的一个重要因素，但我们无法获取空间数据来体现这一点，人为干扰如道路通常与偷猎者的可进入性有关（Bennett and Robinson，2000；Kerley et al.，2002）。所有距离的测量都是通过 ArcGIS 9.0 软件中的 GIS 空间分析软件包完成的。地形和河流变量（海拔、海拔标准差、坡度、坡向和隐蔽级）通过分辨率为 1km 的数字高程图及河流分布的 GIS 矢量数据获得。

1.2.4　空间自相关检测

东北虎出现的分布格局受其空间扩散和行为的影响（Goodrich et al.，2005），这种种群的聚群分布可能存在空间自相关问题（José and Mariana，2002；Haining，2003）。东北虎的分布信息具有局部的聚群性，其分布模式反映了东北虎的移动特征对当前中国境内东北虎分布的影响。在分析这样的空间分布模式时，存在一种风险，即独立取样的假设将被空间自相关所推翻（Koenig，1999），因为相邻生境栅格的变量可能具有相似的值。在选择最佳模型后，使用 SAM 版本 4.0（Rangel et al.，2010）计算模型残差的 Moran's I 统计量（Cliff and Ord，1981），以测试因变量中的空间自相关。为了衡量聚类的大小，创建了一个基于相邻栅格内东北虎的出现与否的空间自协变量（SA）（Beard et al.，1999；Segurado and Araujo，2004；Jiang et al.，2009）。自协变量是为了对模型残差发生空间自相关的距离范围进行可视化模拟（Augustin et al.，1996），是通过计算栅格中心点两两之间到参照栅格的距离的平方的倒数获得的，并将以 Moran's I 值检验确定的显著自相关距离作为半径内的所有栅格的权重总和进行标准化（Davis et al.，2007；Jiang et al.，2009）。

在本研究中，当根据适用性对在不同区域选择的最佳预测模型进行排序和选择时，对拟合结果显示的空间自相关问题进行诊断，然后应用随机检验对残差中的空间结构进行检验（Rangel et al., 2010）。如果残差中存在显著的空间自相关，使用参数 SA 作为没有显著空间结构的残差数据来进行建模，从而，在 196km^2 的栅格尺度下，对无论是否有空间自回归项的基于环境变量的东北虎出现数据都进行了建模，应用这两种模型，分别比较了长白山和完达山区域东北虎的生境差异，并预测了东北虎的空间分布，还对各区域建立的模型在预测东北虎生境适宜性方面的准确性进行了验证。最后，对空间模型（有 SA）与非空间模型（没有 SA）拟合的稳健性进行了比较，同时建立了不同分区生境适宜性的预测等级概率图，并根据近年来记录的东北虎出现数据，对观察到的频率等级图进行了比较与核实。

1.2.5 统计分析

完达山和长白山地区都有着不同的环境、人为干扰和气候条件（陈九屹等，2011；Zhou and Zhang, 2011），这可能导致虎栖息地选择特征的差异（Goodrich et al., 2010）。Davis 等（2007）建议，对这些区域分别进行栖息地建模，可以揭示环境或人为干扰的区域性差异。因此，本研究将完达山和长白山作为独立的研究区域进行建模分析，共在 808 个栅格内记录到东北虎出现信息，其中长白山区域有 549 个，完达山区域有 259 个。该模型的抽样设计是参照 Manly 等（2002）的方法使用二次采样，该设计的局限性在于它在统计上仅给出栖息地适宜性排序的相对概率。没有记录到老虎信息的栅格被视为伪缺失。为了选择出会显著影响东北虎生境选择的变量并放入模型分析，基于出现与伪缺失数据利用 Wilcoxon 秩和检验，对生境变量与东北虎出现之间进行两两相关分析。使用标准化转换方法对高度偏态分布的变量进行标准化。所有检验，在 10 000 次蒙特卡罗采样中，$P<0.05$ 时认为差异显著。随后，将所有显著变量输入 Pearson 相关矩阵以检测变量的共线性（即 $r>0.5$）（Ramsay et al., 2003）。因为广义可加模型（GAM）可以提供与原始数据接近的拟合结果并输出比同等的具备多个多项式的广义线性模型（GLM）更可靠的结果，GAM 中的非线性函数部分可以很好地表现数据特征中物种对环境的真实响应（Suarez-Seoane et al., 2002）。因此，使用非线性广义可加模型模拟东北虎存在与量化的栖息地因子间的函数关系（Hastie and Tibshirani, 1990；Guisan et al., 2002）。为了减少 GAM 中共线性的潜在影响，在模型选择时对具有共线性的变量的数量进行了限制。之后，使用逐步筛选的方法获得多变量 GAM（Hastie and Tibshirani, 1990；Guisan et al., 2002）。对于各分区的数据，栖息地变量的选择都是通过上述方法获得的。所有分析均利用 R 软件 2.12.1 版本中 MGCY 包 1.7-2 版本（Wood，2006；R Development Core Team，2010）完成。通过最小化广义交叉验证值（表示模型对测试样本预测能力的均方差的指数，GCV）来确定平滑曲线的最佳粗糙度（Stige et al.，2006）。按照 GCV 最小化的原则，在所有变量都具有统计学意义（$P<0.05$）的条件下，从候选模型中选择最适合的模型。

通过使用受试者操作特征曲线（receiver operating characteristic curve，ROC 曲线）下的面积（AUC）来评估模型的判别能力（Altman and Bland，1994；Fielding and Bell，1997）。使用拟合模型的 5 重交叉验证来测试模型的稳健性，即将出现和不出现数据分成 5 等份，保留其中 1 份用于测试，并使用剩余的数据进行模型的拟合。如此重复 5 次并计算 5 次试验的平均性能（Davis et al.，2007），分别比较了模型 AUC 和交叉验证 AUC（CV AUC）。此外，基于 AUC 分析，检验了最佳空间模型或非空间模型对不同区域东北虎出现情况预测的分类结果。然后，利用最佳空间模型来量化和绘制东北虎出现对环境与人为干扰的响应曲线，并分别确定了 2 个区域内该非线性响应的生态阈值。通过线性回归确定空间模型或非空间模型预测的东北虎出现的相对概率与实际观测频率的关系，以此检验考虑生物学意义条件下的空间模型是否比非空间模型具有更好的拟合效果（Granadeiro et al.，2004）。

1.3 结　　果

整个调查期间，东北虎在整个地区 12.25%（99/808）的栅格内都有出现，包括长白山 11.66%（64/549）的栅格和中国东北的完达山 13.51%（35/259）的栅格。有东北虎活动的栅格距离中俄边境最远的达 293km。完达山与长白山区域内有东北虎活动的栅格间的最小距离为 60km，小于成年雄虎活动范围的最大直径（Goodrich et al.，2010）。

1.3.1　长白山模型

在长白山区域，坡度、海拔、道路密度、积雪深度（雪深）、海拔标准差、到中俄边界的距离及到铁路的距离等变量都与东北虎的活动显著相关（Wilcoxon Z，$P<0.05$）（表 1.2）。

在这些相关的栖息地变量中，东北虎的出现与林型、阔叶林或农田没有显著关系（Wilcoxon Z，$P=0.143$）。然而，研究发现，坡度与海拔（+）、坡度与海拔标准差（+），以及海拔与海拔标准差（+）（$r>0.5$）之间存在显著的相关性（表 1.3）。

表 1.2 基于 10 000 次蒙特卡罗采样的 Wilcoxon U 检验，东北虎出现与环境因子之间相关性的单变量检验

变量	长白山（$n=64$）		完达山（$n=35$）	
	Wilcoxon Z	P	Wilcoxon Z	P
隐蔽级	−0.342	0.737	**−4.88**	**<0.001**
坡向	−0.989	0.327	−0.575	0.571
坡度	**−1.977**	**0.047**	**−4.811**	**<0.001**
海拔	**−2.023**	**0.043**	**−3.261**	**0.001**
针叶林	−0.678	0.501	**−3.026**	**0.003**
阔叶林	−0.713	0.475	**−4.839**	**<0.001**
林型	−0.47	0.633	**−5.558**	**<0.001**
农田	−0.345	0.73	**−5.24**	**<0.001**
道路密度	**−3.596**	**<0.001**	**−4.668**	**<0.001**
河流密度	−0.165	0.867	**−3.523**	**<0.001**
积雪深度	**−5.272**	**<0.001**	**−5.695**	**<0.001**
海拔标准差	**−2.143**	**0.031**	**−4.991**	**<0.001**
到城市的距离	−0.132	0.895	−0.507	0.618
到中俄边境的距离	**−8.991**	**<0.001**	**−6.789**	**<0.001**
到村庄的距离	−1.777	0.075	**−3.2**	**0.001**
到铁路的距离	**−4.518**	**<0.001**	−1.461	0.142

注：粗体表示显著相关

表 1.3 应用于东北长白山区域东北虎生境模型的生境变量之间的相关性

	隐蔽级	坡向	坡度	海拔	针叶林	阔叶林	林型	农田	道路密度	河流密度	积雪深度	海拔标准差	到城市的距离	到中俄边境的距离	到村庄的距离	到铁路的距离
坡向	0.08	1														
坡度	0.16	−0.08	1													
海拔	0.08	−0.11	**0.66**	1												
针叶林	0.16	0.03	−0.05	0.01	1											
阔叶林	−0.13	−0.02	0.36	0.24	−0.39	1										
林型	0.03	−0.06	0.49	**0.54**	0.17	**0.61**	1									
农田	−0.08	0.06	−0.44	−0.48	−0.18	**−0.55**	**−0.92**	1								
道路密度	−0.07	0.02	−0.23	−0.14	−0.06	−0.24	−0.28	0.29	1							
河流密度	0.17	−0.03	−0.27	−0.24	0.08	−0.19	0.16	0.22	1							
积雪深度	−0.12	0.15	−0.26	−0.12	−0.15	−0.11	−0.22	0.25	0.17	−0.12	1					
海拔标准差	0.21	−0.02	**0.95**	**0.62**	−0.04	0.33	0.47	−0.43	−0.25	−0.26	−0.25	1				
到城市的距离	0.12	−0.06	0.31	0.42	0.03	0.17	0.3	−0.29	−0.21	−0.16	0.08	0.29	1			
到中俄边境的距离	0.13	0.13	−0.19	−0.27	−0.01	−0.17	−0.22	0.2	0.03	−0.03	0.49	−0.16	0.06	1		
到村庄的距离	0.17	−0.01	0.38	0.47	0.01	0.19	0.34	−0.33	−0.25	−0.16	−0.13	0.36	0.25	−0.1	1	
到铁路的距离	0.16	−0.03	−0.12	−0.2	0.15	0.00	−0.01	−0.08	−0.44	0.07	−0.24	−0.11	−0.03	−0.19	0.01	1

注：加粗表示显著相关

选择这些显著相关的变量,然后逐一比较这些两两显著相关的变量对模型的贡献性。从而,将海拔、雪深、到中俄边境的距离和到铁路的距离一并放入 GAM 中,作为最佳模型(偏差解释为 29.96%,GCV 得分为 0.0746)(表 1.4),当空间尺度大于 70km(5 个栅格)时模型残差(观测–预测)表现出显著的正空间自相关(图 1.2a),主要是因为该模型对长白山东部的东北虎出现情况进行了过度预测,产生了负残差(图 1.3d)。添加空间自协变量后,交叉验证 AUC 从 0.834 增加到 0.997,说明空间模型分类能力获得了改善,并且消除了模型残差的显空间自相关(表 1.5,图 1.3e)。

表 1.4 模型选择过程中得到的 5 个最佳模型

地区		组合模型	偏移值	GCV 值
长白山	1	海拔+雪深+到中俄边境的距离+到铁路的距离	0.2996	0.0746
	2	雪深+到中俄边境的距离+到铁路的距离	0.2945	0.0749
	3	到中俄边境的距离+到铁路的距离	0.2820	0.0755
	4	雪深+到中俄边境的距离	0.2787	0.0761
	5	到中俄边境的距离	0.2679	0.0765
完达山	1	隐蔽级+海拔+农田+道路密度+雪深	0.3480	0.0817
	2	隐蔽级+海拔+农田+雪深+到村庄的距离	0.3509	0.0818
	3	隐蔽级+海拔+农田+雪深	0.3327	0.0832
	4	隐蔽级+海拔+道路密度+雪深+到村庄的距离	0.3312	0.0835
	5	隐蔽级+海拔+道路密度+河流密度+雪深	0.3331	0.0843

注:采用模型逐步程序广义交叉验证,用 5 个最佳模型估算了东北虎在中国东北南部(长白山)和北部(完达山)的出现概率

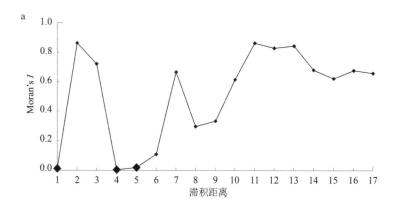

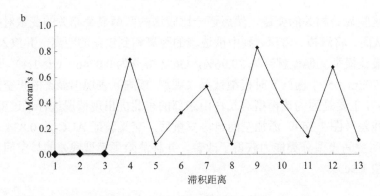

图 1.2 基于非空间模型的广义可加模型（GAM）残差的 Moran's I 空间自相关图

横坐标显示位点两两之间的距离。模型包括长白山（a）、完达山（b）。◆表示 Moran's I 显著（$P<0.05$）。距离用点之间的网格数表示

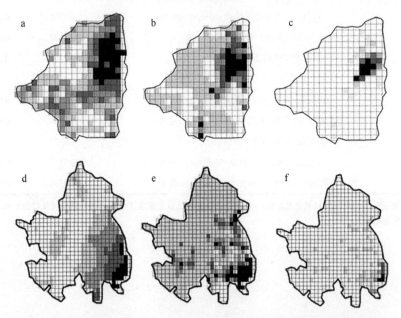

图 1.3 东北虎出现频率和广义可加模型（GAM）预测的出现概率结果图

较暗的方块（14km×14km）表示该区域观测或预测的出现概率较高。a. 基于非空间模型的完达山区域预测概率；b. 基于空间模型的完达山区域预测概率；c. 完达山区域实际观测频率；d. 基于非空间模型的长白山区域预测概率；e. 基于空间模型的长白山区域预测概率；f. 长白山区域实际观测频率

1.3.2 完达山模型

完达山仅坡向、到城市的距离和到铁路的距离与东北虎的出现无显著相关（Wilcoxon Z, $P>0.05$），其他 13 个生境变量则与东北虎是否出现显著相关（表 1.2）。

表 1.5　基于 2000~2012 年监测数据得到的用于预测东北虎出现和不出现的最佳空间广义可加模型与非空间广义可加模型的总结

模型和区域数据	n（比例）	非空间广义可加模型			空间广义可加模型	
		变量	AUC	CV AUC	AUC	CV AUC
长白山	64（11.66%）	海拔、雪深、到中俄边境的距离、到铁路的距离	0.835	0.834	0.998	0.997
完达山	35（13.51%）	隐蔽级、海拔、农田、道路密度、雪深	0.895	0.895	0.998	0.998

注：AUC 表示曲线下面积；CV AUC 表示 5 重交叉验证的平均 AUC 值

研究发现，坡度与海拔（+）、阔叶林（+）、林型（+）及海拔标准差（+）；海拔和阔叶林（+）、林型（+）及海拔标准差（+）；阔叶林和林型（+）、农田（−）及海拔标准差（+）；林型和农田（−）及海拔标准差（+）；到中俄边境的距离和道路密度（+）、雪深（−）；到铁路的距离和道路密度（−）之间存在显著的相关性（$r > 0.5$）（表 1.6）。

表 1.6　应用于东北完达山区域东北虎生境模型的生境变量间的相关性

	隐蔽级	坡向	坡度	海拔	针叶林	阔叶林	林型	农田	道路密度	河流密度	积雪深度	海拔标准差	到城市的距离	到中俄边境的距离	到村庄的距离	到铁路的距离
隐蔽级	1															
坡向	0.08	1														
坡度	0.49	0.14	1													
海拔	0.36	0.16	**0.90**	1												
针叶林	0.09	0.08	0.34	0.40	1											
阔叶林	0.34	0.22	**0.74**	**0.72**	0.13	1										
林型	0.38	0.24	**0.81**	**0.78**	0.30	**0.95**	1									
农田	−0.24	−0.03	−0.49	−0.47	−0.21	**−0.61**	**−0.64**	1								
道路密度	−0.08	0.04	−0.08	0.02	0.07	−0.21	−0.20	0.46	1							
河流密度	−0.13	−0.09	−0.31	−0.25	−0.22	−0.23	−0.28	0.23	0.14	1						
积雪深度	0.19	−0.02	0.09	−0.09	−0.03	0.03	0.06	−0.16	−0.20	−0.04	1					
海拔标准差	0.48	0.13	**0.97**	**0.85**	0.32	**0.74**	**0.81**	−0.45	−0.08	−0.29	0.10	1				
到城市的距离	0.01	0.03	0.26	0.37	0.19	0.27	0.28	−0.35	−0.22	−0.24	0.03	0.22	1			
到中俄边境的距离	−0.06	0.09	0.18	0.41	0.02	0.15	0.17	**0.50**	−0.10	**−0.65**	0.16	0.12	1	1		
到村庄的距离	0.17	−0.12	0.38	0.31	0.05	0.29	0.32	−0.42	−0.31	−0.23	0.24	0.30	0.28	−0.28	1	
到铁路的距离	−0.17	−0.19	−0.10	−0.18	−0.16	0.00	−0.04	−0.30	**−0.68**	−0.05	0.01	−0.10	0.04	−0.43	0.21	1

注：加粗表示显著相关

选择这些两两之间显著相关的变量，逐个比较每个变量对模型的贡献大小。从而，将海拔、针叶林、农田、雪深和到村庄的距离输入 GAM 模型，并最终获得以隐蔽级、海拔、农田、道路密度和雪深为解释变量的最佳模型（偏差解释

34.80%，GCV 得分 0.0817)（表 1.4）。当空间尺度大于 42km（3 个栅格）时模型残差（观测-预测）表现出显著的正空间自相关（图 1.2b），主要是因为模型对完达山西部和南部东北虎出现过度预测，产生了负残差（图 1.3a）。添加空间自相关变量后，交叉验证 AUC 从 0.895 提高到 0.998，分类能力得到提高，模型残差的自相关系数显著降低（表 1.5，图 1.3b）。

1.3.3 分区比较和定量化响应

在单变量检验中发现同一生境变量在不同区域对东北虎出现的影响不同（表 1.2）。例如，隐蔽级在完达山区域与东北虎的出现显著相关，而在长白山区域与东北虎的出现则不相关。有趣的是，东北虎出现的相对概率对雪深的响应在两个分区完全相反：长白山区域东北虎对雪深的响应是正相关的，而在完达山区域东北虎对雪深的响应则是负相关的（图 1.4a、b）。

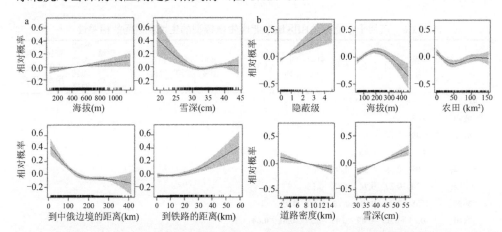

图 1.4 中国东北长白山地区基于地理（海拔）、气候（雪深）、与俄罗斯虎种群的关系（到中俄边境的距离）及到铁路的距离等变量的广义可加模型（GAM）的部分概率相应曲线（a）和中国东北完达山地区基于地理（隐蔽级、海拔）、气候（雪深）、农田、道路密度等变量的 GAM 的部分概率相应曲线（b）

横坐标表示模型中相互独立的变量的值，纵坐标表示变量对非参数 GAM 平滑度方程的可叠加的贡献概率。阴影部分表示估计模型的置信区间

根据模型结果，在完达山区域，东北虎更喜欢隐蔽级高的地区（图 1.4b）。中俄边境东北虎种群向边境线 150km 范围内的长白山区域扩散明显（图 1.4a）。距铁路 15km 以内的地区东北虎的出现明显减少（图 1.4a）。农田侵蚀和道路密度对完达山区域东北虎的出现有消极影响（图 1.4b）。当完达山区域农田面积比例为每个栅格内 50km^2 时，老虎的相对出现概率最低（图 1.4b）。

不同区域检测到的空间自相关距离不同（图 1.2a、b）。当去除这些最佳模型的空间自相关效应时,两个区域的最佳模型的稳健性得到了改善(表1.5,图1.3)。此外,研究发现,空间模型的相对预测概率与观察到的东北虎出现的相对频率呈显著的线性相关(完达山区域,$R^2 = 0.7042$ vs 0.3693;长白山区域,$R^2 = 0.7212$ vs 0.4945)(图1.5a、b)。因此,对于这两个区域,考虑到与实际观测频率的关系,空间模型的预测概率要优于非空间模型。

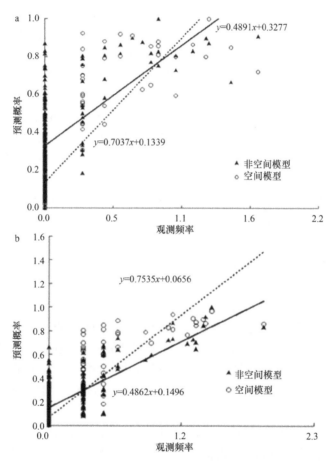

图1.5　完达山区域(a)和长白山区域(b)基于空间模型或非空间模型的预测概率与观测频率间的线性相关性

x轴的观测频率是经过$\ln(n+1)$转换后的数值。空间模型和非空间模型中R^2值较高的模型方程分别在其拟合直线附近进行了显示。虚线表示空间模型

1.4　讨　　论

1.4.1　环境因素的影响

考虑到俄罗斯东北虎种群作为我国东北虎恢复的资源种群,研究发现,在

150km 范围内，到中俄边境的距离与东北虎出现栅格数量之间存在显著的响应关系（图 1.4a，表 1.4）。此外，对中国东北虎雪地足迹的跟踪研究和俄罗斯远东地区信息都显示出，一些野生东北虎个体共享着中俄边境地区的这片栖息地。因此，在今后的保护工作中，在东北虎分布的中俄边境地区，开放或建立国际生态廊道是维持东北虎数量稳定和增长的关键任务。

研究表明，隐蔽级和海拔是影响东北虎出现的两个关键环境因素。东北虎出现的地区有较高的隐蔽级和海拔，这可能与东北虎对有蹄类及其他猎物的捕食策略有关（Karanth and Sunquist，2000；Valeix et al.，2009），也可能是东北虎对人类干扰的一种回避［详见 Jiang 等（2009）中的驼鹿栖息地选择策略］。在完达山区域，东北虎更偏好 150~250m 的低海拔区域。本研究得到的最适海拔的中值低于根据经验认为的长白山区域 400~800m 的最适海拔（Hebblewhite et al.，2012）。

Heptner 和 Sludskii（1992）的研究表明，东北虎无法在深雪中猎食，已有由冬季持续的厚积雪导致东北虎被饿死的案例。雪深和积雪残留的时间都可能会影响老虎的死亡率（Kitchener and Dugmore，2000）。实际上，不同的雪深会直接影响东北虎的猎物（Pauley et al.，1993；Jiang et al.，2008，2010）。例如，当积雪深度超过 40cm 时，白尾鹿（Odocoileus virginianus）的基础代谢率会降低（Pauley et al.，1993）。姜广顺等（2008）认为，相比于马鹿，体型较小的狍可能更喜欢积雪少的区域，因为马鹿体型较大，在较深的积雪中活动更自由，消耗的能量更少。在本研究区，梅花鹿是长白山东部猎物群落中较为丰富的物种（陈九屹等，2011），体型较小，喜欢在雪浅的区域活动。而在完达山地区，马鹿或野猪是丰富度较高的物种，能够适应较深的积雪（Zhou and Zhang，2011）。此外，完达山地区纬度较高，积雪时间较长且积雪较深。我们的研究结果表明，在长白山地区，年最大积雪深度超过 30cm 时，会降低东北虎出现的概率（图 1.4a），但在完达山地区，雪深却对东北虎的出现有积极影响（图 1.4b）。Miquelle 等（1999）的研究表明，在俄罗斯远东地区，没有证据表明老虎会回避深雪区域，但是在北方有蹄类动物群落中，捕食和积雪经常被认为是致死因素（Jędrzejewski et al.，1992）。因此，我们认为积雪影响了不同猎物群落的分布，导致不同区域东北虎对积雪深度的反应不同。然而，中国境内积雪影响下的猎物群落与东北虎的相互作用还有待进一步研究。

1.4.2 人为干扰的影响

人为干扰可能直接导致野生动物栖息地的丧失和回避行为的增加，进而改变其空间分布（Stevens and Boness，2003；Jiang et al.，2007，2009；Proulx and MacKenzie，2012；Mushtaq et al.，2013）。东北虎会回避人为干扰（田瑜等，2009；

Kang et al., 2010），而这种回避使得一些看似适宜的栖息地被放弃而导致栖息地的进一步丧失（Mace and Waller, 1996; Stevens and Boness, 2003; Johnson et al., 2005）。一些专家认为，大多数野生动物会在距离人为干扰 500m 至 5km 的阈值之外的区域活动（Mahoney and Schaefer, 2002; Frid, 2003; Jiang et al., 2009, 2010）。然而，目前的研究大多采用广义线性模型（Mahoney and Schaefer, 2002; Frid, 2003; Jiang et al., 2009, 2010），通过对模型预先定义几个多项式来确定野生动物规避人类干扰的阈值距离（Suarez-Seoane et al., 2002），其结果并不可靠。本研究结果表明，人为干扰因素对老虎出现的影响是非线性的。当每个栅格内的农田面积超过 $50km^2$ 时，东北虎出现的概率急剧下降，表现出明显的回避（图 1.4b）。在长白山地区，在距离铁路 15km 范围内，老虎的出现受到明显抑制（图 1.4a）。这些 GAM 的响应曲线反映了东北虎回避人为干扰的阈值距离。此外，完达山地区道路密度的增加对东北虎的出现具有显著的负线性效应（图 1.4b）。Kerley 等（2002）认为，俄罗斯公路的建设使一个连续分布的东北虎种群变为两个隔离种群，以致无法与其他种群进行遗传交流。对于其他物种，道路引起的交通事故增加了猞猁的非正常死亡率，对整个猞猁种群造成了消极的影响（Kramer-Schadt et al., 2004），另外道路会限制狼在其栖息地间的移动（Whittington et al., 2004）。此外，道路为农民提供了更多的交通和生产上的便利（Linkie et al., 2004），为偷猎者提供了更多进入动物栖息的偏远区域的机会（Bennett and Robinson, 2000）。

1.4.3　空间自相关效应

空间自相关（SA）是生态学家需要注意的一个非常具有挑战性的问题（José and Mariana, 2002; Haining, 2003）。Dormann（2007）认为，应同时考虑空间模型和非空间模型，以帮助比较合并空间自相关对模型预测的影响。虽然 SA 对分布模型的影响仍存在广泛争议（Dormann et al., 2007），但由于东北虎出现数据具有空间自相关，因此它是我们的研究中要考虑的一个重要问题。具有显著空间自相关的距离是否反映了不同区域的差异，或者空间模型和非空间模型预测的栖息地适宜性之间是否存在差异？我们的结果表明，空间模型的预测概率与观测频率的关系比非空间模型更为密切，证明了空间模型能够更好地解释数据的生物学意义（图 1.5）。此外，不同尺度的区域模型反映了焦点区域东北虎出现点的不同聚群模式，这可能是由于我们使用了出现/伪缺失数据对模型进行校准，也可能是由于在许多明显适宜老虎栖息的区域没有收集到老虎出现信息。事实上，当生境适宜性越高，未出现点的绝对残差自然越高，相反，出现点的残差则随预测值增大而减小（Lyet et al., 2013）。研究结果使我们对长白山东部和完达山北部区域生境的预测结果更加可靠。然而，对于长白山西部地区模型的解释需要更加谨慎，那些出现信息频次较少或只出现一次的区域明显有游荡的东北虎家族或个体。因此，由于附近老虎种群的缺失和扩散能力的限制，长白山西部作为东北虎的潜在生境的适宜性可能被低估。

目前，很少有研究关注空间模型偏差是否比非空间模型更小（Dormann，2007），但这个问题确实值得进一步研究，因为模型预测很大程度上依赖于因模型结构引起的偏差（Reineking and Schröder, 2006）。Augustin 等（1996）和 Betts 等（2009）认为，自协变量模型最适合用于生态学中的预测，但在推理研究中的应用并非必要。在本研究中，比较了不同分区尺度下的模型对生境适宜性的预测，发现与非空间模型相比，利用空间模型预测的生境适宜性更符合生物学意义。

1.4.4 保护意义

本研究首先定量分析了中国东北地区环境和人为干扰对东北虎出现概率的影响，这些量化的研究结果对于采取切实措施恢复东北虎栖息地具有重要意义。因此，应该考虑如何通过减少人为干扰来恢复东北虎的核心分布区，如封闭道路、退耕还林，以及为铁路交通修建隧道或桥梁以建设智能绿色的基础设施（Quintero et al., 2010; Wikramanayake et al., 2011）。2012 年 12 月 28 日，对东北虎雪地足迹的跟踪发现，东北虎通过公路隧道向长白山西部地区活动，这表明，生态友好型基础设施的设计可能在协调未来人类发展和东北虎共存方面发挥关键作用（Quintero et al., 2010）。在野外保护实践中，为了减轻人为干扰的负面影响，应注重通过建立保护区网络，构建中俄边境之间及破碎的生境斑块之间的生态廊道（Tian et al., 2011）。

此外，建模工具的使用对于识别威胁因素和新的潜在栖息地至关重要（Li and Wang, 2013）。虽然多尺度建模的应用更普遍（Riitters et al., 1997），但针对大尺度的保护，同样应该考虑空间模型和多分区尺度方法。空间模型和多分区尺度方法可以为濒危物种栖息地预测提供新的见解，应根据目标动物的生态学特征和未来实际保护要求使用。

尽管本研究提供了有用的保护信息，但本研究的数据收集并没有基于系统调查法进行动物真实利用-未利用检测的抽样设计。因此，未来的研究，必须要克服该调查设计中的固有缺陷和弱点，我们应该考虑不同动物种群的分布模式，然后采用不同的方法来收集必要的信息。例如，在出现频率高、生境适宜性高的区域，应选择系统样线调查法进行调查，因为可能存在定居老虎；只有游荡个体信息的大范围区域可以作为信息网络，随机地监测其分布动态，如 Duchamp 等（2012）针对狼的网络监测系统。虽然随机收集的信息有其无法克服的缺陷，但它仍然提供了关于一些稀有、难以发现和游荡的动物种群分布的重要信息，以便开展监测和栖息地评估。

1.5 本章小结

基于 2000～2012 年收集的野生东北虎分布信息数据，以及环境和人为干扰等变量数据，我们建立了中国东北地区东北虎适宜栖息地的分布模型，定量化研究了环境和人为干扰对大型哺乳动物栖息地丧失的影响。不同分区的种群分布模型显示了影响东北虎种群分布格局的生境因子的区域性差异。在农田覆盖面积超过 $50km^2$/栅格（$196km^2$）的区域、长白山地区距离铁路 15km 以内区域，以及完达山区道路密度（每个栅格内长度）增加的情况下，东北虎出现的相对概率呈现明显的下降趋势；然而，距离中俄边境 150km 范围内，东北虎的出现概率相对较高，研究也显示了东北虎分布对海拔、雪深和隐蔽级的回避或偏好性选择。此外，由于不同的种群分布模式，不同分区模型检测到的空间自相关距离也各不相同。结果表明，空间模型的拟合效果比非空间模型有显著的改善，对栖息地的适宜性预测也比非空间模型更准确。这些发现可为今后栖息地保护和管理提供有益的参考。

参 考 文 献

陈九屹, 那顺得力格尔, 孙全辉, 等. 2011. 吉林珲春自然保护区东北虎及其猎物资源调查. 动物学杂志, 46(2): 46-52.

田瑜, 邬建国, 寇晓军, 等. 2009. 东北虎种群的时空动态及其原因分析. 生物多样性, 17(3): 211-225.

Altman D G, Bland M. 1994. Diagnostic tests 2: predictive values. British Medical Journal, 309(6947): 102.

Augustin N H, Mugglestone M A, Buckland S T. 1996. An autologistic model for the spatial distribution of wildlife. Journal of Applied Ecology, 33: 339-347.

Beard K H, Hengartner N, Skelly D K. 1999. Effectiveness of predicting breeding bird distributions using probabilistic models. Conservation Biology, 13: 1108-1116.

Bennett E L, Robinson J G. 2000. Hunting of wildlife in tropical forests: implications for biodiversity and forest peoples//Environment Department working papers, no.76. Biodiversity series. Washington DC: The World Bank.

Betts M G, Ganio L M, Huso M M P, *et al*. 2009. Comment on "Methods to account for spatial autocorrelation in the analysis of species distributional data: a review". Ecography, 30(5): 609-628.

Carroll C, Miquelle D G. 2006. Spatial viability analysis of Amur tiger *Panthera tigris altaica* in the Russian Far East: the role of protected areas and landscape matrix in population persistence. Journal of Applied Ecology, 43: 1056-1068.

Channell R, Lomolino M V. 2000. Dynamic biogeography and conservation of endangered species. Nature, 403: 84-86.

Cliff A D, Ord J K. 1981. Spatial Processes-models and Applications. London: Pion.

Davis F W, Seo C, Zielinski W J. 2007. Regional variation in home range scale habitat models for fisher (*Martes pennanti*) in California. Ecological Application, 17: 2195-2213.

Dormann C F, McPherson J M, Araujo M B, et al. 2007. Methods to account for spatial auto-correlation in the analysis of species distributional data: a review. Ecography, 30: 609-628.

Dormann C F. 2007. Effects of incorporating spatial autocorrelation into the analysis of species distribution data. Global Ecology and Biogeography, 16: 129-138.

Duchamp C, Boyer J, Briaudet P E, et al. 2012. A dual frame survey to assess time-and space-related changes of the colonizing wolf population in France. Hystrix the Italian Journal of Mammalogy, 23: 14-28.

Fielding A H, Bell J F. 1997. A review of methods for the assessment of prediction errors in conservation presence/absence models. Environmental Conservation, 24: 38-49.

Frid A. 2003. Dall's sheep responses to overflights by helicopter and fixed-wing aircraft. Biological Conservation, 110: 387-399.

Goodrich J M, Kerley L L, Miquelle D G, et al. 2005. Social structure of Amur tigers on Sikhote-Alin Biosphere Zapovednik//Miquelle D G, Smirnov E N, Goodrich J M. Tigers in Sikhote-Alin Zapovednik. Vladivostok, Russia (in Russian): PSP: 50-60.

Goodrich J M, Miquelle D G, Smirnov E N, et al. 2010. Spatial structure of Amur (Siberian) tigers (*Panthera tigris altaica*) on Sikhote-Alin Biosphere Zapovednik, Russia. Journal of Mammalogy, 91: 737-748.

Granadeiro J P, Andrade J, Palmeirim J M. 2004. Modelling the distribution of shorebirds in estuarine areas using generalised additive models. Journal of Sea Research, 52: 227-240.

Guisan A, Edwards T C, Hastie T. 2002. Generalized linear and generalized additive models in studies of species distributions: setting the scene. Ecological Modelling, 157: 89-100.

Haining R. 2003. Spatial Data Analysis Theory and Practice. Cambridge: Cambridge University Press.

Hastie T J, Tibshirani R J. 1990. Generalized Additive Models. London: Chapman, Hall/CRC.

Hebblewhite M, Zimmermann F, Li Z, et al. 2012. Is there a future for Amur tigers in a restored tiger conservation landscape in Northeast China. Animal Conservation, 15(6): 1-14.

Hemmer H. 1976. Fossil history of the living Felidae // The Carnivore Research Institute. The World's Cats. Seattle: Burke Museum, 111(2): 1-14.

Heptner V G, Sludskii A A. 1992. Mammals of the Soviet Union. Volume II, Part 2. Carnivora (hyenas and cats). Leiden: Brill.

Hurlbert S H. 1984. Pseudoreplication and the design of ecological field experiments. Wildlife Monographs, 54: 187-211.

Jędrzejewski W, Jędrzejewska B, Okarma H, et al. 1992. Wolf predation and snow cover as mortality factors in the ungulate community of the Białowieża National Park, Poland. Oecologia, 90: 27-36.

Jiang G, Ma J, Zhang M, et al. 2009. Multiple spatial scale resource selection function models in relation to human disturbance for moose in northeastern China. Ecological Research, 24: 423-440.

Jiang G, Ma J, Zhang M, et al. 2010. Multi-scale foraging habitat use and interactions by sympatric cervids in northeastern China. Journal of Wildlife Management, 74: 678-689.

Jiang G, Ma J, Zhang M. 2006. Spatial distribution of distribution of ungulate responses to habitat factors in Wandashan, northeastern China. Journal of Wildlife Management, 70: 1470-1476.

Jiang G, Zhang M, Ma J. 2007. Effects of human disturbance on movement, foraging and bed site selection of red deer *Cervus elaphus xanthopygus* in the Wandashan Mountains, northeastern China. Acta Theriologica, 52: 435-446.

Jiang G, Zhang M, Ma J. 2008. Habitat use and separation between red deer *Cervus elaphus xanthopygus* and roe deer *Capreolus pygargus bedfordi* in relation to human disturbance in the Wandashan Mountains, northeastern China. Wildlife Biology, 14(1): 92-100.

Johnson C J, Boyce M S, Case R L, et al. 2005. Cumulative effects of human developments on Arctic wildlife. Wildlife Monographs, 160: 1-36.

José A F D-F, Mariana O D C T. 2002. Spatial autocorrelation analysis and the identification of operational units for conservation in continuous populations. Conservation Biology, 16: 924-935.

Kang A, Xie Y, Tang J, et al. 2010. Historic distribution and recent loss of tigers in China. Integrative Zoology, 5: 335-341.

Karanth K U, Nichols J D, Kumar N S, et al. 2004. Tigers and their prey: predicting carnivore densities from prey abundance. Proceedings of the National Academy of Sciences of the United States of America, 101: 4854-4858.

Karanth K U, Sunquist M E. 2000. Behavioural correlates of predation by tiger (*Panthera tigris*), leopard (*Panthera pardus*) and dhole (*Cuon alpinus*) in Nagarahole, India. Journal of Zoology, 250: 255-265.

Kerley L L, Goodrich J M, Miquelle D G, et al. 2002. Effects of roads and human disturbance on Amur tigers. Animal Conservation, 16: 97-108.

Kitchener A C, Dugmore A J. 2000. Biogeographical change in the tiger, *Panthera tigris*. Animal Conservation, 3: 113-124.

Kitchener A C, Yamaguchi N. 2010. What is a tiger? Biogeography, morphology, and taxonomy// Tilson R, Nyhus P. Tigers of the World: The Science, Politics and Conservation of *Panthera tigris*. 2 nd ed. Oxford: Elsevier: 53-85.

Koenig W D. 1999. Spatial autocorrelation of ecological phenomena. Trends in Ecology and Evolution, 14: 22-26.

Kramer-Schadt S, Revilla E, Wiegand T, et al. 2004. Fragmented landscapes, road mortality and patch connectivity: modelling influences on the dispersal of *Eurasian lynx*. Journal of Applied Ecology, 41: 711-723.

Latimer A M, Wu S S, Gelf A E, et al. 2006. Building statistical models to analyze species distributions. Ecological Application, 16: 33-50.

Lawton J H. 1993. Range, population abundance and conservation. Trends in Ecology and Evolution, 8: 409-413.

Li B, Zhang E, Liu Z. 2009. Livestock depredation by Amur tigers in Hunchun Nature Reserve, Jilin, China. Acta Theriologica Sinica, 29: 231-238.

Li B, Zhang E, Zhang Z, et al. 2008. Preliminary monitoring of Amur tiger population in Jilin Hunchun National Nature Reserve. Acta Theriologica Sinica, 28: 333-334.

Li T, Jiang J, Wu Z, et al. 2001. Survey on Amur tigers in Jilin Province. Acta Theriologica Sinica, 21: 1-6.

Li X, Wang Y. 2013. Applying various algorithms for species distribution modeling. Integrative Zoology, 8: 124-135.

Li Z, Zimmerman F, Hebblewhite M, et al. 2010. Study on the Potential Tiger Habitat in the Changbaishan Area. Beijing: China Forestry Publishing House.

Linkie M, Smith R J, Leader-Williams N. 2004. Mapping and predicting deforestation patterns in the lowlands of Sumatra. Biodiversity and Conservation, 13: 1809-1818.

Luan X, Qu Y, Li D, et al. 2011. Habitat evaluation of wild Amur tiger (*Panthera tigris altaica*) and conservation priority setting in north-eastern China. Journal of Environmental Management, 92: 31-42.

Luo S J, Kim J H, Johnson W E, et al. 2004. Phylogeography and genetic ancestry of tigers (*Panthera tigris*). PLoS Biology, 2: 2275-2293.

Lyet A, Thuiller W, Cheylan M, et al. 2013. Fine-scale regional distribution modelling of rare and threatened species: bridging GIS Tools and conservation in practice. Diversity and Distributions, 19(7): 651-663.

Ma Y. 2005. Changes in numbers and distribution of the Amur tiger in Northeast China in the past century: a summary report//Zhang E D, Miquelle D G, Wang T H. Recovery of the Wild Amur Tiger Population in China: Process and Prospect. Beijing: China Forestry Publishing House.

Mace R D, Waller J S. 1996. Grizzly bear distribution and human conflicts in Jewel Basin Hiking Area, Swan Mountains, Montana. Wildlife Society, 24(3): 461-467.

Mahoney S P, Schaefer J A. 2002. Hydroelectric development and the disruption of migration in caribou. Biological Conservation, 107: 147-153.

Manly B F J, McDonald L L, Thomas D L, et al. 2002. Resource Selection by Animals: Statistical Analysis and Design for Field Studies. Boston: Kluwer: 2.

Miquelle D G, Smirnov E N, Merrill T W, et al. 1999. Hierarchical spatial analysis of Amur tiger relationships to habitat and prey//Seidensticker J, Christie S, Jackson P. Riding the Tiger: Tiger Conservation in Human-dominated Landscapes. Cambridge: Cambridge University Press: 71-99.

Miquelle D G, Smirnov E N, Quigley H G, et al. 1996. Food habits of Amur tigers in Sikhote-Ali Zapovednik and Russian Far East, and implication for conservation. Journal of Wildlife Research, 1: 138-147.

Mushtaq M, Hussain I, Mian A, et al. 2013. Field evaluation of some bait additives against Indian crested porcupine (*Hystrix indica*) (Rodentia: Hystricidae). Integrative Zoology, 8: 285-292.

Pauley G R, Peek J M, Zager P. 1993. Predicting white-tailed deer habitat use in northern Idaho. Journal of Wildlife Management, 57: 904-913.

Proulx G, MacKenzie N. 2012. Relative abundance of American badger (*Taxidea taxus*) and red fox (*Vulpes vulpes*) in landscapes with high and low rodenticide poisoning levels. Integrative Zoology, 7: 41-47.

Quintero J, Roca R, Morgan A, et al. 2010. Smart Green Infrastructure in Tiger Range Countries. Washington D. C.: The World Bank.

R Development Core Team. 2010. R: A language and environment for statistical computing. Vienna, Austria: R Foundation for Statistical Computing.

Ramsay T, Burnett R, Krewski D. 2003. Exploring bias in a generalized additive model for spatial air pollution data. Environmental Health Perspectives, 111: 1283-1288.

Rangel T F, Diniz-Filho J A F, Bini L M. 2010. SAM: a comprehensive application for spatial analysis in macroecology. Ecography, 33: 46-50.

Reineking B, Schröder B. 2006. Constrain to perform: regularization of habitat models. Ecological Modelling, 193: 675-690.

Riitters K H, O'neill R V, Jones K B. 1997. Assessing habitat suitability at multiple scales: a landscape-level approach. Biological Conservation, 81: 191-202.

Segurado P, Araujo M B. 2004. An evaluation of methods for modelling species distributions. Journal of Biogeography, 31: 1555-1568.

Seidensticker J. 1996. Tigers. Vancouver: Voyageur Press.

Shenko A N, Bien W F, Spotila J R, et al. 2012. Effects of disturbance on small mammal community structure in the New Jersey Pinelands, USA. Integrative Zoology, 7: 16-29.

Stevens M A, Boness D J. 2003. Influences of habitat features and human disturbance on use of breeding sites by a declining population of southern fur seals (*Arctocephalus australis*). Journal of Zoology, 260: 145-152.

Stige L C, Ottersen G, Brander K, et al. 2006. Cod and climate: effect of the North Atlantic Oscillation on recruitment in the North Atlantic. Marine Ecology Progress, 325: 227-241.

Suarez-Seoane S, Osborne P E, Aloneso J C. 2002. Large-scale habitat selection by agricultural steppe birds in Spain: identifying species-habitat responses using generalized additive models. Journal of Applied Ecology, 39: 755-771.

Sunquist M. 2010. What is a tiger ecology and behavior//Tilson R, Nyhus P. Tigers of the World: The Science, Politics and Conservation of *Panthera tigris*. 2nd ed. Oxford: Elsevier: 19-34.

Tian Y, Wu J, Smith A T, et al. 2011. Population viability of the Siberian Tiger in a changing landscape: going, going and gone? Ecological Modelling, 222: 3166-3180.

Valeix M, Loveridge A J, Jammes C, et al. 2009. Behavioral adjustments of African herbivores to predation risk by lions: spatiotemporal variations influence habitat use. Ecology, 90: 23-30.

Whittington J, St Clair C C, Mercer G. 2004. Path tortuosity and the permeability of roads and trails to wolf movement. Ecology and Society, 9(1): 1759-1763.

Wikramanayake E, Dinerstein E, Seidensticker J, et al. 2011. A landscape-based conservation strategy to double the wild tiger population. Conservation Letters, 4: 219-227.

Wood S N. 2006. Generalized Additive Models: An Introduction with R. Boca Raton: CRC Press.

Yackulic C B, Sanderson E W, Uriartea M. 2011. Anthropogenic and environmental drivers of modern range loss in large mammals. Proceedings of the National Academy of Sciences of the United States of America, 108(10): 4024-4029.

Zhang C, Zhang M, Jiang G. 2012. Assessment of monitoring methods for population abundance of Amur tiger in Northeast China. Acta Ecologica Sinica, 32: 5943-5952.

Zhou S, Sun H, Zhang M, et al. 2008. Regional distribution and population size fluctuation of wild Amur tiger (*Panthera tigris altaica*) in Heilongjiang Province. Acta Theriologica Sinica, 28: 165-173.

Zhou S, Zhang M. 2011. An integrated analysis into the causes of ungulate mortality in the Wanda Mountains (Heilongjiang Province, China) and an evaluation of habitat quality. Biological Conservation, 144(10): 2517-2523.

第 2 章　东北虎保护中的生态与社会的协同效应

2.1 引　言

顶级捕食者由于其在生态系统中不可或缺的生态和社会经济作用，一直被作为生物多样性保护中的指示物种、保护伞物种和旗舰物种（Ripple et al., 2014）。然而，全球许多食肉动物都面临着种群数量大幅减少、地理分布范围收缩严重、栖息地丧失和破碎化等威胁（Ceballos and Ehrlich, 2002；Morrison et al., 2007）。对于虎和豹等家域范围大的大型食肉动物，一方面需要保证大面积的可渗透景观来满足其活动的需求，另一方面还要应对高昂的栖息地恢复投资、人类与野生动物冲突等社会经济和政治问题，使得保护其可持续种群要面临很多复杂的难题，具有巨大的挑战（Wikramanayake et al., 2004；Athreya et al., 2013）。近年来，由于森林恢复、有蹄类猎物数量增加及人为干扰（如偷猎、放牧、栖息地破坏）减少，一些大型食肉动物种群数量逐渐开始恢复。无论是在欧洲发达且拥挤的景观格局中通过与人类共存（Chapron et al., 2014），还是在非洲保护区或荒野地区（Packer et al., 2013）及北美洲地区通过与人类隔离分而治之（Gompper et al., 2015），大型食肉动物种群恢复都取得了显著成效。这些不同的成功保护模式反映了在生物多样性保护领域对于土地共享（land sharing）和土地抽取（land sparing）两种不同保护模式的有益探索，这两种模式都是景观管理和保育如何促进关键种种群恢复的探索与深入思考。深入了解每一种模式在促进种群恢复中的作用机制及使用方式（两种模式单一使用或结合使用）都将有助于对受胁物种的保护。

在过去 60 年里，中国政府一直致力于消除贫困，改善十几亿人口的民生问题。特别是 1978 年改革开放以来，中国经济飞速发展，到 2010 年中国已成为世界第二大经济体。20 世纪 90 年代以前，中国林区拥有大量的林业工人，在中国东北地区甚至全国各地进行了长达近 50 年的大规模森林采伐（从中华人民共和国成立直到 1998 年）。因此，森林资源的过度开发，严重影响了区域生物多样性。梅花鹿、马鹿等有蹄类动物在我国东北地区濒临灭绝或局部灭绝，大部分东北虎（Panthera tigris altaica）的栖息地丧失（田瑜等，2009）。森林的破坏，以及干旱、洪水等自然灾害的频发，不仅对国民经济造成了巨大冲击，还危及人民生命财产安全。经调查，我国东北虎的数量从 20 世纪 50 年代的 200 只下降到 1999 年的 14 只（Yu et al., 2009），野生东北豹（Panthera pardus orientalis）数量不足 10 只（Yang et al., 1998）。中国政府认识到森林面积减少、生物多样性丧失导致的社会经济和生态安全问题，于 1998 年启动了天然林保护工程（以下简称天保工程），其目标是解决天然林的休养生息和恢复发展问题，最终实现林区资源、经济、社会的协调发展。该工程是全球迄今为止政府投资最大、空间范围最广的生态环境修复工程之一，其中覆盖了中国东北地区濒危大型猫科动物历史分布区中的所有林区。

最新研究显示，在过去 20 年中，随着天保工程的实施、森林经营管理方式的转变和森林资源的保护，中国东北虎和东北豹的栖息地质量有所改善，其种群数量逐渐回升（Wang et al.，2016；Jiang et al.，2015）。近年来，Wang 等（2016）利用红外相机数据对东北虎和东北豹种群分布进行了评估，Jiang 等（2014，2015）研究了东北虎和东北豹的生境适宜性与潜在生境，他们的研究发现，大型猫科动物的种群增长与栖息地面积的增加具有一定的关联性，但未涉及人类活动和社会经济因素对这两种大型猫科动物保护的长期影响。因此，比较森林经营管理方式、人口变化、管理行动、资金投入随时间变化对大型猫科动物及猎物种群数量的影响，将有助于阐明土地共享和土地抽取两种保护策略的作用机制及其彼此间相互排斥与协同作用的程度。

要检验保护投入的社会效益与生态效益的协同效应，就需要对人类生产生活与生物多样性保护之间的关系进行定量分析（Persha et al.，2011）。生物多样性保护和消除贫困之间的权衡或二者之间的协同效应也可能揭示保护需求与保护机遇（Naughton-Treves et al.，2005）。然而，很少有研究关注顶级捕食者在社会和生态维度中的互作关系（Persha et al.，2011）。本研究调查了中国东北地区东北虎和东北豹核心分布区景观保护投资、森林采伐量、林业工人的数量变化（如改变其生计方式）与东北虎、东北豹栖息地之间的关系。

基于生态学中的捕食者-猎物理论及政府实施的移民安置计划，对中国东北国有林区东北虎和东北豹种群恢复的步骤提出以下假设。假设 1，林业工人的转移（土地抽取）导致森林生物量的增加。假设 2，森林生物量的增加和人类干扰（即偷猎、放牧、非林木产品采摘等）的减少导致有蹄类猎物丰富度增加（支持土地共享或土地抽取）。假设 3，人口密度的降低导致针对大型猫科动物和有蹄类猎物的偷猎活动减少（支持土地抽取）。假设 4，预测大型猫科动物的种群数量和分布范围会随着猎物数量的增加而增加，随人为干扰活动的减少而增加。我们还分析了人类干扰消除后与森林生物量恢复的时间对应关系，从而确定是土地共享方式还是土地抽取方式或两者相结合的方式更适于物种保护。

2.2 研究方法

2.2.1 森林管理数据的收集与分析

为了检验保护成本与移民、人口下降、森林生物量增加之间的关系，本研究选择在中国东北地区东北虎、东北豹现有分布区内的 31 个林业局（69 605km²）和 10 个自然保护区（4050km²）开展数据收集与调查，总计 73 655km² 的调查面积构成了中国境内东北虎、东北豹的整个现有分布区（表 2.1，图 2.1）。多数林业局成立于 20 世纪 50 年代，进行商业性采伐已有 60 余年（表 2.1）。除一个自然保护区外，大部分自然保护区都是在 2000 年之后建立的（表 2.1）。我们整理了 1950～2015 年相关变量的年度记录，这些变量包括保护资金投入、森林采伐量、年森林蓄积量超过采伐量的部分（以下简称森林净蓄积量增长量）、人工林面积、林业工人的数量及密度、人类与野生动物冲突的生态补偿费用，以及各林业局和自然保护区为林业工人的搬迁安置提供的福利费用。

表 2.1　中国东北地区东北虎和东北豹分布区的 31 个林业局和 10 个自然保护区的名称、地理位置、面积和成立年份

序号	林业局名称	省份	E (°)	N (°)	森林面积（km²）	成立年份
1	柴河	黑龙江	129.460 2	45.238 83	3 452.72	1947
2	东京城	黑龙江	129.134 7	43.936 18	4 184.68	1948
3	方正	黑龙江	129.425 9	45.761 73	2 035.82	1958
4	海林	黑龙江	129.110 5	44.783 35	1 566.21	1958
5	山河屯	黑龙江	127.870 7	44.336 3	2 064.09	1948
6	绥阳	黑龙江	130.799 6	44.029 39	5 160.00	1948
7	林口	黑龙江	130.061 8	45.532 36	2 730.25	1963
8	勃利县	黑龙江	130.569 7	45.752 32	785.07	1958
9	鸡东县	黑龙江	131.122 5	45.243 25	1 640.99	1965
10	林口县	黑龙江	130.270 6	45.274 11	1 939.04	1950
11	穆棱市	黑龙江	130.514 4	44.915 73	335.316	1958
12	宁安市	黑龙江	129.477 7	44.344 62	1 487.369	1958
13	安图县	吉林	128.912 7	43.110 6	1 237.65	1962
14	敦化市	吉林	128.226 2	43.369 63	1 743.51	1951
15	安图林管局	吉林	128.912 7	43.110 6	1 092.41	1962
16	白河	吉林	128.220 2	42.405 72	1 904.70	1971
17	大石头	吉林	128.579 8	43.140 93	2 634.38	1952
18	大兴沟	吉林	129.577 8	43.476 87	1 272.00	1960
19	敦化	吉林	127.976 5	43.122 54	2 373.55	1958
20	和龙	吉林	128.664 7	42.213 04	1 704.89	1958
21	和龙市	吉林	129.003 4	42.535 29	2 708.13	1950
22	黄泥河	吉林	128.106 4	43.833 15	1 965.96	1953
23	珲春市	吉林	130.363 2	42.874 27	880.04	1950

续表

序号	林业局名称	省份	E（°）	N（°）	森林面积（km²）	成立年份
24	天桥岭	吉林	130.077 8	43.768 04	1 935.06	1959
25	图们市	吉林	129.844 1	42.966 02	886.36	1974
26	汪清	吉林	130.472 1	43.392 19	3 040.00	1947
27	汪清县	吉林	129.756 4	43.309 12	3 290.11	1956
28	珲春	吉林	130.501 7	43.034 8	4 000.40	1994
29	大海林	黑龙江	128.474 4	44.378 39	2 663.10	1947
30	穆棱	黑龙江	130.212 2	44.266 63	2 675.30	1947
31	东方红	黑龙江	133.627 8	46.523 41	4 216.76	1963
总计					69 605.87	

序号	自然保护区名称	省份	E（°）	N（°）	森林面积（km²）	成立年份
1	大峡谷	黑龙江	127.917 3	44.183 4	249.98	2004
2	蝴蝶岭	黑龙江	129.447 7	45.607 9	108.02	2004
3	老爷岭	黑龙江	131.045 4	43.575 4	781.278	2011
4	穆棱红豆杉	黑龙江	130.153 7	43.969 3	356.48	2007
5	凤凰山	黑龙江	131.163 2	44.977 4	265.7	2006
6	牡丹峰	黑龙江	129.745 1	44.447 8	194.68	1981
7	鸟青山	黑龙江	131.011 1	44.540 4	180.02	2007
8	黄泥河	吉林	128.127	43.975	234.76	2000
9	汪清	吉林	130.894	43.397	674.34	2002
10	珲春	吉林	130.775 6	42.932 5	100 8	2002
总计					4 053.26	

注：经度和纬度代表了林业局和自然保护区的地理中心。林业局名称中有"县""市"字样的属于集体所有，没有"县""市"字样的属于森工集团所有

我们采用线性或非线性回归的方法检验 1998 年天然林保护工程启动后年森林经营投资总额、森林采伐量、森林净蓄积量增长量、林业工人密度与东北虎、东北豹栖息地面积变化之间的关系，通过将偷猎数量进行对数转换，对 1998～2006 年及 2007～2015 年东北虎偷猎的线性增长率进行了量化。所有统计检验都是双尾检验，显著水平为 0.05，利用 Prism（GraphPad Prism, Prism 5.0; www.graphpad.com）对数据进行分析。

2.2.2 东北虎和东北豹种群大小与生境分布

从 20 世纪 50 年代以来中国发表的相关文献（如 Yu *et al.*，2009）中收集东北虎和东北豹的种群丰富度与生境分布数据，并在下文进行了总结。1998～2015

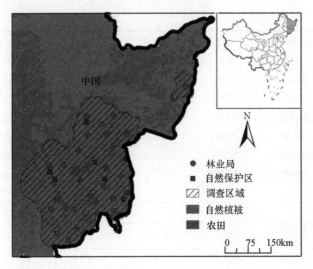

图 2.1 林业局（红色圆点）和自然保护区（蓝色方块）的空间分布（彩图请扫封底二维码）

年东北虎和东北豹历史种群大小是通过虎、豹冬季雪地足迹样线调查进行评估的。

2013 年以来，在老爷岭、张广才岭、完达山的核心栖息地中，我们架设了 1500 台以上的自动相机，覆盖了这些大型猫科动物核心分布区约 2900km² 的区域。自动相机架设密度为 2 台/10km²，2013～2015 年整个监测期间每 3 个月检查一次。

基于通过自动相机监测和历史报道（Yu et al., 2009）获得的种群大小的数据，我们评估东北虎的实际种群增长率为 $R=[\ln(N_{2015})–\ln(N_{1999})]/t$，其中 N_{1999} 为 1999 年东北虎种群数量；N_{2015} 为 2015 年东北虎种群数量；$t = 16$（年）（Berryman and Turchin, 2001）。然后通过公式 $\lambda=e^R$，将年种群增长率 R 转换为周限增长率（λ）。由于东北虎存活率相对较低（如 85%），要确保其种群的长期持续存在，至少需要有 100 只东北虎个体（Chapron et al., 2008）。我们根据计算的增长率预测了还需要多少年中国东北地区东北虎个体才能达到 100 只（Berryman and Turchin, 2001），与此同时，我们也对东北豹进行了类似的统计分析。

1998～2015 年，我们在研究区域共记录到东北虎出现信息 779 条，其中捕食猎物或牲畜 355 次，粪便样本 51 份（通过 DNA 鉴定），足迹链 345 条，自动相机拍摄影像数据 71 张。此外，还记录到东北豹出现信息 643 条，包括 24 次捕食猎物或牲畜信息、36 份粪便样本（通过 DNA 鉴定）、133 条足迹链信息、自动相机拍摄影像数据 459 张。我们还在国家林业和草原局猫科动物研究中心建立了中国东北虎豹监测信息数据库。

栖息地丧失通常是指示种群数量下降的一个有效指标（Dinerstein et al., 2007），为评估该大型猫科动物的栖息地面积，我们首先利用 ArcGIS 软件（Environmental Systems Resource Institute，ArcGIS 10.0；www.esri.com）在整个研究区域内创建两个栅格化的多边形，分辨率分别为 20km×20km 和 10km×10km，这两种空间分辨率是根据雌性东北虎或东北豹的平均家域大小来确定的（Goodrich et al.，2010；Hebblewhite et al.，2011）。然后，我们利用 Hawth 的 ArcGIS 分析工具（http://www.spatialecology.com/htools/download.php），以 400km² 栅格图层、100km² 栅格图层分别统计东北虎和东北豹的出现频次。当一个栅格内的出现频次大于零（>0）时则认为该栅格区域有东北虎或东北豹的分布。利用每年有东北虎或东北豹分布的栅格的总和来评估每种大型猫科动物当年的栖息地总面积。图 2.2a 和图 2.2b 分别代表 1999~2014 年东北虎、东北豹的总分布区域，我们使用线性或非线性回归模型检验了这两种大型猫科动物年栖息地总面积与各种社会经济和森林变量之间的关系。

2010~2014 年，在吉林省汪清国家级自然保护区（以下简称汪清自然保护区）878km² 范围内，利用雪地大样方方法对有蹄类猎物密度进行了调查和统计（Qi et al.，2015）。其中，2010~2011 年完成 33 个大样方的调查，2012~2013 年完成 14 个，2013~2014 年完成 10 个。每个大样方面积约 10km²（即 5km×2km），由 5 条平行的 5km 样线组成，样线间距 500m，只记录 24h 以内的动物足迹来计算样方中每种猎物的个体数量（Qi et al.，2015），进而计算出有蹄类猎物的密度。此外，为了评估偷猎的减少对东北虎、猎物和猎套数量的影响，我们收集了 2009 年、2015 年汪清自然保护区巡护记录中清除钢制猎套数量及其他涉及钢制猎套密度的数据。

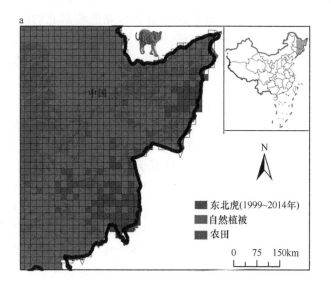

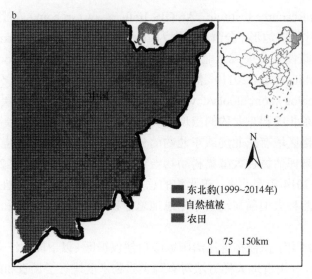

图 2.2　1999~2014 年有东北虎分布的栖息地面积（41 200km²）（a）和有东北豹分布的栖息地面积（10 200km²）（b）（彩图请扫封底二维码）

红色方块代表有大型猫科动物分布的栖息地

2.3　结　　果

中国境内的东北虎大多生活在中俄边境区域，但我们利用自动相机在中俄边境以西 270km 处拍摄到东北虎个体活动的信息，说明东北虎已经成功地向中国东北内陆地区扩散。另外，有关研究还记录到东北虎和东北豹的繁殖家族（Shi et al., 2015; Jiang, 2014）。研究发现，中国东北地区东北豹的现有分布区面积为 48 000km²，包括 37 个适宜栖息地斑块，可保证至少 195 只东北豹个体的生存（Jiang et al., 2015）。

在这两种大型猫科动物现有分布区域内，中国政府投入的资金已达 44.76 亿美元，其中天保工程（图 2.3a）投入 27.23 亿美元，人与野生动物冲突补偿款（图 2.3b）投入 0.13 亿美元，生态移民安置项目中居住环境改善（图 2.3c）投入 17.12 亿美元，自然保护区建设投入 0.27 亿美元。1998~2015 年，天保工程年投资总额稳步增长（$R^2 = 0.78$，$n = 17$，$P < 0.001$；图 2.3a）。2007~2014 年，大型猫科动物捕食牲畜事件增长了 31%。为缓解当地居民生产生活与大型猫科动物或森林恢复之间的冲突（图 2.3b），中国政府于 2008 年启动了林业工人迁移居住地的棚户区改造项目。政府建造福利住房，将林场或村屯的林业工人安置到附近城镇，并帮助大多数移民改行从事与森林采伐无关的其他职业（图 2.3c）。

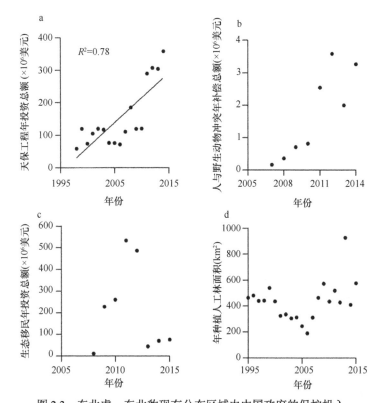

图 2.3　东北虎、东北豹现有分布区域内中国政府的保护投入

a. 1998～2015 年 31 个林业局天保工程年投资总额；b. 2007～2015 年人与野生动物冲突年补偿总额；c. 2008～2015 年林业工人迁移出林场用来改进人居环境年投资总额；d. 1995～2015 年年种植人工林面积

整个天保工程的平均投资额为每年 2927 美元/km², 并且从开始的每年 1098 美元/km² 增长到 3930 美元/km², 呈显著的线性增长趋势（$R^2 = 0.76$, $n = 17$, $P<0.001$；图 2.4a）。与假设 1 和土地抽取一致，1998 年启动的天保工程政策导致人口从 1999 年的 $17.77×10^4$ 人下降到 2015 年的 $7.18×10^4$ 人（图 2.4b, 图 2.5a），在此期间，近 10 万林业工人的职业发生了改变（图 2.5a）。1999～2015 年，年平均林业工人密度从 2.73 人/km² 下降到 1.07 人/km²（$R^2 = 0.95$, $n = 17$, $P<0.001$；图 2.4b）。此外，2008～2015 年，在这两种大型猫科动物分布区域内的 31 个林业局中，有 9 个林业局 2861km² 的林区无人居住。随着人口的减少，1999～2015 年森林采伐总量从 $5.51×10^6 m^3$ 线性下降到 $2.06×10^6 m^3$（$R^2 = 0.88$, $n = 17$, $P<0.001$；图 2.5b）。年平均采伐量从 1998 年的 117m³/km² 下降到 2015 年的 30.3m³/km²（$R^2 = 0.90$, $n = 17$, $P<0.001$；图 2.4c），2015 年 4 月东北林区全面禁止商业性采伐。与假设 1 一致，年平均森林蓄积量净增长从 1999 年的 66.62m³/km² 上升到 2014 年的 232.45m³/km²（$R^2 = 0.61$, $n = 16$, $P = 0.004$；图 2.4d）。1995～2015 年，森林面积以 368km²/年的速度增长，1995～2015 年，种植人工林总面积为 7736km²（图 2.3d）。

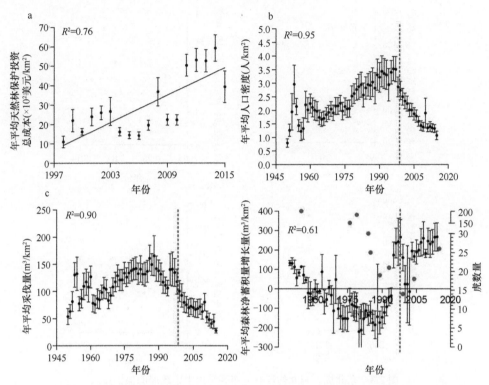

图 2.4　1998~2015 年每平方千米的年平均天然林保护投资总成本（a）、年平均人口密度（b）、年平均采伐量（c）、1950~2015 年东北虎种群动态变化与年平均森林净蓄积量增长量（d）

虚线代表 1998 年实施天然林保护工程。d 图中圆点代表东北虎的数量

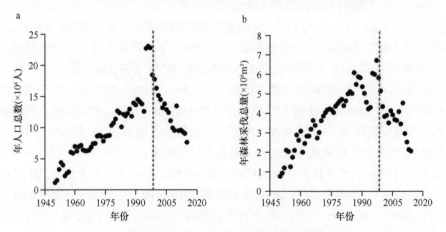

图 2.5　1950~2015 年东北虎栖息地内年人口总数（a）及年森林采伐总量（m³）的动态变化（b）

虚线表示 1998 年启动天然林保护工程

天保工程每年每平方千米的投资金额与每平方千米的林业工人数量（$R^2 = 0.68$，$n = 17$，$P = 0.002$；图 2.6a）和年平均森林采伐量（$R^2 = 0.74$，$n = 17$，$P < 0.001$；图 2.6b）呈负相关，与森林净蓄积量增长量呈正相关（$R^2 = 0.45$，$n = 16$，$P = 0.069$；图 2.6c）。

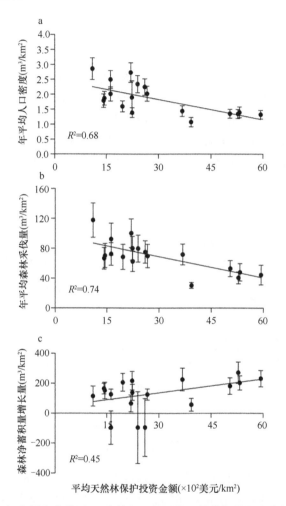

图 2.6 1998～2015 年中国东北地区 31 个林业局的平均天然林保护投资金额（横坐标）与年平均人口密度（a）、年平均森林采伐量（b）和森林净蓄积量增长量（c）之间的关系

与假设 2 和假设 3 一致，1998～2015 年，公开报道的被猎套猎到的东北虎总数为 5 只，其中 1998～2006 年为 4 只，2007～2015 年为 1 只，说明在加强了行政管理后，在一定程度上促进了东北虎种群的增长。此外，巡护记录显示，2009 年吉林汪清自然保护区的猎套密度为 0.4 个/km，而 2015 年的线样调查结果显示，该区域用于捕捉动物的猎套密度仅为 0.1 个/km。

东北虎种群数量下降的速率与人口密度和森林平均采伐量的增加及森林净蓄积量增长量的减少是同步的（图 2.4b~d）。然而，正如假设 4 的预测，我们根据 2013~2015 年的自动相机监测数据发现，东北虎种群数量从 1999 年的 14 只恢复到 27 只，与 Wang 等（2016）报道的 26 只或更多个体的结果相似。此外，东北豹的数量从 1998 年的约 10 只（Yang et al., 1998）增加到 2015 年的 42 只（Wang et al., 2016）。据此，东北虎 1999~2015 年周限增长率（λ）为 1.04，中国境内野生东北虎种群数量增长到 100 只需要到 2050 年。此外，1999~2015 年，东北豹周限增长率（λ）为 1.08，野生东北豹种群数量增长到 100 只可能需要到 2025 年。

对于假设 4，我们证实了人口密度、森林生长、森林采伐量和猎物的变化对预测的两种大型猫科动物栖息地面积与种群大小的影响。1999~2014 年，东北虎分布区域的总面积为 41 200km^2（图 2.2）。1999~2014 年，东北豹分布区总面积为 10 200km^2（图 2.2）。此外，东北虎和东北豹年栖息地面积 1999~2015 年呈线性上升且增加迅速（东北虎：$R^2 = 0.52$，$n = 14$，$P = 0.049$；东北豹：$R^2 = 0.86$，$n = 19$，$P < 0.001$）（图 2.7）。

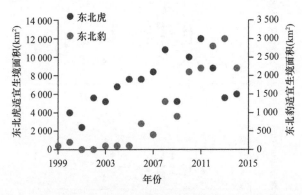

图 2.7　1998 年天然林保护工程实施后中国东北地区有东北虎和东北豹分布的栖息地估算面积
（彩图请扫封底二维码）

东北虎占据的栖息地面积与年平均人口密度呈负相关（$R^2 = 0.55$，$n = 14$，$P = 0.032$；图 2.8a），与年森林净蓄积量增长量呈正相关（$R^2 = 0.63$，$n = 14$，$P = 0.011$；图 2.8b），但与年平均森林采伐量无关（$R^2 = 0.24$，$n = 14$，$P = 0.387$；图 2.8c）。与此类似，东北豹占据的栖息地面积与年平均人口密度也呈负相关（$R^2 = 0.80$，$n = 15$，$P = 0.001$；图 2.8d），与年森林净蓄积量增长量呈正相关（$R^2 = 0.83$，$n = 15$，$P = 0.002$；图 2.8e），与年平均森林采伐量呈负相关（$R^2 = 0.79$，$n = 15$，$P < 0.001$；图 2.8f）。有蹄类猎物调查显示，2010~2014 年吉林汪清国家级自然保护区 878km^2 范围内虎、豹主要猎物种群密度是增加的，支持假设 3。狍（*Capreolus pygargus*）密度从 2010 年的（0.67±0.13）只/km^2 上升到 2013 年的（1.88±0.22）只/km^2，2014

年又上升到（2.33±0.52）只/km²。野猪（*Sus scrofa*）密度从 2010 年的（0.32±0.13）只/km² 增加到 2013 年的（0.40±0.02）只/km²，2014 年增加到（0.93±0.66）只/km²。梅花鹿（*Cervus nippon*）密度由 2010 年的（0.03±0.022）只/km² 上升到 2013 年的（0.06±0.03）只/km²，2014 年上升到（0.34±0.34）只/km²。有蹄类猎物种群数据也为东北虎、东北豹栖息地面积或种群恢复与猎物种群密度增长相关的假说提供了证据。

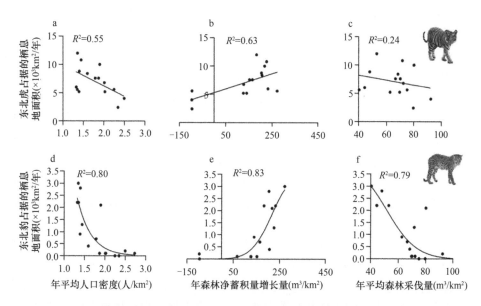

图 2.8　自 1998 年天然林保护工程开始以来，大型猫科动物占据的栖息地面积与年平均人口密度（a、d）、年森林净蓄积量增长量（b、e）和年平均森林采伐量（c、f）之间的线性关系

2.4　讨　　论

中国东北林区的森林管理已经从不可持续的森林采伐向土地抽取等方式进行了转变。在这一转变过程中，区域人口密度、森林采伐量和已知东北虎偷猎数量大幅下降，东北虎豹三种主要的有蹄类猎物密度也在增加，与此同时，两种大型猫科动物种群数量和栖息地面积也呈增长趋势。

我们的研究发现，中国东北林区的森林管理在 1999 年之前缺乏土地共享的理念。东北虎种群数量 1950～1998 年呈线性下降趋势，而 1999 年以后开始逐渐回升（图 2.4d）。与此相对应，森林年采伐量在过去一直呈增长趋势，直到 1998 年天保工程实施以后才逐渐下降，1999～2015 年森林净蓄积量增长量呈上升趋势（图 2.4d）。在过去的 65 年里，森林管理和大型猫科动物种群动态随时间、空间的变化趋势，证实了过去（1999 年以前）我们的森林管理缺乏土地共享的理念。具体来说，对森林资源的过度开发导致森林生物量和有蹄类猎物减少，而对有蹄类猎物和大型猫科动物的盗猎增加。自 1998 年天保工程实施以来，土地抽取促进了森林生物量的恢复，进而促进了猎物的恢复，大型猫科动物栖息地面积和种群丰富度都有所增长。

森林管理应注重森林资源的可持续利用和人文社会发展（Carpenter et al., 2009）。尽管中国的虎、豹等大型猫科动物种群数量仍然维持在较低的水平，但是我们的研究为中国将来的猫科动物保护提供了参考。我们不能简单地将这两种大型猫科动物的种群恢复具体归功于土地共享或土地抽取中的其中一种模式，因为这是生态协同综合作用的结果。在 1998 年天保工程实施之前，由于大型猫科动物和人类之间缺乏土地共享，东北虎豹种群数量急剧下降。但自天保工程实施以来，通过对大部分林区限伐后产生的土地抽取以保持最小的森林采伐量，以及通过其他对森林资源的利用以实现土地共享，增加了这两种大型猫科动物的种群丰富度和栖息地分布，这是东北虎保护向着 Chapron 等（2014）提出的目标迈出的重要一步，即对大型食肉动物的保护应该致力于维持动物在人类主导的景观中及不同破碎化程度的森林生态系统中的生态过程。因此，我们应该利用可衡量的社会指标实时地评价保护投资的社会进程（Mace，2014），否则我们可能无法确定保护投资的生态效益与成效。

另外，由于土地所有者的意愿，土地所有权的类型也会影响保护投资的有效性（Wilcove et al., 1998）。西方国家在生物多样性保护方面面临的一大挑战是说服土地所有者同意保护私有土地上的野生动物（McShane et al., 2011）。但在中国情况有所不同，中国的土地都归国家所有。尽管天保工程这种自上而下的项目非常复杂，但事实证明，土地国有化的性质为项目实施提供了诸多便利。考虑到国家统一推行项目实施及影响范围大等因素，中国政府首先进行了试点，然后根据项目进展情况逐步提高年度投资额度（图 2.4a，图 2.9）。此外，建设福利住房、增加森林蓄积量、实施野生动物危害补偿政策等在解决人虎冲突和保护居民安全方面都发挥了重要作用（Jiang et al., 2014）。就天保工程的投资水平而言，它与非洲保护区成功保护狮子的投资水平（>3000 美元/km^2）非常接近（Packer et al., 2013）。

大型食肉动物对人口密度变化的敏感程度具有物种特异性，从而为它们成功适应人类主导的生态景观提供可能性（Chapron et al., 2014）。我们的研究结果表明，人口密度的下降促进了东北虎和东北豹的分布区域的扩张（图 2.4b，图 2.6a，图 2.8a、d）。大型猫科动物核心分布区内人口密度的控制不仅直接减少了森林采伐量，还减轻了人为干扰，如林产品采集、耕地开垦、基础设施建设、偷猎和其他人类活动（Qi et al., 2015；Giller et al., 2008）。由于盗猎对动物的影响具

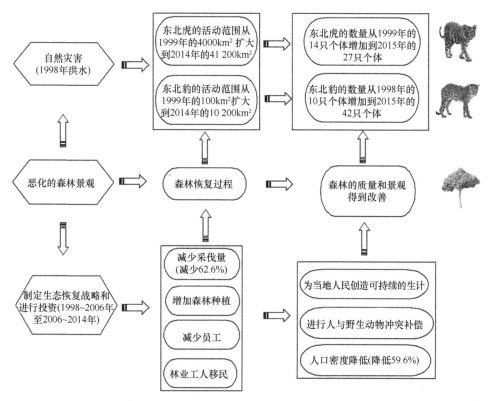

图 2.9 保护方案与行动（1998～2015 年中国发生事件原因与同步行动路线图）

有时滞性，因此要揭示盗猎与人口密度变化之间的关系是比较困难的。例如，一旦猎套布设到野外，它们对野生动物的威胁可能持续多年。随着大型猫科动物出现的频率增加，当地政府已经在东北虎和东北豹的主要栖息地进行了定期巡护及清山清套等工作，以确保东北虎和东北豹的安全。结果表明，2009～2015 年吉林汪清自然保护区猎套密度有所下降。与此同时，2007～2015 年，东北虎被偷猎的案件明显少于 1998～2006 年。虽然现在生活在林区的人越来越少，但对于地方管理部门而言，还是需要定期开展巡护及清山清套等工作。

有证据显示，长期的森林恢复及对非法盗猎和其他人为干扰的严格控制增加了有蹄类动物的数量与分布范围（Jiang et al.，2015）。例如，通过调查，在吉林汪清自然保护区 878km² 范围内，大型有蹄类猎物（狍、野猪、梅花鹿）的密度 2010～2014 年都呈上升趋势。前人研究及本研究结果，为揭示关于猎物驱动大型猫科动物分布的理论提供了实证（Jiang et al.，2015；Qi et al.，2015），我们认为，森林恢复有助于有蹄类等食肉动物猎物的种群恢复，进而扩大了大型猫科动物在中国东北地区的分布范围。我们的研究也证实了林业管理部门实施土地共享和土地抽取的管理方式后，大型猫科动物在种群数量和栖息地面积上都有所增加。另外，充分考虑因移民而引起的社会问题也是十分必要的。例如，要充分考虑移民安置过程中的基础建设和提供新的工作岗位等问题，但这一点，生态保护学家往往容易忽视（Jiang et al., 2014）。我们的研究表明，从保护投资的角度来看，通过控制东北虎和东北豹栖息地的人口密度，可以产生社会效益或生态效益的协同效应（Persha et al., 2011）。

欧洲的一些研究表明，在整个欧洲大陆，通过土地共享模式，大型食肉动物成功地在高密度人口地区与人类实现共存，棕熊（Ursus arctos）、欧亚猞猁（Lynx lynx）、灰狼（Canis lupus）及貂熊（Gulo gulo）等4种大型食肉动物在保护区以外的区域成功地繁衍生息（Chapron et al., 2014）。然而，在较小的地理尺度上，如加纳西南部和印度北部，为了尽量减少粮食生产对树木和鸟类物种多样性的负面影响，土地抽取是一个更有效的保护策略（Phalan et al., 2011）。与欧洲的研究结果相似，我们认为，中国大型食肉动物的保护可能不仅仅依赖于当地的自然保护区系统，如大熊猫的保护就是如此（Liu et al., 2001；Loucks et al., 2001）。长白山自然保护区是20世纪60年代为保护东北虎而建立的，然而，这个保护区已经无法有效地保护老虎种群，因为在过去的30年里这个保护区始终没有再发现过老虎。本研究发现，在大地理尺度上的景观保护对我国东北虎和东北豹种群恢复是有效的。天保工程自1998年实施以来，中国东北地区的林地管理不但在大空间尺度上采用了土地共享模式来降低人口密度，而且在区域尺度上采用了土地抽取模式。例如，通过移民安置等措施，当地政府在9个林业局内划分出861km^2的无人区。因此，在不同的时空尺度上，需要因地制宜地采取不同的保护模式来开展对大型食肉动物的保护（Hurlbert, 1978；Carter et al., 2012；Chapron et al., 2014）。

与其他大型食肉动物一样，大型猫科动物的种群恢复往往受制于较低的年增长率和疾病的威胁。现在中国东北虎种群的周限增长率与俄罗斯东北虎种群过去41年的增长率（1.046）相似（Miquelle et al., 2015）。由于缺乏足够敏感的生态指示数据，以及物种自然寿命较长，种群变化过程缓慢，短期监测评估无法进行准确评估。因此，科学地评估东北虎和东北豹种群恢复状况就需要长期监测（Miquelle et al., 2015）。在大型猫科动物保护的初期阶段，可以利用间接的社会效益、经济效益、初级生产力（即栖息地植被改善等）或猎物指标来衡量保护过程成功与否。保护工程实施较长时间后，我们才能将东北虎和东北豹栖息地面积或种群丰富度的变化与天保工程的保护投资联系起来。到目前为止，中国政府还没有公布天保工程项目的具体结束日期。在中国东北地区已全面禁止采伐的情况下，中国政府仍在继续为森林保护提供资金支持。此外，中国政府计划在东北虎和东北豹栖息地景观中逐步建立一系列由中央直接管辖的大型国家公园（14 600km^2）（Li et al., 2016）。新的国家公园计划将确保对中国东北地区这两种大型猫科动物的长期持续保护（Kathleen, 2016）。除东北虎豹国家公园体制试点外，中国政府正计划建立其他 9 个国家公园，包括三江源国家公园体制试点（123 100km^2）、大熊猫（*Ailuropoda melanoleuca*）国家公园体制试点（27 000km^2）、湖北省神农架国家公园体制试点（1170km^2）、浙江省钱江源国家公园体制试点（252km^2）、湖南南山国家公园体制试点（635.94km^2）、福建省武夷山国家公园体制试点（982.59km^2）、祁连山国家公园体制试点（50 200km^2）、北京市长城国家公园体制试点（59.91km^2）、云南普达措国家公园体制试点（602.1km^2）。目前，在国家公园范围内，科学评估维持东北虎和东北豹生存所需的可渗透景观面积，以及连接中俄边境和国内分布区的生态廊道的面积是东北虎和东北豹国家公园体制试点迫切需要解决的问题（Chapron *et al.*，2014；Ekroos *et al.*，2016）。从历史上看，大型猫科动物种群数量下降与人类扩张、森林采伐及自然灾害密切相关，最后为挽救受胁物种才开始制定相应的保护策略（图 2.9）。随着生态保护的不断深入，必然会出现一些新的社会或生态问题。人虎冲突（主要是虎捕食牲畜）随着东北虎种群的恢复将不断加剧（图 2.3b）。因此，未来应在宏观尺度上关注大型猫科动物或其他大型食肉动物与人类生存的相互影响，通过可持续的森林资源管理来增强保护成效及种群恢复潜力，同时也要注意两种大型猫科动物在栖息地和猎物需求上的差异。东北豹在与同域分布的东北虎的种间竞争中处于劣势（Jiang *et al.*，2015），因此，如何平衡同域分布的大型食肉动物的种间关系可能成为大型猫科动物保护人士未来面临的新挑战。本研究关于在物种保护过程中的土地管理模式的分析不仅为大型猫科动物的保护提供了实证，还为全球其他复杂社会和生态因素协同效应下的大型食肉动物集团及其栖息地的保护提供了有益的参考。

2.5　本章小结

目前全球生物多样性保护开始关注土地共享和土地抽取两种土地管理策略在保护中的作用。然而，很少有研究关注社会和生态因素对顶级捕食者保护过程中土地管理策略的交互作用。我们利用中国东北林区近 65 年的生态与社会经济数据，评估了政府的社会政策在解决濒危物种保护、人类与野生动物冲突和改善人居生活方面的作用。研究表明，1998～2015 年，中国东北虎和东北豹分布区内人口密度下降了 59.6%，森林砍伐量减少了 62.6%，东北虎和东北豹的种群数量、分布区面积、猎物丰富度均显著增加。尽管东北虎和东北豹的种群数量仍然维持在较低水平，但在天保工程实施后的 1999～2015 年，东北虎和东北豹的种群数量分别以每年 1.04、1.08 的周限增长率增长，两种大型猫科动物的分布区面积也显著增加。随着天保工程的启动、新的土地抽取利用模式的应用，过去由不可持续的土地利用方式导致的森林资源过度开发与东北虎和东北豹种群下降的局面正在好转。大量的经济投入和移民安置工程，再加上非法盗猎活动的减少，彰显了社会和生态的综合效应在中国大型猫科动物保护工作中发挥的重要作用。

参 考 文 献

田瑜, 邬建国, 寇晓军, 等. 2009. 东北虎种群的时空动态及其原因分析. 生物多样性, 17(3): 211-225.

Athreya V, Odden M, Linnell J D, et al. 2013. Big cats in our backyards: persistence of large carnivores in a human dominated landscape in India. PLoS ONE, 8: e57872.

Berryman A, Turchin P. 2001. Identifying the density-dependent structure underlying ecological time series. Oikos, 92: 265-270.

Carpenter S R, Mooney H A, Agard J, et al. 2009. Science for managing ecosystem services: beyond the millennium ecosystem assessment. Proceedings of the National Academy of Sciences of the United States of America, 106: 1305-1312.

Carter N H, Shrestha B K, Karki J B, et al. 2012. Coexistence between wildlife and humans at fine spatial scales. Proceedings of the National Academy of Sciences of the United States of America, 109: 15360-15365.

Ceballos G, Ehrlich P R. 2002. Mammal population losses and the extinction crisis. Science, 296: 904-907.

Chapron G, Kaczensky P, Linnell J D C, et al. 2014. Recovery of large carnivores in Europe's modern human-dominated landscapes. Science, 346: 1517-1519.

Chapron G, Miquelle D G, Lambert A, et al. 2008. The impact on tigers of poaching versus prey depletion. Journal of Applied Ecology, 45: 1667-1674.

Dinerstein E, Loucks C, Wikramanayake E, et al. 2007. The fate of wild tigers. BioScience, 57: 508-514.

Ekroos J, Ödman A M, Andersson G K S, et al. 2016. Sparing land for biodiversity at multiple spatial scales. Frontiers in Ecology and Evolution, 3: 145.

Giller K E, Leeuwis C, Andersson J A, et al. 2008. Competing claims on natural resources: what role for science? Ecology and Society, 13(2): 34.

Gompper M E, Belant J L, Kays R. 2015. Carnivore coexistence: America's recovery. Science, 347: 382-383.

Goodrich J M, Miquelle D G, Smirnov E N, et al. 2010. Spatial structure of Amur (Siberian) tigers (*Panthera tigris altaica*) on Sikhote-Alin biosphere Zapovednik, Russia. Journal of Mammalogy, 91: 737-748.

Hebblewhite M, Miquelle D G, Murzin A A, et al. 2011. Predicting potential habitat and population size for reintroduction of the Far Eastern leopards in the Russian Far East. Biological Conservation, 144: 2403-2413.

Hurlbert S H. 1978. The measurement of niche overlap and some relatives. Ecology, 59: 67-77.

Hyde W F, Belcher B M, Xu J. 2003. China's Forests: Global Lessons from Market Reforms. Washington DC: Resource for the Future Press.

Jiang G, Qi J, Wang G, et al. 2015. New hope for the survival of the Amur leopard in China. Scientific Reports, 5.

Jiang G, Sun H, Lang J, et al. 2014. Effects of environmental and anthropogenic drivers on Amur tiger distribution in northeastern China. Ecological Research, 29: 801-813.

Jiang G. 2014. New evidence of wild Amur tigers and leopards breeding in China. Oryx, 48: 326.

Kathleen M. 2016. Can a new park save China's big cats? Science. http: //dx.doi.org/10.1126/science.

aah7201[2017-3-6].

Li J, Wang W, Axmacher J C, et al. 2016. Streamlining China's protected areas. Science, 351: 1160.

Liu J, Linderman M, Ouyang Z, et al. 2001. Ecological degradation in protected areas: the case of Wolong Nature Reserve for giant pandas. Science, 292: 98-101.

Loucks C J, Lü Z, Dinerstein E, et al. 2001. Giant pandas in a changing landscape. Science, 294: 1465.

Ma Z, Melville D S, Liu J, et al. 2014. Rethinking China's new great wall. Science, 346(6212): 912-914.

Mace G M. 2014. Whose conservation? Science, 345: 1558-1560.

McShane T O, Hirsch P D, Trung T C, et al. 2011. Hard choices: making trade-offs between biodiversity conservation and human well-being. Biological Conservation, 144: 966-972.

Miquelle D G, Smirnov E N, Zaumyslova O Y, et al. 2015. Population dynamics of Amur tigers (*Panthera tigris altaica*) in Sikhote-Alin Biosphere Zapovednik: 1966-2012. Integrative Zoology, 10: 315-328.

Morrison J C, Sechrest W, Dinerstein E, et al. 2007. Persistence of large mammal faunas as indicators of global human impacts. Journal of Mammalogy, 88: 1363-1380.

Naughton-Treves L, Holl M B, Brandon K. 2005. The role of protected areas in conserving biodiversity and sustaining local livelihoods. Annual Review of Environment and Resources, 30: 219-252.

Packer C, Loveridge A, Canney S, et al. 2013. Conserving large carnivores: dollars and fence. Ecology Letters, 16: 635-641.

Persha L, Agrawal A, Chhatre A. 2011. Social and ecological synergy: local rule making, forest livelihoods, and biodiversity conservation. Science, 331: 1606-1608.

Phalan B, Onial M, Balmford A, et al. 2011. Reconciling food production and biodiversity conservation: land sharing and land sparing compared. Science, 333: 1289-1291.

Qi J, Shi Q, Wang G, et al. 2015. Spatial distribution drivers of Amur leopard density in northeast China. Biological Conservation, 191: 258-265.

Ripple W J, Estes J A, Beschta R L, et al. 2014. Status and ecological effects of the world's largest carnivores. Science, 343: 1241484.

Shi Q, Li Q, Zhang M. 2015. First camera-trap video evidence of the Amur tiger breeding in China. Oryx, 49: 205-206.

Sikes R S, Gannon W L. 2011. Guidelines of the American Society of Mammalogists for the use of wild mammals in research. Journal of Mammalogy, 92: 235-253.

Wang T, Feng L, Mou P, et al. 2016. Amur tigers and leopards returning to China: direct evidence and a landscape conservation plan. Landscape Ecology, 31: 491-503.

Wikramanayake E, McKnight M, Dinerstein E, et al. 2004. Designing a conservation landscape for tigers in human-dominated environments. Conservation Biology, 18: 839-844.

Wilcove D S, Rothstein D, Dubow J. 1998. Quantifying threats to imperiled species in the United States. BioScience, 48: 607-615.

Xu J, Yin R, Li Z, et al. 2006. China's ecological rehabilitation: Unprecedented efforts, dramatic impacts, and requisite policies. Ecological Economics, 57: 595-607.

Yang S, Jiang J, Wu Z, et al. 1998. Report on the Sino-Russian joint survey of Far Eastern leopards and Siberian tigers and their habitat in the Sino-Russian boundary area, eastern Jilin Province, China, winter 1998// A Final Report to the UNDP and the Wildlife Conservation Society.

第3章 有蹄类动物死亡原因和生境质量评价综合分析

3.1 引 言

大型有蹄类动物在世界许多地区都是重要的经济资源（Loibooki et al., 2003; Corlett, 2007）。然而，人类在利用动物资源的过程中，非法盗猎已经成为大型有蹄类动物生存最主要的威胁之一（IUCN, 2010），使许多大型有蹄类动物面临区域灭绝的风险（Milner-Gulland and Bennett, 2003; Baillie et al., 2004; Sodhi et al., 2004; Corlett, 2007），这种现象在中国东北完达山东部林区尤其突出。一方面，当地政府大力加强对森林资源的保护，但另一方面，地方经济的发展还需要依靠森林采伐创收，通过建设公路网改善民生，这就使资源保护与经济发展的矛盾难以调和。此外，公路网的建设也为盗猎分子到达更偏远地区盗猎提供了便利条件（Blake et al., 2007）。因此，为了找到一种能科学、切实可行地保护野生动物资源的管理措施（Zhang and Liu, 2008），研究大型有蹄类动物的生境选择和空间分布规律极其重要（Jiang, 2008, 2011），这将为有蹄类动物的保护计划提供有益的参考。然而，盗猎引起的是动物繁殖和生存需求之外的非正常死亡，在进行生境选择和质量评价研究时，若不考虑人类盗猎引起的动物个体死亡，获得的结果通常与实际情况不符，且用于保护计划的可参考价值较小（Garshelis, 2000）。

许多研究表明，在进行动物栖息生境评价时，通过野外收集动物的出现点建立的生境评价模型计算出高质量生境区域，这些区域因动物的繁殖和生存率较低且死亡率较高而成为沦陷生境（attractive sink-like habitat），因此这些沦陷区域很难维持动物种群的生存。但是，生活在这类生境中的动物个体并不会意识到人类活动给它们带来的潜在死亡风险，也不了解这种生境对它们的生存繁衍不利（Delibes et al., 2001; Naves et al., 2010）。在动物保护和管理工作中，如果将沦陷生境作为优质生境，这将导致保护管理的错误，使保护的目标物种逐渐走向濒危甚至灭绝。在动物沦陷生境研究中，Naves 等（2010）最初建议用 2 个尺度的生境模型分析沦陷生境。Nielsen 等（2008）采用动物出现点和人类引起的动物死亡点及与之相关的变量信息用于沦陷生境分析。Falcucci 等（2009）利用动物出现点和人类引起的动物死亡点信息通过生态位分析方法建立动物出现生境评价模型和动物死亡危险生境评价模型，然后将 2 个模型进行综合分析、重分类，并基于生境适宜性等级和死亡危险等级的交互作用将综合分析后的模型获得的生境质量划分为 7 个等级：不适宜、沦陷生境（3 个等级的沦陷生境，具有中或高等级死亡危险的生境）和资源类生境（3 个等级的适宜生境，具有无或低等级死亡危险的生境）。在动物生境分析中，虽然建立动物出现–死亡生境评价模型能够获得很好的结果，但是这种方法不能很好地说明单个生境因子对动物的影响，也不能明确解释人类盗猎动物引起动物死亡的原因。

在物种保护研究工作中，野外调查和监测收集动物死亡及生存现状资料，对深入分析动物死亡的原因并采取有效方法减少动物死亡以形成科学的保护计划极其重要（Groom et al.，2006）。在动物生境选择和质量评价研究中，通过收集动物个体出现点数据分析动物个体对特定生境成分的偏好（Duberstein et al.，2008；Guisan and Thuiller，2010；Jiang et al.，2011），然后利用生境适宜性指数（HSI）评价模型评价生境质量已被广泛运用（Johnson，1980；Underwood，2014）。此外，基于动物出现点信息建立的生境评价模型和基于动物死亡点信息建立的动物死亡危险评价模型，经综合分析而形成的动物出现–死亡生境评价模型不仅能更好地说明生境适宜性现状，还能更好地服务于动物保护和管理（Falcucci et al.，2009）。

本研究通过对有蹄类动物个体出现点和死亡点位置等数据资料进行分析，其目的为：①分析野猪（*Sus scrofa*）、马鹿（*Cervus elaphus*）和狍（*Capreolus pygargus*）的死亡原因；②分析影响野猪、马鹿和狍出现（生境选择）及死亡（与死亡有关）的生境因子；③建立一个能够识别沦陷生境和资源类生境的综合有蹄类动物（野猪、马鹿和狍）出现–死亡生境评价模型，为完达山东部林区有蹄类动物的保护提供基础科学依据。

3.2 研究地区概况

研究地区总面积 3692.06km^2，位于 46°07′55″N～47°01′41″N、132°48′52″E～133°56′55″E。研究地区与俄罗斯远东地区相邻，多山，平均海拔约 200m。植被类型包括针叶林、阔叶林、针阔混交林、灌丛、林间湿地和农田。年平均降雨量为 500～800mm。气温年内变化较大，月最高气温达 34.6℃，最低气温达–34.8℃。一般 11 月下雪，雪深可达 110cm，积雪融化一般在 4 月末 5 月初。人口密度为 32 人/km^2（村屯水平）。

生境多样性孕育了生物的多样性。该区域主要分布的野生动物有野猪、马鹿、狍、斑羚（*Naemorhedus goral*）、雪兔（*Lepus timidus*）及濒危顶级捕食动物东北虎（*Panthera tigris altaica*）（李言阔等，2008；马逸清等，1986）等。但是，由于多年来的森林采伐和非法盗猎活动，有蹄类动物（野猪、马鹿和狍）种群数量已经急剧下降（周绍春等，2010）。为了保护和恢复野生动物种群，动物保护管理部门已经颁布了多个法律法规去阻止盗猎，如猎枪在 20 世纪 90 年代末被严格禁止。

3.3　数据收集和分析方法

3.3.1　有蹄类动物死亡点数据来源

在本研究中，使用了 2 种方法收集数据：① 2008～2009 年东北虎猎物种群样线调查收集的动物个体死亡数据；② 2002～2009 年东北虎野外种群监测收集的动物个体死亡数据（周绍春等，2008）。2 种方法共收集到 58 只野猪、40 只马鹿和 62 只狍死亡的相关信息。在进行数据分析时，我们考虑了数据在时间和空间方面的独立性，即选用的死亡个体信息必须满足出现点相距 500m，且时间间隔大于 24h。最终 151 个信息（野猪 56 个、马鹿 37 个和狍 58 个）满足数据分析要求。此外，为了使研究结果更加准确，将样线调查和监测 2 种方法收集的动物死亡个体信息整合到一起用于有蹄类动物死亡建模分析。

3.3.2　死亡原因分析

本研究将动物死亡原因划分为 4 类：①自然死亡；②猎套致死；③毒药致死；④狗围猎杀。此外，为了进一步分析动物死亡与季节的关系，本研究将 1 年时间划分为 2 个时期：①11 月至翌年 4 月积雪覆盖期；②5～10 月非积雪覆盖期。

3.3.3　有蹄类动物出现点数据来源

在本研究中，使用了 2 个时期收集的有蹄类动物出现点数据：①2003 年、2006 年晚冬季节和 2005 年早春季节开展的 105 条样线（每条长 5km）调查收集的 558 个动物活动点（135 个野猪出现点、165 个马鹿出现点、258 个狍出现点）；②2008 年冬季和 2009 年早春积雪覆盖期野外调查的 240 条样线收集的 234 个动物活动点（63 个野猪出现点、54 个马鹿出现点、117 个狍出现点）。

3.3.4　生境选择分析

3.3.4.1　基于有蹄类动物活动点的生境选择指数和生境适宜性评价

在本研究中，"生境类型"代表生境变量在"生境等级"中的分类。在利用有蹄类动物出现点数据分析生境选择时，综合参考以前的研究，最终选择 6 个生境因子用于生境选择分析（Jiang et al., 2011；Zhou et al., 2006；周绍春等，2010）

（表 3.1）。生境选择指数（E_i^*）分析方法用于分析 3 种有蹄类动物对单个生境因子不同等级的选择程度（韩宗先等，2004）。生境选择指数公式为

$$E_i^* = \frac{W_i - 1/n}{W_i + 1/n} \tag{3.1}$$

$$W_i = \frac{r_i / p_i}{\sum (r_i / p_i)} \tag{3.2}$$

式（3.1）中，n 为一种生境因子的等级；W_i 为选择系数。式（3.2）中，p_i 为动物可获得的第 i 类等级特征的某类生境的面积占整个区域面积的百分比；r_i 为动物在第 i 类等级特征生境中活动点的数量占整个研究地区收集到的该动物活动点数量的百分比；E_i^* 是生境选择指数值，范围为[-1, +1]，表示动物对不同生境的利用程度。当 E_i^*=-1，表示动物强烈回避；E_i^*=1，表示动物强烈利用；E_i^*=0，表示动物随机选择。为了探究动物对某一生境因子第 i 类等级生境的利用程度是否具有统计显著性，通过卡方检验比较第 i 类等级特征的区域内动物活动点的数量和基于随机选择理论的数量是否存在差异，只有具有统计显著性差异的生境因子才被用于动物生境适宜性评价（表 3.2）。拥有统计显著性差异且 E_i^*>0 的生境被划分为适宜生境；拥有统计显著性差异且 E_i^*<0 的生境被划分为不适宜生境；E_i^*>0 且没有统计显著性差异的生境被划分为次适宜生境；E_i^*<0 且没有统计显著性差异的生境被划分为一般适宜生境（蒋志刚，2006）。我们将 0～3 分别赋值给相应等级的生境（0 为不适宜生境，1 为一般适宜生境，2 为次适宜生境，3 为适宜生境）。最终，本研究利用坡向、海拔、植被和到居民点的距离 4 个生境因子建立了多个生境因子的生境适宜性指数（HSI）评价模型。

表 3.1 生境因子的范围与等级

序号	坡向	坡度（°）	海拔（m）	植被类型	到居民点的距离（m）	到公路的距离（m）
1	平	<5	0～150	针叶林	0～1500	0～1000
2	北	5～15	150～300	针阔混交林	1500～3000	1000～2000
3	东北	15～25	300～450	阔叶混交林	3000～4500	2000～3000
4	东	>25	450～830	灌丛	4500～6000	3000～4000
5	东南			其他	>6000	>4000
6	南					
7	西南					
8	西					
9	西北					

表 3.2 基于活动点建立的野猪、马鹿和狍的生境选择指数及利用

序号	坡向	海拔	植被	到居民点的距离
1	-0.22^a, 0.18^b, -0.10^c	-0.08^a, -0.05^b, -0.16^c	-0.16^a, -0.15^b, -0.13^c	-0.16^a, -0.23^b, -0.11^c
2	-0.12^a, -0.25^b, -0.29^c	0.32^{a*}, 0.18^b, 0.41^{c*}	0.45^{a*}, 0.25^b, 0.21^c	0.15^a, 0.05^b, 0.01^c
3	0.08^a, 0.10^b, 0.31^c	0.11^a, 0.36^{b*}, 0.06^c	0.36^a, 0.38^{b*}, 0.57^{c*}	0.43^{a*}, 0.22^b, 0.36^{c*}
4	0.40^a, 0.32^b, 0.45^{c*}	-0.15^a, -1^{b*}, -1^{c*}	-0.29^a, -0.12^b, -0.28^c	0.19^a, 0.47^{b*}, 0.18^c
5	0.19^a, 0.13^b, 0.12^c		-1^{a*}, -1^{b*}, -0.31^c	0.02^a, 0.06^b, -0.11^c
6	-0.23^a, 0.19^b, -0.22^c			
7	-0.14^a, -0.07^b, -0.11^c			
8	-0.65^{a*}, -0.37^b, -0.58^{c*}			
9	-0.15^a, -0.55^{b*}, -0.26^c			

注：a. 野猪；b. 狍；c. 马鹿。*表示显著性（$P<0.05$）

如果模型中任何一个生境成分值 $V_i=0$，则 HSI=0，表明某一生境斑块为不适宜生境。在分析中，我们假定坡向、海拔、植被和到居民点的距离 4 个生境因子在 3 种有蹄类动物生境选择中扮演着同等重要的角色（Thomasma，1981；Jiang et al.，2011）。基于此，用 4 个生境因子的几何平均数 $HSI_{geometric}$ 为每种动物建立生境评价模型，模型公式为

$$HSI_{geometric} = (V_1 V_2 V_3 V_4)^{1/4} \quad (3.3)$$

V_1、V_2、V_3 和 V_4 均被赋予生境因子相应的适宜性等级值。HSI 模型值为 0~3（0 为不适宜生境，0~1 为一般适宜生境，1~2 为次适宜生境，2~3 为适宜生境）。最终建立野猪、马鹿和狍的生境评价模型，并进一步合并 3 个物种的生境评价模型图去形成一个多物种（野猪、马鹿和狍）出现的综合生境评价模型图。

此外，虽然水源是动物生存最重要的三要素之一（Simcharoen et al.，2008），但在本研究中，由于完达山东部林区冬季寒冷，积雪期长，河流封冻，分布在该地区的动物对水源的需求主要来自雪被。因此，动物与水源距离之间的关系不被考虑进入生境选择模型。

3.3.4.2 有蹄类动物死亡危险、综合的有蹄类动物死亡危险评价模型

本研究采用与通过有蹄类动物出现点进行生境选择和质量评价相同的分析方法来探究各生境因子与有蹄类动物死亡之间的关系，并建立每个物种的死亡危险评价模型，然后将 3 个物种的死亡危险评价模型进行综合分析，形成一个 3 个物种的综合死亡危险评价模型，并对死亡危险等级进行分类。

本研究采用 Falcucci 等（2009）的分析方法，根据适宜生境和死亡危险等级的分类与交互作用，将研究地区有蹄类动物的生境划分为 7 个亚类。

3.4 研究结果

3.4.1 有蹄类动物死亡原因

研究结果表明，人类盗猎引起的有蹄类动物死亡占 87.42%，自然死亡仅占 12.58%。对单个物种死亡原因的分析表明，40.07%的野猪死亡源于毒药致死，27.79% 由猎套致死；51.35%的马鹿死亡源于毒药致死，40.54%由猎套致死；29.31%的狍死亡源于毒药致死，56.90%由猎套致死，见图 3.1。对死亡时间的分析表明，冬季和早春季节是林区盗猎最猖獗的时期，该时期 3 种动物死亡数占死亡总数的 86.00%。

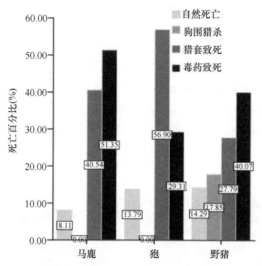

图 3.1 2007～2009 年完达山区域 56 头野猪、37 头马鹿、58 头狍死亡原因分析

3.4.2 生境选择分析

用动物出现点分析生境选择时，剔除了统计不显著的坡度和公路因子，利用统计显著的 4 个因子进行生境选择建模分析（表 3.2）。在进行有蹄类动物死亡危险建模分析时，剔除了不具有统计显著性差异的坡度和居民点，利用具有统计显著性差异的 4 个生境因子建立生境选择模型（表 3.3）。

3.4.3 生境评价

通过物种出现模型创建的模型图计算结果表明（图 3.2a），研究地区野猪不

表 3.3 基于死亡点建立的野猪、马鹿和狍生境选择指数与利用显著性

序号	坡向	海拔	植被	公路
1	-0.11^a, 0.05^b, -0.10^c	-0.04^a, -0.06^b, -0.05^c	-0.06^a, -0.08^b, -0.15^c	0.36^{a*}, 0.31^{b*}, 0.42^{c*}
2	-0.13^a, -0.18^b, -0.16^c	0.02^a, 0.05^b, 0.09^c	0.38^{a*}, 0.16^b, 0.11^c	0.21^a, 0.22^b, 0.21^c
3	0.23^a, 0.17^b, 0.13^c	0.21^{a*}, 0.39^{b*}, 0.28^{c*}	0.15^a, 0.43^{b*}, 0.45^{c*}	0.06^a, 0.09^b, 0.06^c
4	0.30^{a*}, 0.25^{b*}, 0.33^{c*}	-0.15^a, -1^{b*}, -1^{c*}	-1^{a*}, -1^{b*}, -1^{c*}	-0.08^a, -0.05^b, -0.10^c
5	0.09^a, 0.02^b, 0.02^c		-1^{a*}, -1^{b*}, -1^{c*}	-0.21^a, -0.18^b, -0.05^c
6	-0.14^a, 0.06^b, -0.18^c			
7	-0.34^a, -0.07^b, -0.11^c			
8	-0.18^a, -0.37^b, -0.15^c			
9	-0.11^a, -0.10^b, -0.15^c			

注：a. 野猪；b. 狍；c. 马鹿。*表示显著性（$P<0.05$）

同等级的生境为：①不适宜生境面积 573.38km²；②一般适宜生境面积 852.25km²；③次适宜生境面积 1101.50km²；④适宜生境面积 1164.93km²。马鹿不同等级的生境为：①不适宜生境面积 954.06km²；②一般适宜生境面积 954.87km²；③次适宜生境面积 1177.95km²；④适宜生境面积 605.18km²。狍不同等级的生境为：①不适宜生境面积 303.49km²；②一般适宜生境面积 374.06km²；③次适宜生境面积 1596.07km²；④适宜生境面积 1418.44km²。

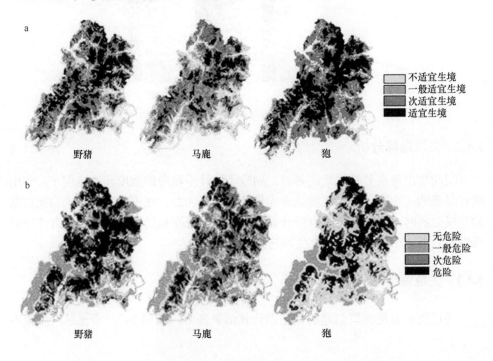

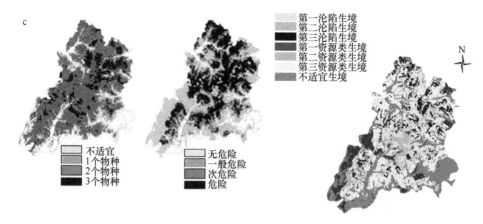

图3.2 野猪、马鹿和狍出现模型（a）、死亡模型（b）和出现-死亡模型（c）

利用具有统计显著性的坡向、海拔、植被和公路4个生境因子分别建立野猪、马鹿、狍的死亡危险生境评价模型。模型分析结果表明（图3.2b），野猪分布的不同危险等级的生境为：①无危险生境面积511.35km^2；②一般危险生境面积957.35km^2；③次危险生境面积846.59km^2；④危险生境面积1376.77km^2。马鹿分布的不同危险等级的生境为：①无危险生境面积557.87km^2；②一般危险生境面积1247.92km^2；③次危险生境面积1149.33km^2；④危险生境面积736.94km^2。狍分布的不同危险等级的生境为：①无危险生境面积1667.19km^2；②一般危险生境面积663.46km^2；③次危险生境面积285.39km^2；④危险生境面积1076.02km^2。

基于出现点的综合生境适宜性评价划分不同质量等级的生境为：①不适宜生境（没有物种出现）面积297.49km^2，占总面积的8.06%；②一般适宜生境面积528.06 km^2，占总面积的14.30%（1个物种出现）；③次适宜生境面积2337.07km^2，占总面积的63.30%（2个物种同时出现）；④适宜生境（3个物种同时出现）面积529.44km^2，占总面积的14.34%，见图3.2c。基于综合死亡危险生境评价图划分不同危险等级的生境：①无危险生境面积448.06km^2，占总面积的12.13%；②一般危险生境面积1531.00km^2，占总面积的41.47%；③次危险生境面积997.30km^2，占总面积的27.01%；④危险生境面积715.70km^2，占总面积的19.39%，见图3.2c。

模型图划分不同等级的生境为：①297.49km^{2}为不适宜3个物种栖息的生境，占总面积的8.06%；②200.44km^{2}为第一沦陷生境，占总面积的5.43%；③1209.85km^{2}为第二沦陷生境，占总面积的32.76%；④302.77km^{2}为第三沦陷生境，占总面积的8.20%；⑤329.00km^{2}为第一资源类生境，占总面积的8.91%；⑥1127.22km^{2}为第二资源类生境，占总面积的30.53%；⑦225.29km^{2}为第三资源类生境，占总面积的6.11%，见图3.2c。

3.5 讨 论

3.5.1 有蹄类动物死亡原因及比例

由于盗猎属于非法行为，隐蔽性强，很难被发现，因此3种有蹄类动物死亡数据主要来自长期对东北虎的监测收集（周绍春等，2010）。死亡数据分析表明，自然死亡仅占12.58%，人类盗猎引起的有蹄类动物死亡占87.42%。在自然死亡的个体中，3只野猪可能是由于冬季严寒，雪被较厚，觅食困难饥饿致死；1只马鹿和1只狍是因为年老体衰自然死亡。本研究表明，乱捕滥猎是导致有蹄类动物死亡的主要原因。钢丝套和投毒是盗猎者经常采用的2种偷猎方式，这2种方式给有蹄类动物生存带来严重威胁。我们的研究说明了有蹄类动物种群数量在研究地区为什么下降如此之快（Baillie et al., 2004；Kühl et al., 2009；张常智和张明海，2011），林区经济不发达、生产力低、贫困和失业人口增加是导致林区居民在近几年增加盗猎有蹄类动物的主要原因（Kühl et al., 2009）。本研究也表明，部分林区居民为了家庭获得更多的经济收入，铤而走险盗猎有蹄类动物（野猪、马鹿和狍）。此外，已有研究表明，在人类活动频繁的区域，人们对野生动物的保护态度与有蹄类动物的生存息息相关（Linnell et al., 2001）。因此，在完达山东部林区，要保护和恢复有蹄类动物种群，需要我们多开展能引导公众参与具有教育意义的动物保护活动，以及增加就业机会，提高林区居民经济收入。如果公众对有蹄类动物的保护态度向积极方面发展，这将有利于有蹄类动物的生存，促进种群恢复（Linnell et al., 2001）。此外，由于盗猎主要发生在晚冬和早春季节，且盗猎行为隐秘，也需要在政府的支持下成立专门的反盗猎巡逻队伍。

3.5.2 生境选择和质量评价

深入研究掌握野生动物的生态需求在动物保护管理工作中极其重要（Freitas et al., 2010）。然而，环境资源怎样影响动物的生境选择行为是野生动物研究的焦点问题之一。在生境选择研究中，科研人员提出2个假设：①野外记录的资料能够被用于推测动物对生境的选择；②进行的生境选择分析是否与目标物种的适合度和种群增长相联系（Alldredge and Ratti, 1986; Porter and Church, 1987）。此外，在研究环境资源怎样影响动物个体生存时，最重要的就是研究动物个体怎样选择利用不同的生境因子（Ratcliffe and Crowe, 2001; Simcharoen et al., 2008）。

研究发现,各生境因子对有蹄类动物生境选择的影响程度存在差异(Podchong et al., 2009)。一些研究表明,东坡向较暖,食物资源丰富,3 种动物喜欢在这类生境中活动 (Singer et al., 1981; 高中信等, 1995)。此外,植被类型也是影响生境选择的一个主要因子。在冬季,针叶林中食物资源(如松子)丰富,野猪常在针叶林中活动 (Grodzinski and Sawicka-Kapusta, 1970)。但我们的研究发现,人类活动干扰对有蹄类动物生境选择和生存均产生显著的不利影响。首先,在研究地区,由于林区居民收集针叶林中的松子,野猪回避针叶林和远离居民区;其次,虽然林间公路窄,车流量相对较少,不会对有蹄类动物的生境选择产生直接的显著影响,但是林间公路方便盗猎者前往动物频繁活动的区域偷猎;最后,其他的研究还表明,当鹿类动物与人相遇时,它们会产生心理压力而影响机体健康 (Sauerwein et al., 2004),从而产生回避行为 (Bechet et al., 2004)。在冬季,农田、林间湿地是两类缺乏食物资源和使有蹄类动物暴露无遗的特殊生境,3 种有蹄类动物均回避农田和林间湿地生境,偏好选择针阔混交林和阔叶林生境 (周绍春等, 2010)。

我们的研究发现有蹄类动物经常在海拔 150~450m 的区域活动,因此对这些区域应该长期保护。该海拔区域森林植被以针阔混交林和阔叶林为主,食物资源丰富,隐蔽条件好,人类活动干扰小 (周绍春等, 2010)。

有蹄类动物死亡数据主要来自长期的东北虎野外种群监测和猎物资源(野猪、马鹿和狍)调查收集。通过分析收集的死亡个体位置信息,发现死亡事件主要发生在东坡向针阔混交林和阔叶林,这表明动物的出现与死亡存在潜在的联系。本研究还发现,森林采伐增加了公路密度,为居民伐木提供了更方便的条件,但是人类活动干扰给有蹄类动物的生存带来潜在威胁,居民点和公路建设增加了有蹄类动物的死亡风险 (Stankowich, 2008; Laurance et al., 2010)。

在人类活动频繁的林区,人类活动对目标动物生存的影响是生境评价的一个关键因素 (Naves et al., 2010; Thomas and Kunin, 2010)。虽然通过生境评价可以认识动物的生态需求,然后形成有效的动物保护管理计划 (Underwood, 2014),但是仅用动物出现点位置信息开展生境评价获得的高质量生境评价模型存在问题,在野生动物管理中运用该研究成果开展动物保护工作容易形成错误的保护计划 (Falcucci et al., 2009)。本研究利用环境因子和人类活动干扰因子建立生境评价模型,这将避免在制订保护计划时产生潜在的错误。在研究地区,有研究认为生境丧失是有蹄类动物生存的主要威胁 (李言阔等, 2008),但在本研究中,对有蹄类动物的出现−死亡模型的分析表明,生境的可获得性并不是有蹄类动物生存的主要限制因子 (Zhang and Liu, 2008)。对综合评价模型的分析表明,46.39%的适宜生境存在有蹄类动物死亡风险,这也说明传统的生境评价模型分析结果用于制订动物保护计划时存在缺点。例如,神顶峰周围林区远离公路、农田和居民区,是有蹄类动物栖息的沦陷生境，我们认为保护该区域的有蹄类动物首先应该有效控制人类乱捕滥猎和破坏森林资源。

有蹄类动物的出现–死亡模型是基于一个假设建立的，即人类活动引起的动物死亡信息能够用于建模分析人类活动对有蹄类动物生存的影响，有蹄类动物的出现信息也能够用于建模分析有蹄类动物的生境适宜性。

在我们的研究中，有蹄类动物出现模型分析的一个缺点在于收集的数据仅来自冬季积雪覆盖期，但是冬季的雪地足迹信息比其他季节收集的数据更加精确（Zhang and Liu, 2008；周绍春等, 2010；Alexander et al., 2011）。此外，另一个缺点是有蹄类动物死亡模型分析的数据来源于全年收集。不过，86.00%的有蹄类动物死亡发生在冬季，这就使用于有蹄类动物出现模型和死亡模型分析的两部分数据资料在时间上趋于统一。

综合的有蹄类动物出现–死亡模型分析表明，许多适宜生境都因为人类盗猎成为沦陷生境（图3.2c）。用有蹄类动物死亡危险生境建模，然后在有蹄类动物出现–死亡模型建模分析中对重分类的生境分析表明，盗猎是导致有蹄类动物死亡的主要威胁因素。此外，公路建设减少了生境适宜性并增加了沦陷生境面积。

3.6 本章小结

分析引起有蹄类动物死亡的原因及其生存的生境是有蹄类动物保护的重要组成部分。为了保护有蹄类动物资源，本研究分析了有蹄类动物死亡原因，建立了野猪、马鹿和狍综合的有蹄类动物出现–死亡危险生境评价模型。对死亡原因的分析表明，猎套和毒药盗猎是导致有蹄类动物死亡的2种主要威胁方式。综合的有蹄类动物出现–死亡危险生境评价模型分析表明，研究地区大面积的栖息生境存在人类盗猎引起有蹄类动物死亡的潜在危险。因此，研究认为，要保护好有蹄类动物资源，资源类生境应该被保护，生境退化应该被阻止，对沦陷生境应该加强管理，消除动物死亡危险隐患。管理部门应该采取有效措施，如增加巡逻减少盗猎，以及为当地居民提供更多的就业机会，增加其经济收入，提供更多的教育机会，加强保护动物的思想意识，尽可能减少人为因素引起的动物死亡。

参考文献

高中信, 张明海, 胡瑞滨. 1995. 小兴安岭地区野猪冬季卧息地选择的初步研究. 兽类学报, 15(1): 25-30.

韩宗先, 魏辅文, 张泽钧. 2004. 蜂桶寨自然保护区小熊猫对生境的选择. 兽类学报, 24(3): 185-192.

蒋志刚. 2006. 陕西老县城自然保护区的生物多样性. 北京: 清华大学出版社: 78-79.
李言阔, 张明海, 蒋志刚. 2008. 基于生境可获得性的完达山地区马鹿(*Cervus elaphus xanthopygus*)冬季生境选择. 生态学报, 28(10): 4619-4628.
马逸清, 等. 1986. 黑龙江省兽类志. 哈尔滨: 黑龙江科学技术出版社.
张常智, 张明海. 2011. 黑龙江省东完达山地区东北虎猎物种群现状及动态趋势. 生态学报, 31(21): 6481-6487.
周绍春, 孙海义, 张明海, 等. 2008. 黑龙江省东北虎分布区域及其数量动态. 兽类学报, 28(1): 165-173.
周绍春, 张明海, 孙海义, 等. 2010. 完达山东部林区野猪种群数量和栖息地特征的初步分析. 兽类学报, 30(1): 28-34.
Alexander S M, Paquet P C, Logan T B, et al. 2011. Snow-tracking versus radiotelemetry for predicting wolf-environment relationships in the Rocky Mountains of Canada. Wildl Soc Bull, 33(4): 1216-1224.
Alldredge J R, Ratti J T. 1986. Comparison of some statistical techniques for analysis of resource selection. The Journal of Wildlife Management, 50: 157-165.
Baillie J E M, Hiltontaylor C, Stuart S N. 2004. 2004 IUCN Red List of Threatened Species.
Bechet A, Giroux J, Gauthier G, et al. 2004. The effects of disturbance on behaviour, habitat use and energy of spring staging snow geese. Journal of Applied Ecology, 41(4): 689-700.
Blake S, Strindberg S, Boudjan P, et al. 2007. Forest elephant crisis in the Congo Basin. PLoS Biology, 5(4): e111.
Corlett R T. 2007. The impact of hunting on the mammalian fauna of tropical Asian forests. Biotropica, 39(3): 292-303.
Delibes M, Gaona P, Ferreras P. 2001. Effects of an attractive sink leading into maladaptive habitat selection. American Naturalist, 158(3): 277-285.
Duberstein C A, Simmons M A, Sackschewsky M R, et al. 2008. Development of a habitat suitability index model for the sage sparrow on the Hanford Site. Revista Colombiana De Entomología, 37(2): 228-233.
Falcucci A, Ciucci P, Maiorano L, et al. 2009. Assessing habitat quality for conservation using an integrated occurrence-mortality model. Journal of Applied Ecology, 46(3): 600-609.
Freitas C, Kovacs K, Lydersen C, et al. 2010. A novel method for quantifying habitat selection and predicting habitat use. Journal of Applied Ecology, 45(4): 1213-1220.
Garshelis D. 2000. Delusions in Habitat Evaluation: Measuring Use, Selection, and Importance. New York: Colomia University Press: 111-164.
Grodzinski W, Sawicka-Kapusta K. 1970. Energy values of tree-seeds eaten by small mammals. Oikos, 21(1): 52-58.
Groom M J, Meffe G K, Carroll C R, et al. 2006. Principles of Conservation Biology. Sunderland: Sinauer Associates: 63-109.
Guisan A, Thuiller W. 2010. Predicting species distribution: offering more than simple habitat models. Ecology Letters, 8(9): 993-1009.
IUCN. 2010. IUCN Red List of Threatened Species. http://www.iucnredlist.org[2017-3-6].
Jiang G S, Ma J Z, Zhang M H. 2011. Spatial distribution of ungulate responses to habitat factors in Wandashan forest region, Northeastern China. Journal of Wildlife Management, 70(5): 1470-1476
Jiang G S, Zhang M H, Ma J Z. 2008. Habitat use and separation between red deer *Cervus elaphus xanthopygus* and roe deer *Capreolus pygargus bedfordi* in relation to human disturbance in the

Wandashan Mountains, northeastern China. Wildlife Biology, 14(1): 92-100.
Johnson D H. 1980. The comparison of usage and availability measurements for evaluating resource preference. Ecology, 61(1): 65-71.
Kühl A, Balinova N, Bykova E, et al. 2009. The role of saiga poaching in rural communities: linkages between attitudes, socio-economic circumstances and behaviour. Biological Conservation, 142(7): 1442-1449.
Laurance W F, Croes B M, Guissouegou N, et al. 2010. Impacts of roads, hunting, and habitat alteration on nocturnal mammals in African rainforests. Conservation Biology the Journal of the Society for Conservation Biology, 22(3): 721-732.
Linnell J D C, Swenson J E, Anderson R. 2001. Predators and people: conservation of large carnivores is possible at high human densities if management policy is favourable. Animal Conservation, 4: 345-350.
Loibooki M, Hofer H, Campbell K L I, et al. 2003. Bushmeat hunting by communities adjacent to the Serengeti National Park, Tanzania: the importance of livestock ownership and alternative sources of protein and income. Environmental Conservation, 29(3): 391-398.
Milner-Gulland E J, Bennett E L. 2003. Wild meat: the bigger picture. Trends in Ecology & Evolution, 18(7): 351-357.
Naves J, Wiegand T, Revilla E, et al. 2010. Endangered species constrained by natural and human factors: the case of brown bears in Northern Spain. Conservation Biology, 17(5): 1276-1289.
Nielsen S E, Stenhouse G B, Beyer H L, et al. 2008. Can natural disturbance-based forestry rescue a declining population of grizzly bears? Biological Conservation, 141(9): 2193-2207.
Podchong S, Schmidt-Vogt D, Honda K. 2009. An improved approach for identifying suitable habitat of Sambar Deer (*Cervus unicolor* Kerr) using ecological niche analysis and environmental categorization: Case study at Phu-Khieo Wildlife Sanctuary, Thailand. Ecological Modelling, 220(17): 2103-2114.
Porter W F, Church K E. 1987. Effects of environmental pattern on habitat preference analysis. Journal of Wildlife Management, 51(3): 681-685.
Ratcliffe C S, Crowe T M. 2001. Habitat utilisation and home range size of helmeted guineafowl (*Numida meleagris*) in the Midlands of KwaZulu-Natal province, South Africa. Biological Conservation, 98(3): 333-345.
Sauerwein H, Müller U, Brüssel H, et al. 2004. Establishing baseline values of parameters potentially indicative of chronic stress in red deer (*Cervus elaphus*) from different habitats in western Germany. European Journal of Wildlife Research, 50(4): 168-172.
Simcharoen S, Barlow A C D, Simcharoen A, et al. 2008. Home range size and daytime habitat selection of leopards in Huai Kha Khaeng Wildlife Sanctuary, Thailand. Biological Conservation, 141(9): 2242-2250.
Singer F J, Otto D K, Tipton A R, et al. 1981. Home ranges, movements, and habitat use of European wild boar in Tennessee. Journal of Wildlife Management, 45(2): 343-353.
Sodhi N S, Pin K L, Brook B W, et al. 2004. Southeast Asian biodiversity: an impending disaster. Trends in Ecology & Evolution, 19(12): 654-660.
Stankowich T. 2008. Ungulate flight responses to human disturbance: a review and meta-analysis. Biological Conservation, 141(9): 2159-2173.
Thomas C D, Kunin W E. 2010. The spatial structure of populations. Journal of Animal Ecology, 68(4): 647-657.
Thomasma L E. 1981. Standards for the development of habitat suitability index Models. Wildlife Society Bulletin, 19: 1-171.

Underwood F. 2014. Resource selection by animals: statistical design and analysis for field studies Journal of Animal Ecology, 63(3): 351.

Zhang M H, Liu Q X. 2008. Estimation of winter carrying capacity of wapiti in the Eastern Wandashan Mountains, Heilongjiang Province, China. Acta Theriologica Sinica, 28(1): 56-64.

Zhou S C, Zhang M H, Wang S L. 2006. Habitat selection of red deer (*Cervus elaphus*) and roe deer (*Capreolus capreolus*) in winter in logged and unlogged forest of the Wandashan Mountains, Heilongjiang. Zoological Research, 27(6): 575-580.

第4章 人为干扰对森林植被、东北虎及其猎物的影响

4.1 引　　言

在以人为主导的景观中，人为干扰对植被和野生动物的影响是野生动物管理中的一个关键问题。近几十年来，随着人口增长及人类活动区域的扩展，人类与野生动物之间的冲突变得更加频繁（Woodroffe，2000；Conover，2001；Pettigrew et al.，2012）。人为干扰导致栖息地丧失，直接或间接地导致野生动物产生回避行为（Mace and Waller，1996）。人为干扰可能导致野生动物栖息地利用的改变和降低食物质量（Hernandez and Laundre，2005）。大型食肉动物（>40kg）更容易受到由人为干扰和生境破碎化甚至丧失等栖息地配置与连通性改变的影响（Dusit et al.，2007）。东北虎（*Panthera tigris altaica* Linnaeus，1758）栖息地的主要人类干扰包括道路、定居点、农田、伐木、偷猎、放牧和采石（Kerley et al.，2002；Barbermeyer et al.，2013；Bhattarai and Kindlmann，2013）。一些研究表明，猎物数量和人为干扰是老虎栖息地质量最重要的参数，严重的人为干扰会导致猎物减少甚至老虎的区域性灭绝（Bhattarai and Kindlmann，2013）。然而，大多数关于人为干扰对老虎影响的研究都是描述性的，只描述了其中受干扰的原因，对其机制却鲜见报道。大多数野生老虎仍然受到人为干扰的严重影响，自然保护区内也不例外（Linkie et al.，2003）。另外，很少有研究关注不同的人为干扰对野生动物及其栖息地中的植被的影响。在自然界中，不同的因素对物种的生存往往有不同的作用、贡献（Tisseuil et al.，2013），许多干扰因素往往相互作用共同影响目标类群（Soh et al.，2014），仅仅关注特定的干扰可能会导致分析问题的片面性，因此比较不同类型的人类干扰作用有助于我们理解对野生动物构成威胁的因素的深层复杂机制。

珲春东北虎国家级自然保护区（简称珲春保护区）是我国少数几个同时有野生东北虎和东北豹分布的区域之一，是我国东北虎豹保护的重点区域。前人的研究发现，在珲春保护区内虎、豹及有蹄类动物的主要人为干扰包括道路、放牧、有蹄类动物偷猎（猎套）和作物种植（人参地和农田）（Li et al.，2006；Li，2009；Chen et al.，2011）。由于历史遗留等，珲春保护区内仍有许多牧场存在，放牧是典型的人为干扰。以前的放牧研究主要集中在由家畜掠夺栖息地导致野生动物生境选择变化方面（Hernandez and Laundre，2005；Li et al.，2009），很少评估干扰对东北虎及其栖息地的实际影响。Soh等（2014）的研究表明，在有猎套分布的

区域有蹄类动物分布密度较低,且家牛被东北虎捕食的概率会随着靠近猎套分布区而增加。除放牧活动外,人参种植也是珲春地区典型的人为干扰类型,不同于农田主要分布在山脚等平原地区,参地多选择在山坡上。在开垦参地时,需要清除山坡上所有的乔木与灌木丛,并搭建大量的覆盖蓝色塑料薄膜的参棚。参地的开垦使得大量野生动物栖息地遭到破坏,并且在人参种植期间为防止虫害需要施加大量农药,可能对植被及土壤造成负面影响。研究表明,道路会对虎的活动造成负面影响(Smirnov and Miquelle,1999)。在俄罗斯远东地区,道路会显著降低东北虎的存活率与生殖成功率(Kerley et al.,2002)。虽然多年前珲春保护区已停止伐木,但许多伐木道路仍被用于国防、非林木产品采集和旅游。珲春保护区及其周边有三种道路类型,分别是三级公路、二级公路和一级公路(图4.1)。大部分三级公路为当地居民使用的废弃伐木道路,路面崎岖,土质地面仅供农用车辆使用。二级公路用于连接珲春保护区周围的村庄、水泥地面;珲春保护区内没有柏油铺成的一级公路,但保护区周边有,主要用于连接吉林和黑龙江两省,允许车辆高速行驶(图4.1)。

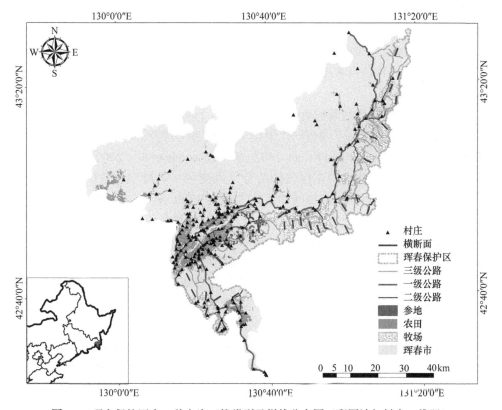

图4.1 珲春保护区内7种人为干扰类型及样线分布图(彩图请扫封底二维码)

统计模型的目的是提供一个基于数学的解释，对解释、参数进行拟合（评估预测是否充分解释响应），加强关联（响应和预测之间的关系是否显著），并确定不同变量的贡献和角色（Guisan et al.，2002）。不同类型的模型为统计建模在生态学中的作用提供了不同的见解（Guisan et al.，2002），并且为产生模式的生态过程提供解读（Austin et al.，1990）。作为线性建模（linear modeling，LM）的数学扩展，GLM 不强制数据进入非自然的尺度，从而允许数据中的非线性和非恒定方差结构（Hastie and Tibshirani，1990）。假设数据来自几个概率分布，包括正态分布、二项分布、泊松分布、负二项分布或伽马分布，其中许多生态数据更符合非正态误差结构（Guisan et al.，2002）。GLM 更灵活，更适合于分析生态关系，而经典高斯分布不能很好地展示生态关系（Austin，1987）。GAM 是 GLM 的半参数扩展（Hastie and Tibshirani，1986，1990）；唯一的基本假设是，函数是可加性的，组件是平滑的。GAM 的优点是能够处理响应变量和解释变量之间高度非线性与非单调的关系，而 GLM 对这些关系没有很强的处理能力（Guisan et al.，2002）。GAM 代表了物种对环境数据特征和非线性函数的真实响应（Suarez and Case，2002）。结构方程模型（structural equation model，SEM）的优点之一是可以估计和测试结构之间的关系，而 GAM 和 GLM 只能交替地显示响应变量和解释变量之间的相关性。与其他一般线性模型相比，SEM 通过研究路径关系来估计因果效应（Grace et al.，2010），它允许使用多个变量来展示结构并解决特定变量的误差问题（Weston and Gore，2006）。

作为捕食者与猎物之间的典型作用方式，下行效应（top-down effect）及上行效应（bottom-up effect）的作用理论受到野生动物生态学家的持续关注（Sabatier，1986；Hunter and Price，1992；Suarezseoane et al.，2002；Aryal et al.，2014）。目前，上行效应的作用理论结合气候变化、病毒学及恐惧生态学理论可以更好地揭示捕食者与猎物间的互作关系（Wilmers et al.，2006；Laundre et al.，2014）。上行效应作用理论的引用可以更好地解释食物链中低等生物对于顶级捕食者的作用机制（Frederiksen et al.，2006）。捕食者与猎物在不同的自然条件下所表现出来的作用机制是不同的：积极的关系表示二者之间是强的上行效应的作用关系，消极的关系表示二者之间是强的下行效应的作用关系，若猎物与捕食者之间无显著关系则表明二者之间可能不存在营养关系（Worm and Myers，2003）。

本部分的目标是确定影响东北虎豹生存的主要人为干扰因子，并为消除或减轻人为干扰对东北虎豹的影响提供建议。我们提出的研究理论假设包括：①人为干扰通过上行效应对珲春保护区中的植被、东北虎及其猎物产生影响；②不同的干扰因子对上行效应网络中不同营养级的影响强度不同。

4.2 研究地区与方法

4.2.1 研究区域

吉林省珲春市面积 5145km², 位于中国东北部。珲春保护区面积 1087km², 东与俄罗斯接壤, 西南与朝鲜接壤, 包括 4 个管理区 (核心区、实验区、缓冲区和社区共同管理区) (Li, 2009)。珲春保护区与俄罗斯 3 个东北虎、东北豹保护区相连, 是中、俄、朝之间东北虎和东北豹活动的重要通道。珲春保护区地处长白山山脉, 属于温带气候, 7~9 月平均降雨量 661mm (占年降雨量的 50%)。珲春保护区主要植被类型为针阔混交林和次生蒙古栎, 主要野生动物有东北虎、东北豹 (*Panthera pardus orientalis*)、梅花鹿 (*Cervus nippon*)、野猪 (*Sus scrofa*)、狍 (*Capreolus pygargus*) 和原麝 (*Moschus moschiferus*)。珲春保护区成立前共有 29 个村庄 14 953 人 (Li, 2009); 2001 年珲春保护区成立, 目前保护区内有 4 个城镇 98 个村庄 (Soh *et al.*, 2014)。

4.2.2 数据收集

为了确定牧场分布和放牧强度, 我们采用实地调查和问卷调查两种方法 (Hasselfinnegan *et al.*, 2013; Zimmermann *et al.*, 2013), 采访了珲春保护区内的每一位牧场主, 调查了牧场分布和牛的数量情况。对于较大的牧场, 由于难以精确地获取其范围和位置, 因此使用 ArcGIS 计算它们的位置和面积; 对于小型的私人牧场, 则使用 HOLUX M241 型轨迹记录仪绕牧场一周来获得其精确位置与面积。

样线设计时采用分层抽样的原则, 按照 10%的抽样强度进行样线设计 (即每 100km² 设计 10km 的样线)。本次调查设计样线 32 条共计 96km, 在样线设计时参考以下因素: ①每条样线长度为 3km, 相邻两条样线的间距不少于 3km; ②林型, 本次调查我们选取了针叶林、阔叶林、针阔混交林及灌木草丛这 4 种主要的生境进行调查。因区内这 4 种生境的面积比为 22∶127∶15∶1, 因此其中调查样线的长度分别为 11.4km、70.3km、10.7km 和 3.6km; ③保护区各功能区, 即珲春保护区内缓冲区、核心区与实验区的面积之比为 6∶5∶1.7, 因此其中调查样线的长度分别为 47km、37km 和 12km。

为了调查保护区内植被的结构与质量, 沿样线每隔 200m 做一个 10m×10m 的大样方, 并记录大样方内所有乔木的种类、棵数、平均胸径及隐蔽级等信息。同时在大样方的 4 角处做 4 个 2m×2m 的小样方, 并记录其中灌木的种类、棵数、可被有蹄类动物采食的枝条数等信息。本次调查共计做了 512 个大样方、2560 个小样方。

采用样线调查法测定野生有蹄类动物和牛的相对丰富度。我们将放牧强度定义为每千米牛的足迹数，对于野生有蹄类动物，计算每个大样方足迹的出现或不出现（1/0）和每千米的足迹数。考虑到无雪被覆盖的调查季节和区域，结合足迹、食物痕迹、粪便等多个参数确定猎物的出现或不出现的计算，也以同样方式计算每个大样方的猎物占有情况。

使用老虎出现记录数据库来评估老虎的分布。该数据库包括2000~2012年在珲春保护区内发现的老虎足迹、捕食地点、粪便等。我们期望根据被确定的老虎出现点数据探测猎物水平对东北虎分布的影响。考虑到雌性成年东北虎的日活动距离平均为7km（Yudakov and Nikolaev，1987），因此以14km为边长制作单位面积为196km^{2}的栅格。使用ArcGIS 9.0的空间分析分区统计，结合东北虎出现点数据计算出每个栅格内东北虎出现的频次。最后把每个栅格内东北虎出现的频次属性值附到该栅格内部所有10m×10m大样方的中心点属性上。

对于人为干扰因素，我们测量了每个大样方到每一种干扰的最近距离，具体包括放牧、人参种植、农田、一级公路（柏油路）、二级公路（水泥路）、三级公路（土路）、居民点等7种人为干扰类型。利用ArcGIS 9.3，我们获得了人参地、农田、一级公路、二级公路、三级公路、居民点的分布矢量数据。

4.2.3 数据分析

使用皮尔逊相关系数（Pearson correlation coefficient，R）来对各种人为干扰等自变量进行共线性检验，如树种、密度、胸径、灌木密度、新生可食用的灌木树枝密度、香农-维纳指数（Shannon-Wiener index）、放牧强度、每个小样方内有蹄类动物的出现或不出现（1/0），以及每个大样方距离人参土地、农田、村庄、一级公路、二级公路、三级公路的最近距离。

如果皮尔逊相关系数大于0.5，则认为变量之间存在很强的相关性，在这种情况下，为了确保进入模型的所有变量都具有低相关性，删除了其中一个自变量（Ramsay *et al.*，2003）。

为探索与人为活动相关的上行效应的作用机制，我们认为干扰分为三个层次，分别为人为干扰—植被、人为干扰—虎豹主要猎物、人为干扰—东北虎。在植被层次结构中，首先只考虑人为干扰；其次利用人为干扰和植被参数对有蹄类动物的影响进行试验；最后利用人为干扰、植被和猎物参数共同确定人类活动对东北虎的影响。在每个层次中，使用了广义可加模型（GAM）、广义线性模型（GLM）和结构方程模型（SEM）来测试具体的效应机制。然后，我们建立了人类干扰和三个层次之间的相互作用网络，以验证我们的假设：人类干扰是否沿上行效应的生态作用关系链条影响植被、东北虎及其猎物。特别值得注意的是，由于有蹄类动物足迹非常少，我们在每个大样方中使用有蹄类动物出现或不出现（1/0）作为它们在 GLM 中的相对丰富度。对于 GAM，我们将每千米的有蹄类足迹数作为它们的相对丰富度，因为这个模型不允许包含许多零值的响应变量。我们采用贡献分割模型（HPM）(Chevan and Sutherland，1991) 分别从三个层次量化每个生物学参数在其对植被、猎物和老虎的影响方面的解释力。HPM 通过 R 语言中的 hier.part 包实现（R Development Core Team，2010），其他分析在 Rv2.12.1 中使用 mgcv 和 hier 进行（R Development Core Team，2010；Li and Wang，2013）。所有模型参数均进行标准化。

4.3 结　　果

通过走访问卷调查，共计在珲春保护区及其周边地区调查到 63 个牧场（共计 3066 头牛）。其中 42 个牧场分布于珲春保护区内，总面积达 355km^2，占保护区总面积的 32.7%。其中有 19 个牧场全部或部分位于保护区核心区内，占核心区总面积的 53.3%。调查到所有牧场内的平均放牧密度为 11 头/km^2。Pearson 相关系数表明，农田和定居点与二级公路、一级公路的相关性较高，放牧、人参种植、一级公路和二级公路在 7 个人为干扰中处于最后（相关性最低）。从 GAM、GLM 及 SEM 探测的作用关系网络中我们可以发现，人为干扰已经影响东北虎食物链的各个层级并形成一种上行效应的作用方式（图 4.2）。贡献分割模型（HPM）显示不同的人为干扰类型对于不同的生物层级有着不同的作用强度（图 4.3）。

4.3.1　人为干扰对植被的影响

在人为干扰对于植被的 16 种作用途径中，GAM 探测出 11 种显著作用关系，

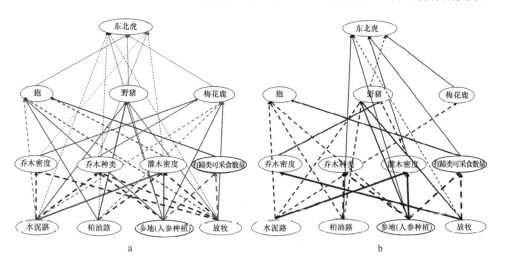

a　　　　　　　　　　　　　　　　　　　b

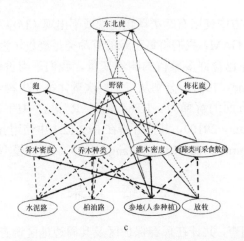

图4.2 GAM（a）、GLM（b）、SEM（c）分别检验水泥路、柏油路、人参种植及放牧作用对于植被、东北虎及其猎物的影响作用（彩图请扫封底二维码）

图a中红色实线表示积极作用，红色虚线表示消极作用，黑色实线表示"n"形作用，黑色虚线表示"u"形作用，蓝线表示波浪形作用。图b和c中，黑色实线表示积极作用，黑色虚线表示消极作用

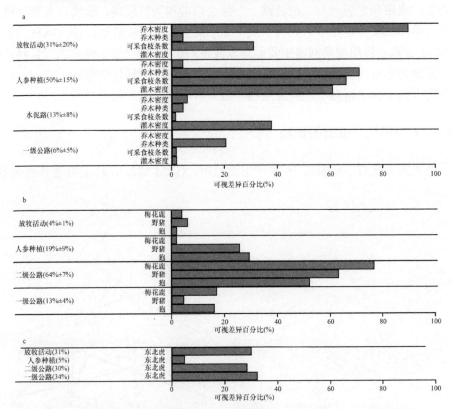

图4.3 贡献分割模型量化的放牧活动、人参种植、二级公路及一级公路对植被、东北虎及其猎物的影响

GLM 探测出 7 种显著作用关系，SEM 探测出 6 种显著作用关系（图 4.2，表 4.1）。所有模型均显示人参种植和放牧活动对有蹄类可采食枝条数有负作用（图 4.2，表 4.1），模型贡献率分别为 65.4% 和 30.9%（图 4.3）。GAM 在 n、u 和 w（波）形状上表现出更多的非单调显著效应，而 GLM 和 SEM 只探测到线性关系。HPM 表明，相比二级公路（13%±8%）和主要道路（6%±5%），放牧活动（31%±20%）和人参种植（50%±15%）有更大的贡献率。放牧活动主要影响乔木密度（u 形，GAM；正作用，GLM）和可采食枝条数，人参种植对乔木种类（u 形，GAM；负作用，GLM 和 SEM）、可采食枝条数（负作用，GAM、GLM 和 SEM）、灌木密度（波形，GAM；正作用，GLM 和 SEM）也有类似的主要贡献（图 4.2，表 4.1）。

表 4.1 GAM、GLM、SEM 模型结果

编号	作用关系	GAM	GLM	SEM	作用结果是否相同
1	水泥路 - 乔木密度	u			×
2	水泥路 - 乔木种类	u			×
3	水泥路 - 灌木密度	w	+	+	×
4	水泥路 - 有蹄类可采食枝条数				
5	水泥路 - 狍出现概率	−	−	−	√
6	水泥路 - 野猪出现概率	w			×
7	水泥路 - 梅花鹿出现概率	−	−	−	√
8	水泥路 - 东北虎出现频率	u			×
9	柏油路 - 乔木密度				
10	柏油路 - 乔木种类		−	−	×
11	柏油路 - 灌木密度	u	+		×
12	柏油路 - 有蹄类可采食枝条数		−		×
13	柏油路 - 狍出现概率	w	−	−	×
14	柏油路 - 野猪出现概率	n	−	−	×
15	柏油路 - 梅花鹿出现概率				
16	柏油路 - 东北虎出现频率	u	+	+	×
17	人参种植 - 乔木密度				
18	人参种植 - 乔木种类	u			×
19	人参种植 - 灌木密度	w	+	+	×
20	人参种植 - 有蹄类可采食枝条数	−	−	−	√
21	人参种植 - 狍出现概率	w	−	−	×
22	人参种植 - 野猪出现概率	w	+	+	×
23	人参种植 - 梅花鹿出现概率	w			×

续表

编号	作用关系	GAM	GLM	SEM	作用结果是否相同
24	人参种植 - 东北虎出现频率	w	+	+	×
25	放牧活动 - 乔木密度	u	+	+	×
26	放牧活动 - 乔木种类	u			×
27	放牧活动 - 灌木密度	u			×
28	放牧活动 - 有蹄类可采食枝条数	−	−	−	√
29	放牧活动 - 狍出现概率	w			×
30	放牧活动 - 野猪出现概率	+	+	+	√
31	放牧活动 - 梅花鹿出现概率	w			×
32	放牧活动 - 东北虎出现频率	u	−	−	×
33	乔木密度 - 狍出现概率				
34	乔木密度 - 野猪出现概率		+	+	×
35	乔木密度 - 梅花鹿出现概率	w			×
36	乔木密度 - 东北虎出现频率	−			
37	乔木种类 - 狍出现概率			+	×
38	乔木种类 - 野猪出现概率				
39	乔木种类 - 梅花鹿出现概率	+			×
40	乔木种类 - 东北虎出现频率	+		+	×
41	灌木密度 - 狍出现概率	u			×
42	灌木密度 - 野猪出现概率		−	−	×
43	灌木密度 - 梅花鹿出现概率				
44	灌木密度 - 东北虎出现频率	u	+	+	×
45	有蹄类可采食枝条数 - 狍出现概率	n	+	+	×
46	有蹄类可采食枝条数 - 野猪出现概率				
47	有蹄类可采食枝条数 - 梅花鹿出现概率				
48	有蹄类可采食枝条数 - 东北虎出现频率		+		×
49	狍出现概率 - 东北虎出现频率	w			×
50	野猪出现概率 - 东北虎出现频率	w	−	−	×
51	梅花鹿出现概率 - 东北虎出现频率	+	+	+	√

注：+、−、n、u、w 分别表示正、负、n 形、u 形、波形相关。"√"表示 3 种模型显示相同的作用结果，否则为"×"。在 3 个生物层级共 51 种作用关系中，广义可加模型、广义线性模型及结构方程模型分别监测出 37 种、27 种、28 种显著作用关系，共有 6 种作用在 3 个模型中显示出相同的结果

4.3.2 人为干扰对猎物的影响

在人为干扰对猎物影响的模型分析中，从 12 种作用途径中 GAM、GLM、SEM 分别显示了 12 个、8 个、9 个显著的人为干扰因子对猎物的直接影响（图 4.2，表 4.1）。HPM 显示二级公路对于有蹄类干扰有 64%±7%的贡献率，大于其他对猎物有影响的人为干扰因子（图 4.3）；GAM 表明，这 4 种干扰对梅花鹿、野猪和狍均有显著影响（图 4.2，表 4.1）。二级公路和一级公路对梅花鹿有负影响，三个模型均表明二级公路对狍和梅花鹿的影响呈负相关，分别为 51.9%和 76.3%的贡献率。人类干扰也以间接的方式影响猎物。以 GAM 为例，二级公路和人参种植对树种有 u 形影响，树种对梅花鹿的存在有正影响（图 4.2）。

4.3.3 人为干扰对东北虎的影响

在人为干扰对东北虎影响的模型分析中，从 4 种作用途径中 GAM、GLM、SEM 分别显示了 4 个、3 个、3 个人为干扰因子对猎物层次结构的直接显著影响（图 4.2，表 4.1），GLM 和 SEM 表明，一级公路和人参种植对东北虎有积极的影响，但放牧有负影响，而 GAM 表明，这些影响可能不是线性的（图 4.2）。HPM 显示，放牧（31%）、二级公路（30%）和一级公路（34%）在影响东北虎出现方面有几乎相同的主要贡献率（图 4.3）。

我们发现人为干扰可能通过影响植被和猎物来影响东北虎的出现（图 4.2），GAM 和 SEM 表明，一级公路和二级公路对梅花鹿的出现有负影响（梅花鹿是东北虎捕食的首选对象）。三种模型均显示二级公路对狍出现概率有负影响。GAM 认为，东北虎的出现频率受狍的影响较大。三种模型均表明，放牧和人参种植均可减少有蹄类动物喜食的可采食枝条数；根据 GLM 和 SEM 分析，可采食枝条数对狍的出现概率有正影响。狍的出现对东北虎的出现有显著的 n 形影响。

4.4 讨 论

4.4.1 放牧的影响

放牧对东北虎出现的影响是多层次的。放牧的间接作用表现为上行效应，即放牧活动通过植被和捕食层次影响东北虎。最重要的发现之一是，放牧活动可以减少有蹄类动物偏好的可采食枝条数（图 4.2b、c），以致有蹄类猎物回避放牧地区，类似的研究也证实了我们的发现（Bhattarai and Kindlmann, 2013）。这种现象可以从狍的行为生态学方面解释，中小型有蹄类动物偏好茂密的灌木生境（Jiang et al., 2008, 2010）。狍的减少可能会影响东北虎的出现频率。Soh 等（2014）

指出盗猎是老虎猎物丰富度下降的重要原因，盗猎对猎物的负面影响加剧了老虎对牛的捕食。然而，Soh 等（2014）未能深入探讨牧场与有蹄类动物分布和栖息地之间的关系。我们认为，放牧通过资源型竞争造成的猎物损失是对东北虎的另一种严重威胁。GAM 表明，放牧对猎物有直接的非线性影响，但 HPM 表明，放牧对植被的影响贡献率为 31%±20%，放牧对猎物的贡献只是 4%±1%。因此，我们认为，与对猎物的影响相比，放牧可能主要通过对植被的影响来影响猎物。

放牧对东北虎生存的直接影响是通过猎物分布的变化来实现的。珲春保护区超过 30% 的区域是牧场，平均放牧密度为 11 头/km²。在珲春保护区中，马鹿、梅花鹿、野猪和狍的预测密度分别为每平方千米 0~0.16 只、0~0.19 只、0~2.9 只和 0~5.5 只（Soh *et al*., 2014）。由于野生和家养有蹄类猎物的选择主要受当地食物丰富度与可获得性的影响（Meriggi and Lovari, 1996），家畜和野生有蹄类动物之间密度的巨大差异意味着老虎遇到并捕食家畜的概率更高，导致当地人与野生动物之间的冲突和农业损失，而且在野生动物与家畜之间可能发生双向疾病感染（Bengis *et al*., 2002）。例如，在珲春保护区中，口蹄疫是家畜的主要疾病（Bengis *et al*., 2002），东北虎可能患有威胁家畜的疾病。放牧是当地居民重要的经济收入来源之一，但放牧对东北虎出现的贡献率为 31%（图 4.3），正确处理该地区的放牧问题对野生动物保护和当地农业发展至关重要（图 4.4）。

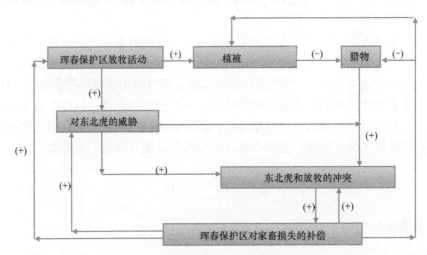

图 4.4　中国政府实施的野生动物造成牲畜损害的补偿政策导致了植被、东北虎及其猎物不断受到威胁的恶性循环

(+) 表示促进效应，(−) 表示抑制效应

4.4.2　道路的影响

俄罗斯远东地区的东北虎项圈跟踪研究结果显示，相较于道路密度低和无道

路的地区，东北虎幼崽在道路密集的区域具有较低的成活率；未受到人为干扰的东北虎相较于受到干扰的东北虎会在捕食场停留更长的时间(Kerley et al., 2002)。与以前的研究结果相比，本次研究中 GLM 与 SEM 都显示柏油路对于东北虎的出现频率有一个正相关作用，而两种模型都未探测到水泥路会对东北虎的出现频率具有显著影响。而 GAM 表明，水泥路与柏油路都会对东北虎的出现频率具有非线性的影响作用。GAM 显示，道路也会对有蹄类动物的出现概率具有显著的非线性作用。在以往道路对野生动物的影响作用的研究中，大约有59%的研究显示道路对于动物丰富度或分布具有负作用（Fahrig and Rytwinski, 2009），但同时也有一些道路对于动物的影响呈现中性作用或正作用（Fahrig and Rytwinski, 2009；Rytwinski and Fahrig, 2012）。通过对 GAM、GLM 及 SEM 结果的比较，我们发现很多在 GLM 与 SEM 中显示线性的作用在 GAM 中显示为非线性。因为 GAM 的一个优势就是能够处理响应变量与一系列解释变量之间高度非线性和非单调的作用关系，以此来增强我们对生态学关系的影响（Guisan et al., 2002）。Rytwinski 和 Fahrig（2013）的研究显示，在捕食压力下的大型有蹄类动物随着道路密度的增加，其受到道路的影响作用也是非线性的，因为道路可能会对有蹄类动物产生两方面作用，一方面是道路存在导致捕食者数量减少而对有蹄类动物产生吸引作用，另一方面是道路上的机动车对有蹄类动物产生排斥作用。在这两种作用的共同影响下，有蹄类动物的出现概率与到道路的距离会呈现一种非线性关系（Rytwinski and Fahrig, 2013）。

HPM 显示，二级公路与一级公路对于东北虎的作用贡献率分别为30%、34%，几乎占据总贡献率的2/3。此外，二级公路与一级公路对于有蹄类动物的作用贡献率也达77%。但是二级公路与一级公路对于植被的作用贡献率只有19%（图4.3）。GLM 和 GAM 都显示二级公路对于梅花鹿的出现概率具有负相关作用，而梅花鹿的出现概率对于东北虎的出现频率具有正相关作用，这就形成了上行效应（图4.2）。

4.4.3 人参种植的影响

HPM 结果显示，相较于其他种类的人为干扰，人参种植对于东北虎与有蹄类动物的作用贡献率较低，只有5%与19%（图4.3）。但是人参种植对于植被的作用贡献率却达到50%。我们认为，通过栖息地的丧失和杀虫剂的使用，人参种植可能仍然是老虎潜在的间接威胁。三种模型均显示人参种植对可采食枝条数有负影响，对灌木密度和树种均有显著的非线性影响（图4.2，表4.1）。研究表明，农药可以导致蛋白质变性、失活，细胞膜受损失活及抑制植被生长（Ciccotelli et al., 1998）。人参种植过程中使用的农药可能进入东北虎的食物链并通过富集作用对东北虎产生潜在伤害。

4.4.4 GAM、GLM、SEM 的模型比较

在我们的研究结果中，GAM 和 GLM 显示出 22 种显著的作用关系（表 4.1）。其中只有 6 种相关作用关系相同，另外 16 种在 GLM 中为线性作用，在 GAM 中为非线性作用。本研究在对 GLM 与 GAM 的分层回归结果进行整理后得出人为干扰系统作用网络（图 4.2a、b），但是使用 SEM 时则直接得到了人为干扰系统作用网络（图 4.2c）。结构方程模型的特点之一就是其可以对整个作用网络之间的各种作用关系进行整体的回归而不是像 GAM 及 GLM 逐一地对每一个响应变量进行回归。相较其他的回归方式，SEM 更注重对每一个影响作用进行路径分析（Grace et al., 2010），并且可以通过多重测量来对作用结构进行展示（Weston and Gore, 2006）。我们的研究结果显示，SEM 中 90%的结果与 GLM 相同，GAM 比 GLM 有更高的预测能力。

4.5 结　　论

人为干扰的影响已经深入东北虎食物链的各个层级，并对东北虎形成上行效应生态关系链的影响。放牧、二级公路和一级公路是主要干扰因素，有直接干扰和间接干扰两种形式；人参种植对东北虎的潜在威胁不容忽视。研究结果还揭示了 4 种人为干扰对 3 个生态关系层级的影响效应。然而，本研究仅涉及保护区内的主要人为干扰，其他干扰的具体影响还有待进一步研究。

4.6 本章小结

多种人为干扰活动影响着中国东北地区珲春保护区内的东北虎、有蹄类动物和植被。为了解不同人为干扰活动对植被、东北虎及其猎物的影响作用，2013 年 8~10 月，应用样线和样方调查了珲春保护区范围内的人为干扰活动。应用广义可加模型、广义线性模型和结构方程模型探索人为干扰活动对植被、东北虎及其猎物的影响。然后，应用贡献分割模型量化并筛选出 4 种主要人为干扰活动。研究结果表明，三种模型结果均表明人为干扰活动直接或间接地通过"上行效应链"（bottom up chain）影响着东北虎及其猎物。在人为干扰活动中，放牧和人参地垦殖比公路对植被的影响更大；次级公路对猎物的影响最大；放牧、次级公路和主要公路是干扰东北虎分布的主要因子。广义可加模型对人为干扰的预测比广义线性模型和结构方程模型具有更强的探测能力。广义可加模型能够探测到更多、

更复杂的非线性捕食者和猎物,以及猎物和栖息地因子等的互作关系。减少或消除一定类型的人为干扰对东北虎及其栖息地的恢复至关重要。

参 考 文 献

Aryal A, Brunton D H, Weihong J I, et al. 2014. Blue sheep in the Annapurna Conservation Area, Nepal: habitat use, population biomass and their contribution to the carrying capacity of snow leopards. Integrative Zoology, 9(1): 34-45.

Austin M P, Nicholls A O, Margules C R. 1990. Measurement of the realized qualitative niche: environmental niches of five *Eucalyptus* species. Ecological Monographs, 60(2): 161-177.

Austin M P. 1987. Models for the analysis of species' response to environmental gradients. Plant Ecology, 69: 35-45.

Barbermeyer S M, Jnawali S R, Karki J B, et al. 2013. Influence of prey depletion and human disturbance on tiger occupancy in Nepal. Journal of Zoology, 289(1): 10-18.

Bengis R G, Kock R A, Fischer J. 2002. Infectious animal diseases: the wildlife/livestock interface. Revue Scientifique et Technique (International Office of Epizootics), 21(1): 53-65.

Bhattarai B P, Kindlmann P. 2013. Effect of human disturbance on the prey of tiger in the Chitwan National Park – Implications for park management. Journal of Environmental Management, 131: 343-350.

Chen J Y, Nasen D, Sun Q H, et al. 2011. Amur tiger and prey in Jilin Hunchun National Nature Reserve. Chinese Journal of Zoology, 46: 46-52.

Chevan A, Sutherland M. 1991. Hierarchical partitioning. The American Statistician, 45(2): 90-96.

Ciccotelli M, Crippa S, Colombo A. 1998. Bioindicators for toxicity assesment of effluents from a wastewater treatment plant. Chemosphere, 37: 2823-2832.

Conover M R. 2002. Resolving Human-wildlife Conflicts: the Science of Wildlife Damage Management. New York: CRC Press.

Dusit N, Antony J L, George A G. 2007. Human disturbance affects habitat use and behaviour of Asiatic leopard *Panthera pardus* in Kaeng Krachan National Park, Thailand. Oryx, 41(3): 343-351.

Fahrig L, Rytwinski T. 2009. Effects of roads on animal abundance: an empirical review and synthesis. Ecology and Society, 14(1): 21.

Frederiksen M, Edwards M, Richardson A J, et al. 2006. From plankton to top predators: bottom-up control of a marine food web across four trophic levels. Journal of Animal Ecology, 75(6): 1259-1268.

Grace J B, Anderson T M, Olff H, et al. 2010. On the specification of structural equation models for ecological systems. Ecological Monographs, 80(1): 67-87.

Guisan A, Edwards T C, Hastie T. 2002. Generalized linear and generalized additive models in studies of species distributions: setting the scene. Ecological Modelling, 157(2): 89-100.

Hasselfinnegan H M, Borries C, Zhao Q, et al. 2013. Southeast Asian primate communities: the effects of ecology and Pleistocene refuges on species richness. Integrative Zoology, 8(4): 417-426.

Hastie T, Tibshirani R. 1986. Gencralized additive models. Statistical Science, 1(3): 297-310.

Hastie T, Tibshirani R. 1990. Generalized Additive Models. London: Champion & Hall: 137-173.

Hernandez L, Laundre J W. 2005. Foraging in the 'landscape of fear' and its implications for habitat use and diet quality of elk *Cervus elaphus* and bison *Bison bison*. Wildlife Biology, 11(3):

215-220.

Hunter M D, Price P W. 1992. Playing chutes and ladders: heterogeneity and the relative roles of bottom-up and top-down forces in natural communities. Ecology, 73(3): 724-732.

Jiang G, Ma J, Zhang M, et al. 2010. Multi-scale foraging habitat use and interactions by sympatric cervids in Northeastern China. Journal of Wildlife Management, 74(4): 678-689.

Jiang G, Zhang M, Ma J. 2008. Habitat use and separation between red deer *Cervus elaphus xanthopygus* and roe deer *Capreolus pygargus bedfordi* in relation to human disturbance in the Wandashan Mountains, northeastern China. Wildlife Biology, 14(1): 92-100.

Kerley L L, Goodrich J M, Miquelle D G, et al. 2002. Effects of roads and human disturbance on Amur tigers. Conservation Biology, 16(1): 97-108.

Laundre J W, Hernandez L, Medina P L, et al. 2014. The landscape of fear: the missing link to understand top-down and bottom-up controls of prey abundance? Ecology, 95(5): 1141-1152.

Li B. 2009. Status of Amur Tiger and Prey Population in Hunchun Nature Reserve, China and Conservation Research. Shanghai: East China Normal University.

Li B B, Wu Y, Zhang E D. 2006. Estimating carrying capacity of red deer (*Cervus elaphus*) at Qinglongtai Forestry in Hunchun Nature Reserve, Jilin. Sich J Zool, 25: 519-523.

Li B, Zhang E D, Liu Z. 2009. Livestock depredation by Amur tigers in Hunchun Nature Reserve, Jilin, China. Acta Theriol Sin, 29: 231-238.

Li X, Wang Y. 2013. Applying various algorithms for species distribution modeling. Integr Zool, (8): 124-135.

Linkie M, Martyr D J, Holden J, et al. 2003. Habitat destruction and poaching threaten the Sumatran tiger in Kerinci Seblat National Park, Sumatra. Oryx, 37(1): 41-48.

Mace R D, Waller J S. 1996. Grizzly bear distribution and human conflicts in Jewel Basin Hiking Area, Swan Mountains, Montana. Wildlife Society Bulletin, 24(3): 461-467.

Meriggi A, Lovari S. 1996. A review of wolf predation in southern Europe : does the wolf prefer wild prey to livestock? Journal of Applied Ecology, 33(6): 1561-1571.

Pettigrew M, Xie Y, Kang A, et al. 2012. Human-carnivore conflict in China: a review of current approaches with recommendations for improved management. Integrative Zoology, 7(2): 210-226.

Ramsay T, Burnett R T, Krewski D. 2003. Exploring bias in a generalized additive model for spatial air pollution data. Environmental Health Perspectives, 111(10): 1283-1288.

R Development Core Team. 2010. R: A language and environment for statistical computing. Vienna, Austria: R Foundation for Statistical Computing.

Rytwinski T, Fahrig L. 2012. Do species life history traits explain population responses to roads? A meta-analysis. Biological Conservation, 147(1): 87-98.

Rytwinski T, Fahrig L. 2013. Why are some animal populations unaffected or positively affected by roads? Oecologia, (173): 1143-1156.

Sabatier P A. 1986. Top-down and bottom-up approaches to implementation research: a critical analysis and suggested synthesis. Journal of Public Policy, 6(1): 21-48.

Smirnov E N, Miquelle D G. 1999. Population dynamics of the Amur tiger in Sikhote-Alin State Biosphere Reserve // Christie J S, Jackson P. Riding the Tiger: Meeting the Needs of People and Wildlife in Asia. Cambridge: Cambridge University Press: 61-70.

Soh Y H, Carrasco L R, Miquelle D G, et al. 2014. Spatial correlates of livestock depredation by Amur tigers in Hunchun, China: Relevance of prey density and implications for protected area management. Biological Conservation, 169: 117-127.

Suarez A V, Case T J. 2002. Bottom-up effects on persistence of a specialist predator: ant invasions and horned lizards. Ecological Applications, 12(1): 291-298.

Suarezseoane S, Osborne P E, Alonso J C. 2002. Large-scale habitat selection by agricultural steppe birds in Spain: identifying species–habitat responses using generalized additive models. Journal of Applied Ecology, 39(5): 755-771.

Tisseuil C, Cornu J, Beauchard O, et al. 2013. Global diversity patterns and cross-taxa convergence in freshwater systems. Journal of Animal Ecology, 82(2): 365-376.

Weston R, Gore P A. 2006. A brief guide to structural equation modeling. The Counseling Psychologist, 34(5): 719-751.

Wilmers C C, Post E, Peterson R O, et al. 2006. Predator disease out-break modulates top-down, bottom-up and climatic effects on herbivore population dynamics. Ecology Letters, 9(4): 383-389.

Woodroffe R. 2000. Predators and people: using human densities to interpret declines of large carnivores. Animal Conservation, 3(2): 165-173.

Worm B, Myers R A. 2003. Meta-analysis of cod-shrimp interactions reveals top-down control in oceanic food webs. Ecology, 84(1): 162-173.

Yudakov A G, Nikolaev I G. 1987. Ecology of the Amur Tiger: Winter Observations During 1970–1973 in the Western Section of Central Sikhote-Alin. Moscow: Nauka Press.

Zimmermann F, Breitenmoserwursten C, Molinarijobin A, et al. 2013. Optimizing the size of the area surveyed for monitoring a Eurasian lynx (*Lynx lynx*) population in the Swiss Alps by means of photographic capture-recapture. Integrative Zoology, 8(3): 232-243.

第 5 章 中国珲春地区与俄罗斯 Primorskii Krai 西南部东北虎食性分析的比较

5.1 引　　言

近 10 年来,随着红外相机和分子遗传分析等先进技术被用于中国野生东北虎的种群监测,这一濒危物种的整体情况变得更加清晰。东北虎现在分布于两个隔离的区域:一个区域为与中国完达山相连的俄罗斯锡霍特-阿林(Sikhote-Alin)山脉,有接近全球 90%的东北虎在此活动(Miquelle et al.,2006;Tian et al.,2009;Jiang et al.,2014);另一个区域在俄罗斯 Primorskii Krai 西南部与中国珲春交界处,此区域存在一个孤立小规模的种群(Miquelle et al.,2006;Henry et al.,2009;Sugimoto et al.,2016b)。这两个种群因城市发展和湿地被长期分隔(Hebblewhite et al.,2011,2014)。从遗传、种群数量和环境的容纳量来看,南部地区小而孤立的东北虎面临着灭绝的危险(Uphyrkina et al.,2002;Henry et al.,2009;Sugimoto et al.,2014)。俄罗斯于 2012 年在该地区建立了豹地国家公园(简称豹地公园),为了提高保护效率,该公园采取了单一管理体系。根据东北亚次区域环境合作项目(NEASPEC)"基于自动相机监测和分子遗传分析方法的东北虎豹跨境移动研究",豹地公园和中国国家林业局猫科动物研究中心共享了 2013～2015 年的红外相机数据。本研究期间,至少记录了 45 只东北虎(仅成体),其中 19 只东北虎(成体)在两国(中国珲春地区与俄罗斯 Primorskii Krai 西南部)均被记录到(http://www.wwf.ru/ resources/news/article/eng/14752)。如此小规模的东北虎种群,特别是在被孤立 20～30 年后,如果栖息地在未来 1～2 代人的努力下不能稳定地增加,种群可能会下降到一个无法恢复的阈值水平(Kenney et al.,2014)。由于 Primorskii Krai 西南部的东南侧是日本海,无法继续扩展栖息地,对于这个小种群来说,栖息地扩张的唯一可能只有在中国一侧(Hebblewhite et al.,2011)。为了改善栖息地条件,中国政府在吉林省和黑龙江省毗邻豹地公园的边境地区建立了一个 150 万 hm^2 的东北虎豹国家公园。

在对东北虎的种群分布和结构有了更清晰的了解之后,我们想知道它们是如何在这一地区生存下来的。捕食是每一种食肉动物生存的基本要素,而猎物的选择对于理解任何食肉动物的生存策略是至关重要的(Miquelle et al.,1996)。Kerley 等(2015)、Sugimoto 等(2016a)分别于 2008～2012 年和 2001～2003 年对俄罗斯 Primorskii Krai 西南部东北虎的食性与捕食偏好进行了研究。由于多只东北虎在两国境内均有活动,因此了解东北虎栖息地向中国长白山腹地的扩张情况,对于研究中国境内东北虎的保护现状是非常有价值的。

我们的研究区域是与俄罗斯豹地公园毗邻的珲春地区,研究该地区东北虎的食性和冬季捕食偏好,并与毗邻的俄罗斯东北虎食性进行比较,以更好地了解这个东北虎小种群的食物利用情况,为我国东北虎栖息地的恢复提供参考。

5.2 材料与方法

5.2.1 研究区域

珲春市面积 5145km^2,位于中国东北吉林省,境内有占地 1087km^2 的珲春东北虎国家级自然保护区(以下简称珲春保护区)(图 5.1)。珲春东部与俄罗斯接壤,西南部与朝鲜接壤,并与俄罗斯的豹地公园相连。对于东北虎、东北豹来说,该

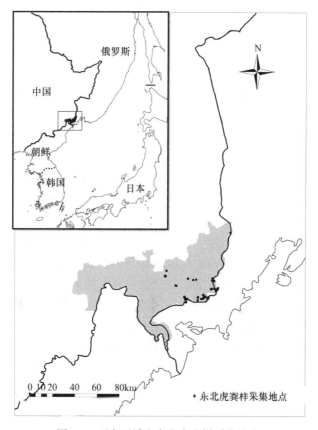

图 5.1　研究区域和东北虎粪样采集地点

地区在中、俄、朝 3 国之间起到了重要的廊道作用。珲春地区是长白山的一部分，属温带气候，平均降水量 661mm，集中在 7～9 月（年降水量的 50%）。植被主要为混交林和次生蒙古栎（*Quercus mongolica*）、野生动物主要包括东北虎、东北豹（*Panthera pardus orientalis*）、马鹿（*Cervus elaphus*）、梅花鹿（*Cervus nippon*）、野猪（*Sus scrofa*）、狍（*Capreolus pygargus*）和原麝（*Moschus moschiferus*）。

5.2.2 野外研究方法

2011~2016 年我们采集了野生东北虎粪便样本，取样区域覆盖了大部分珲春保护区及保护区周边区域（图 5.1）。我们除随机收集了东北虎常见道路上的粪便（Sunquist，1981；Karanth and Sunquist，1995）外，还在冬季通过雪地追踪（Yudakov et al.，1988）东北虎个体时采集了粪便，也收集了靠近其捕食残骸的粪便，这些信息通常为雪地追踪东北虎时发现的，或者为当地居民提供的信息。粪样被收集在采样袋中，并在冰箱中冷冻储存备用。因为同域分布的大型猫科动物东北豹的粪便与东北虎的粪便相似，所以在进行食性分析前，需要通过 DNA 的方法确定粪便的物种来源（Sugimoto et al.，2006）。

2015 年通过冬季样线调查估算猎物的相对丰富度，调查过程中所记录的标准动物雪地足迹应是物种穿过足够降雪后形成的新鲜踪迹（Stephens et al.，2006）。本次调查以记录到的动物足迹及有蹄类动物的日活动距离为基础，通过 18 条路线（总长度约 90km）对珲春大部分东北虎栖息地内的有蹄类动物进行了丰富度评估。

5.2.3 粪便分析

在食肉动物粪便中，猎物的毛发相对完好，因此可以根据粪便中未损伤的毛发来识别被捕食的猎物种类。每一个粪样中发现的任何一种猎物毛发都作为一个独立事件记录。用蒸馏水对收集到的粪便样品进行清洗，用 1.5mm 的筛网过滤，得到的剩余残留物包括毛发、骨头、蹄、羽毛和猎物的牙齿等，用于后续物种鉴定（Sunquist，1981；Karanth and Sunquist，1995）。在我们的样本中，并没有发现骨头、蹄、羽毛及牙齿，所以主要靠粪便中残留的猎物毛发进行物种鉴定。在每个粪样中至少挑选 10 根完整的毛发，用温水脱脂，双氧水脱色，然后经 100% 乙醇脱水。在 400× 的显微镜下观察，主要包括毛色、毛长度、毛细度、髓质形态和表皮鳞片形态并计算髓质指数(髓质细度/毛细度)（Moore et al.，1974；Mukherjee et al.，1994），同时与东北林业大学毛皮标本室的数据库进行对比，得到粪便中残余毛发的物种信息。

鹿科动物毛发形态特征非常类似，难于依靠毛发形态进行物种鉴别（Rozhnov et al.，2011），我们用 DNA 分子技术对鹿科动物毛发进行鉴定。这些鹿科动物毛发样本严格按照 TIAN 微量样本试剂盒（北京）提取步骤提取毛发 DNA。将 10μg 粪样中的毛发 DNA 提取物加入 20μL 反应体系，反应体系包括引物各 0.6μL、10μL 缓冲液、4μL dNTP、2.5μL ddH$_2$O、0.3μL KOD FX Neo DNA 聚合酶（TOYOBO）。PCR 程序的热循环条件为：94℃持续 2min、35 个循环（94℃/15s、55℃/40s、68℃/30s），接着 68℃持续 20min，通过使用引物扩增的 PCR 产物的直接测序，获得来自猎物物种的线粒体 DNA 基因的序列；其中引物采用 mcb398 和 mcb869 引物对、L14841 和 H15149 引物对（Kocher et al.，1989；Ficetola et al.，2010）。

5.2.4 数据分析

我们以猎物在粪样中的出现频率和生物量贡献率估计各猎物对东北虎食物组成的贡献。本研究采用 Ackerman 等（1984）对美洲狮（*Puma concolor*）食性分析时计算相对生物量的方法，来计算各猎物在东北虎食物组成中的生物贡献量：

$$y = 1.98 + 0.035x$$

式中，y 为东北虎实际消耗的某种猎物的质量；x 是猎物的活体质量（Ackerman et al.，1984）。我们计算每种猎物的 y 值，并将 y 值乘以该物种在东北虎粪便中出现的次数，从而计算出该物种被东北虎消耗的生物量。其中被捕食猎物的活体质量根据俄罗斯远东地区的东北虎猎物进行估计（Bromley and Kucherenko，1983；Dalnikin，1999；Prikhodko，2003）。最后，计算每种猎物在东北虎食物组成中的生物量贡献率（某种猎物的生物量消耗/总生物量消耗×100）。

猎物相对丰富度的评估，通过 2015 年在研究区域的冬季样线调查中记录到的新鲜动物足迹结果来计算（Stephens et al.，2006）。依据有蹄类动物的日活动距离，使用 FMP 公式，以 95%的置信区间分析猎物的种群密度（Stephens et al.，2006），然后将密度更改为比例数据（相对丰富度）获得 Jacobs 指数（Jacobs，1974）。该指数范围值从+1（强烈偏好）到-1（强烈回避）。由于马鹿出现在东北虎猎物中的证据仅有一次，因此我们对东北虎冬季猎物偏好的分析仅限于 3 个主要捕食猎物：野猪、梅花鹿和狍。

本研究粪样采集时间为 2011~2016 年，其中 2013~2015 年采集了 77%的样本。在此期间，2012 年、2014 年和 2015 年分别在珲春进行了 3 次冬季有蹄类动物调查。2012 年和 2014 年使用样方法调查有蹄类动物的密度，但俄罗斯的调查方法与此不同，只有 2015 年冬季的调查方法与俄罗斯相同，所以本研究中只使用 2015 年的调查结果，2012 年及 2014 年的调查结果未在此使用。然而，这两年与 2015 年关于有蹄类动物物种多样性和优势种种群密度的调查结果相似，说明该地区有蹄类动物的相对丰富度（百分比）顺序是稳定的，但每年的种群密度可能有所差异。因此，我们用 2015 年关于有蹄类动物的研究结果与俄罗斯数据（相同调查方法）进行对比。

将珲春东北虎各猎物的生物贡献量、猎物种群丰富度及东北虎捕食偏好的研究结果与 Kerley 等（2015）和 Sugimoto 等（2016a）在俄罗斯 Primorskii Krai 西南部区域开展的东北虎食性研究结果进行对比，以便更好地了解这一小规模东北虎种群的食物资源利用状况。

5.3 结 果

2011～2016 年，我们在珲春共收集 68 份东北虎粪便样本，其中 13 份内含有无法用来进行猎物物种识别的残留物；8 份样本为来源于同只虎的足迹追踪或同一捕食地点周围的重复样本。上述这两种样本不列入食性分析，以避免可能带来的误差。最后，珲春地区共有 47 份东北虎粪样用于食性及捕食偏好分析，其中冬季 35 份（11 月至翌年 4 月）、夏季 12 份（5～10 月）（Hojnowski et al., 2012）。在这 47 份东北虎粪样中，共鉴定出 53 次猎物，其中冬季 38 次、夏季 15 次（表 5.1）。因为夏季的样本有限，故没有进行季节间的比较。

表 5.1 识别的东北虎猎物组成种类

猎物种类	出现次数		合计
	冬季（$n=35$）	夏季（$n=12$）	
梅花鹿	4	3	7
狍	8	—	8
野猪	15	5	20
马鹿	1	—	1
牛	3	2	5
家狗	2	—	2
马	1	3	4
豹猫	1	—	1
黑熊	1	—	1
狗獾	1	—	1
羊	1	—	1
猞猁	—	2	2
合计	38	15	53

我们在珲春地区的东北虎冬季粪样中共鉴定出 11 种猎物（表 5.2），而在俄罗斯境内关于东北虎食性的研究中并未发现马鹿、牛、马及羊。在冬季和夏季的样本中，共有 16 份含有鹿科动物毛发（表 5.1）。对于这 16 份样本，我们成功提取了毛囊 DNA 进行物种鉴定。另外，利用传统毛发形态学的方法对 16 份鹿科动物

毛发进行物种鉴别，结果有 3 份样本鉴定结果与分子生物学结果不符。因此，在我们的研究中，对于鹿科动物而言，基于毛发形态学的物种鉴别准确率为 81.25%（13/16）。在东北虎粪便中出现频率最高的前 3 个物种为野猪、狍和梅花鹿。由于家养牛比野生有蹄类猎物重很多，因此牛与野猪、狍和梅花鹿排在生物贡献量的前 4 位，这一结果与俄罗斯一侧的研究结果差异明显。中国境内一侧 4 种家畜占东北虎总生物贡献量的 33.85%，而俄罗斯一侧，家畜（只有家狗）的生物贡献量仅为 0.44%和 6.3%（表 5.2）。

表 5.2 中俄两国东北虎食性研究中猎物的出现频率和生物贡献量

猎物种类	猎物体重（kg）	出现频率（%）	生物贡献量（%）		
			中方（2011~2016 年）$n=35$	俄方†（2008~2012 年）$n=152$	俄方‡（2001~2003 年）$n=63$
梅花鹿	95	11.43	9.28	25.09（18.12~32.06）	23.86
狍	37	22.86	11.45	8.31（4.89~12.22）	41.3
野猪	103	42.86	36.63	58.60（49.47~67.74）	54.0
马鹿	187	2.86	4.19	—	—
牛	418	8.57	21.79	—	—
家狗	31	5.71	2.68	0.44（0.00~1.31）	6.30
马	450	2.78	7.75	—	—
豹猫	4	2.86	0.93	—	1.60
黑熊	150	2.86	3.16	3.11（0.00~7.25）	—
狗獾	6	2.86	0.96	0.34（0.00~1.01）	6.30
羊	50	2.86	1.63	—	—
赤狐	5	—	—	—	3.20
东北兔	1.8	—	—	—	3.20
马鹿	—	—	—	0.37（0.00~1.10）	—
鼬	—	—	—	0.94（0.00~2.19）	—
东北虎				2.81（0.70~5.62）	

注：†引用 Kerley 等（2015）2008~2012 年的研究结果；"$n=152$" 是俄罗斯 Primorskii Krai 西南部冬季和夏季的总粪样数量；Kerley 等（2015）没有提及冬季的粪样数量。‡引用 Sugimoto 等（2016a）2001~2003 年的研究结果

中俄双方在这一交接区域存在两个重要差异（图 5.2）。其一，Kerley 等（2015）和 Sugimoto 等（2016a）在冬季样线调查中并没有发现马鹿的足迹，但是在中国境内一侧 5 条调查线上发现了马鹿的足迹。其二，4 种有蹄类猎物的相对丰富度不同，Kerley 等（2015）和 Sugimoto 等（2016a）关于这 4 种猎物相对丰富度的研究结果是一致的：梅花鹿>狍>野猪（马鹿= 0），但我们的研究结果为：狍>梅花鹿>野猪>马鹿。

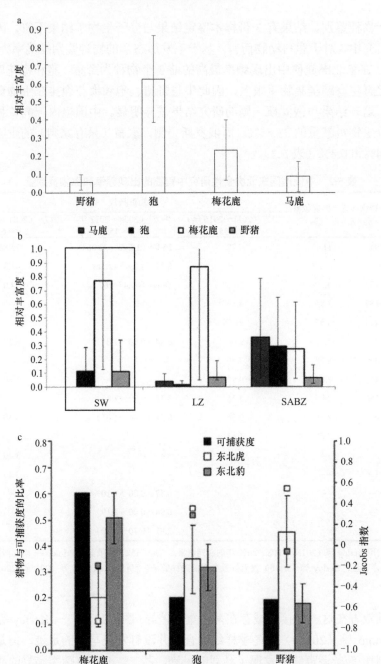

图 5.2 珲春地区和俄罗斯 Primorskii Krai 西南部 4 种有蹄类动物的相对丰富度比较
a. 2015 年我们的调查结果；b. 引用 Kerley 等（2015）2008～2012 年的结果，其中 SW 代表俄罗斯 Primorskii Krai 西南部；c. 引用 Sugimoto 等（2016a）2001～2003 年的研究结果
LZ. 拉佐夫斯基自然保护区；SABZ. 锡霍特-阿林生物圈保护区

Jacobs 指数结果显示，珲春地区的东北虎对野猪表现出强烈的偏好（+0.849），这与俄罗斯的结果一致，但对狍表现出强烈的回避（−0.693），而俄罗斯的数据显示东北虎对梅花鹿具有强烈的回避（−0.698、−0.717）。除狍外，东北虎还会根据其可捕获度回避梅花鹿（−0.495）（表 5.3）。

表 5.3 中俄两国 4 种有蹄类动物 Jacobs 指数 95%置信区间的比较结果

物种	中方 2011～2016 年	俄方†2008～2012 年	俄方‡2001～2003 年
野猪	0.849（0.747～0.957）	0.790（0.618～0.962）	0.547
马鹿	—	−0.326（−0.791～−0.139）	—
梅花鹿	−0.495（−0.672～−0.272）	−0.698（−0.917～−0.268）	−0.717
狍	−0.693（−0.906～−0.355）	−0.368（−1.000～−0.275）	0.353

注：†引用 Kerley 等（2015），但结果来自 3 个研究地点，不只是代表俄罗斯 Primorskii Krai 西南部。‡引用 Sugimoto 等（2016a）

5.4 讨 论

对比中国珲春地区与俄罗斯 Primorskii Krai 西南部地区东北虎的食性，两者在食性及捕食偏好上存在非常大的差异，主要体现在对家畜的捕食上。由于俄罗斯一侧居民较少，饲养的家畜少，东北虎冬季捕食的家畜仅见狗，且占比很低。而在中国境内一侧，东北虎不但在冬季捕食狗，而且全年都有捕食牛、马、羊等家畜的记录。在我们的研究区域内有 4 个镇 98 个村庄（Yi *et al*.，2014）。经问卷调查，研究区域内共有 63 个牧场 3066 头牛，其中 355km^2 范围内牧场竟高达 42 个（Li *et al*.，2016）。Li 等（2016）发现放牧活动主要影响树密度和当年生的新枝数，放牧使新枝数减少 30.9%，这是有蹄类动物最偏好的食物。在冬季，尽管大多数家畜被带回村庄附近，但它们对植被的破坏作用已经产生，从而导致有蹄类猎物的数量减少。尽管如此，东北虎最偏好的猎物仍然是野猪，东北虎的主要食物仍然是大中型有蹄类动物，这与之前对东北虎食性的研究结果是一致的（Karanth and Sunquist，1995；Biswas and Sankar，2002；Miquelle *et al*.，2007，2010；Kapfer *et al*.，2011）。由于中国境内一侧有蹄类动物的密度较低（尤其是野猪），迫于生存压力，与俄罗斯一侧不同，中国境内的东北虎并不明显回避捕食梅花鹿。

本研究对牛和马的生物贡献量可能被高估，在捕食地点我们通过研究捕食残骸发现，东北虎通常只吃牛或马的小部分，不像野猪几乎全部被吃掉。此外，东北虎一般会捕杀小牛犊而不是成年牛。家畜在东北虎猎物中的出现频率也可能被高估，因为当东北虎捕食家畜后，主人会向当地政府提交信息申请生态补偿，所以研究人员对东北虎捕食家畜的情况了解较为全面；而当地居民无法统计东北虎捕食野生动物的事件，所以在统计东北虎的捕食情况时，家畜在东北虎猎物中的出现频率被高估，而野生有蹄类猎物被低估。在冬季调查中，我们虽然没有在森林中发现新鲜的牛或马的足迹，但东北虎在村庄附近搜寻猎物时家畜的足迹并不难被发现。

我们关于东北虎猎物偏好的研究表明，珲春地区东北虎对狍（−0.693）有强烈回避行为，但在俄罗斯则是对梅花鹿（−0.698，−0.717）有强烈回避。与同域分布的东北豹相比，东北虎对特定有蹄类动物有明显的偏好或回避倾向，东北虎更偏好野猪，而东北豹主要捕食梅花鹿和狍（Miquelle et al., 1996；Sugimoto et al., 2016a）。这可能是营养生态位分离以促进同域分布的两种大型食肉动物共存。

另一个值得关注的方面是东北虎对马鹿的偏好（Zhivotchenko，1981；Miquelle et al., 2010）。在 Kerley 等（2015）和 Sugimoto 等（2016a）的研究中，俄罗斯 Primorskii Krai 西南部没有发现东北虎捕食马鹿的记录，但我们通过毛发 DNA 分子技术发现东北虎在我国至少捕食过一次马鹿（表 5.1，图 5.2a）。2012～2013 年，WWF（中国）在珲春以北、距中俄边境 50～60km 的汪清地区投放了 67 头马鹿和梅花鹿，以期恢复东北虎、东北豹猎物资源（http://www.wwfchina.org/pressdetail.php?id=1485）。目前还不清楚这一地区是否有野生马鹿种群存活，也不知道这些新引入者是否会扰乱当地有蹄类动物的现有状况。这些都需要未来进一步的调查研究。

2013 年 5 月 24 日和 10 月 18 日，我们采集了 2 个东北虎粪便样本，后续研究发现这两个样本中含有猞猁的毛发（表 5.1），时间早于 Petrunenko 等（2016）宣布的第一次发现东北虎捕食猞猁的记录日期。Petrunenko 等（2016）报道，研究小组在 2014 年 3 月 4 日位于俄罗斯远东地区犹太州的 Bastak 自然保护区（48°56′37N、133°07′13E）内发现一具被遍布东北虎足迹包围的猞猁遗骸。在珲春地区，我们通过红外相机捕捉到了猞猁影像，说明东北虎、东北豹及猞猁在该区域共存。目前还不清楚在中俄边境分布的东北虎是否会像其他研究提及的大型猫科动物那样抑制小型猫科动物的数量（Harihar et al., 2011），东北虎和猞猁之间的关系更是鲜为人知。虽然关于东北虎和猞猁之间直接竞争这一事件的记录是令人信服的，但是对于生态系统中大型食肉动物之间复杂的相互关系仍待详细研究。

在粪便分析过程中发现，绝大多数重复样本为鹿科动物（Kerley et al., 2015；Sugimoto et al., 2016a）。鹿科动物毛发的外形和微观形态结构非常类似，因此利用光学显微镜比较毛发的髓质类型和表皮鳞片形态，并不能准确地鉴定到物种。简单去除假重复样本可能降低鹿科动物相对生物量的百分比，提高其他猎物种类相对生物量的百分比，因此强烈建议在对鹿科动物的粪便分析中使用 DNA 技术检测以避免假重复样本现象。以我们的研究为例，如果我们没有使用 DNA 鉴定技术，我们就不会发现东北虎捕食马鹿的案例。

我们的结果表明，在中国珲春地区与俄罗斯 Primorskii Krai 西南部的东北虎具有不同的食物组成和捕食偏好，这可能是因为中国境内一侧具有严重的人为干扰，如放牧影响了当地的植被，因此造成依赖这些植被生存的大型有蹄类动物种群密度降低。随着东北虎向中国一侧扩张的趋势越来越明显，减少或消除人为干扰对于东北虎猎物和生境的恢复至关重要。中俄关于有蹄类动物的年度联合调查对于估计这一孤立、小规模的东北虎种群的猎物数量也是必不可少的。

5.5 本章小结

本研究利用粪便分析法并结合4种猎物的丰富度数据研究了珲春地区东北虎的食性及捕食偏好。在2011~2016年我们共采集68份东北虎粪样，结果表明东北虎共捕食12个物种（冬季11种），其中包括4种家畜，其生物贡献量为33.85%，并在此区域第一次记录到东北虎捕食猞猁。野猪是东北虎最常捕食的猎物，东北虎对野猪表现出强烈的偏好（Jacobs指数：+0.849），对狍表现出强烈的回避（Jacobs指数：-0.693）。而对比俄罗斯的研究结果发现，俄方东北虎粪便中很少发现家畜（只有狗），也不是对狍而是对梅花鹿表现出强烈的回避。我们还在冬季调查过程中发现了马鹿的足迹，在中国东北虎会捕食马鹿，但在俄罗斯未发现东北虎捕食马鹿的记录。

参 考 文 献

Ackerman B B, Lindzey F G, Hemker T P. 1984. Cougar food habits in southern Utah. The Journal of Wildlife Management, 48(1): 147.

Biswas S, Sankar K. 2002. Prey abundance and food habit of tigers (*Panthera tigris tigris*) in Pench National Park, Madhya Pradesh, India. Proceedings of the Zoological Society of London, 256(3): 411-420.

Bromley G F, Kucherenko S P. 1983. Ungulates of the Southern Far East USSR. Moscow: Nauka Press.

Dalnikin A A. 1999. Mammals of Russia and Adjacent Regions: Deer(Cervidae). Moscow: GEOC.

Ficetola G F, Coissac E, Zundel S, *et al*. 2010. An in silico approach for the evaluation of dna barcodes. BMC Genomics, 11(1): 434.

Harihar A, Pandav B, Goyal S P. 2011. Responses of leopard *Panthera pardus* to the recovery of a tiger *Panthera tigris* population. Journal of Applied Ecology, 48: 806-814.

Hebblewhite M, Miquelle D G, Murzin A A, *et al*. 2011. Predicting potential habitat and population size for reintroduction of the Far Eastern leopards in the Russian Far East. Biological Conservation, 144: 2403-2413.

Hebblewhite M, Miquelle D G, Robinson H, *et al*. 2014. Including biotic interactions with ungulate prey and humans improves habitat conservation modeling for endangered Amur tigers in the Russian Far East. Biological Conservation, 178: 50-64.

Henry P, Miquelle D, Sugimoto T, *et al*. 2009. *In situ* population structure and *ex situ* representation of the endangered Amur tiger. Molecular Ecology, 18: 3173-3184.

Hojnowski C E, Miquelle D G, Myslenkov A I, et al. 2012. Why do Amur tigers maintain exclusive home ranges? Relating ungulate seasonal movements to tiger spatial organization in the Russian Far East. Journal of Zoology, 287: 276-282.

Jacobs J. 1974. International Association for Ecology Quantitative Measurement of Food Selection: A modification of the forage ratio and Ivlev's electivity index. Oecologia, 14: 413-417.

Jiang G, Sun H, Lang J, et al. 2014. Effects of environmental and anthropogenic drivers on Amur tiger distribution in northeastern China. Ecological Research, 29: 801-813.

Kapfer P M, Streby H M, Gurung B, et al. 2011. Fine-scale spatio-temporal variation in tiger *Panthera tigris* diet: Effect of study duration and extent on estimates of tiger diet in Chitwan National Park, Nepal. Wildlife Biology, 17: 277-285.

Karanth K U, Sunquist M E. 1995. Prey selection by tiger, leopard and dhole in tropical forests. Journal of Animal Ecology, 64: 439.

Kenney J, Allendorf F W, McDougal C, et al. 2014. How much gene flow is needed to avoid inbreeding depression in wild tiger populations? Proceedings of the Royal Society B Biological Sciences, 281 (1789): 20133337.

Kerley L L, Mukhacheva A S, Matyukhina D S, et al. 2015. A comparison of food habits and prey preference of Amur tiger (*Panthera tigris altaica* Timminck, 1884) at three sites in the Russian Far East. Integrative Zoology, 10: 354-364.

Kocher T D, Thomas W K, Meyer A, et al. 1989. Dynamics of mitochondrial DNA evolution in animals: amplification and sequencing with conserved primers. Proceedings of the National Academy of Sciences of the United States of America, 86: 6196-6200.

Li Z, Kang A, Gu J, et al. 2016. Effects of human disturbance on vegetation, prey and Amur tigers in Hunchun Nature Reserve, China. Ecological Modelling, 353: 28-36.

Melville H I A S. 2004. Behavioural ecology of the caracal in the Kgalagadi Transfrontier Park, and its impact on adjacent small stock production units (MSc dissertation). Pretoria: University of Pretoria.

Miquelle D G, Goodrich J M, Smirnov E N, et al. 2010. Amur tiger: A case study of living on the edge. *In*: Macdonald D W, Loveridge A J. Biology and Conservation of Wild Felids. Oxford: University of Oxford Press.

Miquelle D G, Pikunov D G, Dunishenko Y M, et al. 2007. 2005 Amur Tiger Census. Cat News, 46: 14-16.

Miquelle D G, Smirnov E N, Quigley H G, et al. 1996. Food habits of Amur tigers in Sikhote-Alin Zapovednik and the Russian Far East, and implications for conservation. Journal of Wildlife Research, 1: 138-147.

Moore T M, Spence L E, Dugnolle C E, et al. 1974. Identification of the dorsal hairs of some mammals of Wyoming. Cheyenne: Issue 14 of Bulletin, Wyoming Game and Fish Department.

Mukherjee S, Goyal S P, Chellam R. 1994. Standardisation of scat analysis techniques for leopard (*Panthera pardus*) in Gir National Park, Western India. Mammalia, 58(1): 139-143.

Petrunenko Y K, Polkovnikov I L, Gilbert M, et al. 2016. First recorded case of tiger killing Eurasian lynx. European Journal of Wildlife Research, 62: 373-375.

Prikhodko V I. 2003. Musk Deer: Distribution, Systematics, Ecology, Behaviour, and Communication. Moscow: GEOS.

Rozhnov V V, Chernova O F, Perfilova T B. 2011. A Guide to Deer Species in the Diet of Amur Tigers (Microstructure of Deer Species Guard Hairs Found in Amur Tiger Excrement). Moscow: Scientific Press. (in Russian)

Stephens P A, Zaumyslova O Y, Miquelle G D, et al. 2006. Estimating population density from indirect sign: track counts and the Formozov-Malyshev-Pereleshin formula. Animal Conservation, 9: 339-348.

Sugimoto T, Aramilev V V, Kerley L L, et al. 2014. Noninvasive genetic analyses for estimating population size and genetic diversity of the remaining Far Eastern leopard (*Panthera pardus orientalis*) population. Conservation Genetics, 15: 521-532.

Sugimoto T, Aramilev V V, Nagata J, et al. 2016a. Winter food habits of sympatric carnivores, Amur tigers and Far Eastern leopards, in the Russian Far East. Mammalian Biology - Zeitschrift für Säugetierkunde, 81(2): 214-218.

Sugimoto T, Nagata J, Aramilev V V, et al. 2006. Species and sex identification from faecal samples of sympatric carnivores, Amur leopard and Siberian tiger, in the Russian Far East. Conservation Genetics, 7: 799-802.

Sugimoto T, Nagata J, Aramilev V V, et al. 2016b. Population size estimation of Amur tigers in Russian Far East using noninvasive genetic samples. Journal of Mammalogy, 93: 93-101.

Sunquist M E. 1981. The social organization of tigers (*Panthera Tigris*) in Royal Chitawan National Park, Nepal. Journal of Endocrinology, 336: 345-359.

The Amur Tiger Program. 2016. Land of the leopard, Chinese nature reserves exchange tiger and leopard data. http://programmes.putin.kremlin.ru/en/tiger/news/25404[2016-10-20].

Tian Y, Wu J G, Kou X J, et al. 2009. Spatiotemporal pattern and major causes of the Amur tiger population dynamics. Biodiversity Science, 17: 211.

Uphyrkina O, Miquelle D, Quigley H, et al. 2002. Conservation Genetics of the Far Eastern Leopard (*Panthera pardus orientalis*). Journal of Heredity, 93: 303.

WWF-China. 2016. WWF released deer to the wild again. http://www.wwfchina.org/pressdetail.php?id=1485[2016-10-26].

WWF-Russia. 2016. Leopards and tigers freely cross the border of Russia and China. http://www.wwf.ru/resources/news/article/eng/14752[2016-12-5].

Yi H S, Carrasco L R, Miquelle D G, et al. 2014. Spatial correlates of livestock depredation by Amur tigers in Hunchun, China: Relevance of prey density and implications for protected area management. Biological Conservation, 169: 117-127.

Yudakov A G, Nikolayev E G, Olson E C. 1988. Ecology of the Amur Tiger. The Quarterly Review of Biology, 63(4): 472.

Zhivotchenko V E. 1981. Food habits of the Amur tiger. *In*: Predatory Mammals. Moscow: TsNIL Glavokhoty RSFSR: 64-75. (in Russian)

第6章 野生和圈养雌性东北虎繁殖参数的比较

6.1 引 言

东北虎（*Panthera tigris altaica*）是地理分布最北的虎亚种。目前，绝大多数野生东北虎种群分布于俄罗斯，据估计，俄罗斯野生东北虎有430~502只（Miquelle et al.，2007），它们是从20世纪40年代的20~30只的一个小种群经过长期恢复形成现在的规模的（Kaplanov，1948；Siberia Tiger Park，2003）。20世纪末在中国境内还分布有约20只野生东北虎（Zhang et al.，2013），而在韩国境内已绝迹（Kerley et al.，2003；Miquelle et al.，2007）。

在俄罗斯，东北虎研究和保护工作的重点是恢复其野生种群。Kerley等（2003）研究了11只野生雌性东北虎的繁殖参数，发现每年每只雌虎的后代仅有0.7只能存活至12个月，研究表明（Miquelle et al.，2015），由于繁殖率较低，野生东北虎种群数量可能不会像预期的那样得到迅速恢复。Russello等（2004）在从俄罗斯收集的27只东北虎遗传样本检测中发现了极低水平的单型多样性，而Henry等（2009）用更大的样本（$n=95$）估计这些东北虎的有效种群大小只有27~35只。遗传多样性的丧失会导致近亲繁殖，并会降低生殖能力（Roelke et al.，1993；Reed and Frankham，2003）。鉴于雌性老虎的繁殖参数对于评估其种群动态具有重要作用，因此，有必要对圈养和野生老虎种群之间的繁殖参数进行比较，以寻找两者间的差异，往往这些差异可能是潜在遗传缺陷的重要指标。

在中国，圈养东北虎已经成功地饲养繁育30多年。最初，中国横道河子猫科动物饲养繁育中心的奠基种群数量和俄罗斯20世纪40年代野生种群数量的低谷接近。经过30多年的人工繁育，到2010年横道河子猫科动物饲养繁育中心东北虎种群数量达到350只，这些老虎的基础种群都是1986年来自中国动物园和美国的27只奠基个体，通过谱系记录证实这是一个纯种的圈养东北虎种群（Zheng et al.，2005）。因此，对圈养东北虎繁殖参数进行研究，以期为野生东北虎的繁育管理与保护工作提供有益的参考。

如果能有足够数量的圈养东北虎个体通过野外放归开展虎的迁地保护工作，将有助于中国野生东北虎种群数量的恢复与稳定，并提高野外小种群的遗传多样性（Russello et al.，2004）。人工圈养繁育的老虎是否能够在不威胁人类的情况下成功适应野外生活，目前尚不清楚，但俄罗斯近期开展的将人工喂养长大的野生虎野外放归的尝试表明（Shaer，2015），这种方法是可行的。但是，东北虎经过几十年的人工圈养繁育，繁殖参数可能发生了变化，以致我们无法科学地评估圈养虎是否适合放归。因此，我们将中国横道河子猫科动物饲养繁育中心圈养的雌性东北虎的繁殖特征与 Kerley 等（2003）报道的野生雌性东北虎的繁殖参数进行比较。Kerley 等（2003）报道了一只野生雌性东北虎不明原因失去 1 周大的幼崽后，经过 7 个月又产下一胎，因此我们将属于正常生育周期的人工圈养雌虎的繁殖参数与过早失去幼崽又产仔的野生雌虎进行比较，然后我们通过比较 16 年来人工圈养东北虎产仔期的变化，来确定东北虎圈养种群产仔期是否发生了变化。

6.2 材料与方法

6.2.1 研究种群

中国横道河子猫科动物饲养繁育中心（以下简称繁育中心）位于中国黑龙江省牡丹江市（44°47′38″N、129°6′39″E），该区域属于野生东北虎的历史分布区（Yu et al.，2009）。自 1986 年繁育中心成立以来，雌虎的繁殖记录一直得到妥善的保存，目前已形成了一个包含 350 多只个体的繁殖数据库。每只雌虎均有单独的谱系档案，记录其亲本、配偶和后代。2003 年，该中心与国家林业局野生动植物检测中心合作，开始重建该种群精确的分子谱系，并开发了一套管理系统，以尽量减少遗传多样性的损失（Siberia Tiger Park，2003）。该中心所有老虎均按照《野生动物饲养管理技术规范——东北虎》（LY/T 2199—2013）（State Forestry Administration，2013）进行饲养与管理。老虎一周喂 6 天，雄性每天消耗 6~8kg 肉类，雌性每天消耗 5~7kg 肉类，老虎的食物包括牛肉、鸡肉和猪肉等。

同年出生的幼虎与母虎一起生活在户外围栏中，直到母虎再次发情，母虎会被转移到有室外和室内间隔的独立围场，根据预先制订的繁殖计划，雄虎在发情期被引入围场。室内圈舍冬季供暖，每天用水管冲洗圈舍并进行火焰消毒。大多数幼崽在大约 6 个月龄时（有时可达 1 岁）与母虎分离，以促进母虎再次发情，缩短繁殖周期。

我们测量了 7 种繁殖参数，包括初产年龄、交配日期、交配持续期（导致怀孕的第一次和最后一次交配的时间间隔）、产仔期、妊娠期的长度（从受孕日期到产仔期）、胎产仔数和胎产间隔。受孕日期定义为交配成功的第一次交配的日期。

我们测量了 2004~2010 年新生幼崽死亡率（从出生到 7 天）和出生至 6 个月龄的幼崽死亡率，将新生幼崽死亡率分为 4 类：早产、死胎、先天性畸形和遗弃（或母虎照顾不足）。先天性畸形被定义为外部可见的异常情况。

6.2.2 数据分析

计算初产年龄、产仔间隔、妊娠时间、产仔数、幼崽死亡率的均值和标准差，并与野生雌虎对应的繁殖参数进行比较。通过 KS 检验（Kolmogorov-Smirnov test）比较了圈养和野生虎的平均产仔数（Lilliefors，1969）。

我们估计了圈养雌虎每月的交配频率和出生日期，将数据分为 4 个季节：初冬（11 月至翌年 1 月）、晚冬（2~4 月）、初夏（5~7 月）和晚夏（8~10 月）（Kerley et al.，2003）。由于数据不呈正态分布，使用 Kruskal-Wallis 检验来比较圈养老虎和野生老虎的交配与出生高峰期。

为了探究雌虎生育时间是如何随圈养时间变化的，我们将出生日期（从年初）与出生年份（1995~2010 年）进行了回归分析，以寻找 16 年来繁殖规律的变化。

最后，我们将幼崽过早死亡后雌虎再次妊娠后的生育特征与正常生育周期的雌虎进行比较，以确定幼崽过早死亡后繁殖参数是否发生变化。

6.3 结　果

1995~2010 年，繁育中心共有 68 只雌性东北虎产仔 252 窝，共计 724 只。在这 252 次产仔中，只有 123 次交配事件被记录下来。

6.3.1 圈养东北虎和野生东北虎种群的繁殖参数

圈养东北虎首次产仔年龄（4.10 岁±1.12 岁）与野生东北虎非常类似（表 6.1）。妊娠期长度为 108.27 天±3.64 天（$n=123$），与其他研究报道的结果接近（103 天）（Kitchener，1991；Sankhala，1978）。妊娠前的交配平均持续时间为 6.56 天±2.85 天（野外无法观测到这个繁殖参数）。圈养雌虎平均每胎产仔数（2.91 个±0.97 个）显著高于野生东北虎（$P=0.034$）（表 6.1）。圈养东北虎的产仔间隔（384.96 天±198.84 天）要显著短于野生东北虎（642 天±132 天）。

表 6.1　圈养东北虎和野生东北虎繁殖参数对比（平均值±标准差）

均值	圈养	n	野生	n	P
首次产仔年龄	4.10 岁±1.12 岁	68	4 岁±0.4 岁	2	—
产仔间隔	384.96 天±198.84 天	184	21.4 月±4.4 月（642 天±132 天）	7	—
交配期长度	6.56 天±2.85 天	87	—		
妊娠期长度	108.27 天±3.64 天†	126	103 天		
胎死亡率	41.99%±40.96%	252	41%~47%	19	
胎产仔数	2.91 个±0.97 个	252	2.4 个±0.6 个	16	0.034

注：†妊娠期长度是指从最近一次交配期的第一次交配到分娩的天数。除妊娠期长度外的野生虎数据来源于 Kerley 等（2003），妊娠期长度数据来源于 Kitchener（1991）

6.3.2 幼崽死亡率

圈养东北虎每胎幼崽死亡率为 41.99%±40.96%,与野生东北虎类似(表 6.1)。我们还记录到 2004~2010 年的新幼崽死亡率为 13%,共记录 76 例新生幼崽死亡(表 6.2),死胎和先天性畸形是导致新生幼崽死亡的两个最主要的原因。特别是在 2008 年,一些缺乏经验和年龄较大的雌虎照顾不足或遗弃而导致幼崽死亡。2004 年和 2010 年的新生幼崽死亡率较低。

表 6.2 4 种新生幼崽死亡率的发生次数

	2004 年	2005 年	2006 年	2007 年	2008 年	2009 年	2010 年	合计
早产					1	4		5
死胎	1	5	3	9	7	5	1	31
先天性畸形		1	7	4	1	6	2	21
遗弃/照顾不足	3	4	2	2	7	1		19
n	4	10	12	15	16	16	3	76
n/所有幼崽数	0.07	0.17	0.12	0.14	0.19	0.13	0.04	0.13

6.3.3 繁殖的季节性

除 9 月和 10 月外,繁育中心的雌虎在一年中的其他月份都会受孕,但主要集中在冬末(2~4 月)(x^2=8.996,df=3,P=0.004)(图 6.1a)。野生虎也在一年绝大多数月份受孕,但受孕最集中的月份为 3~5 月(图 6.1a)。除 1 月外,繁育中心的圈养雌虎在全年其他月份均有分娩的记录,但在初夏(5~7 月)达到产仔高峰(x^2=8.728,df=3,P=0.006),圈养虎产仔高峰早于野生虎(8~10 月)。

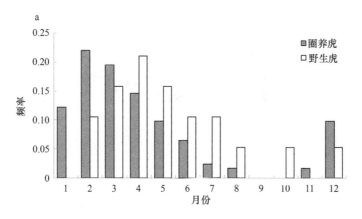

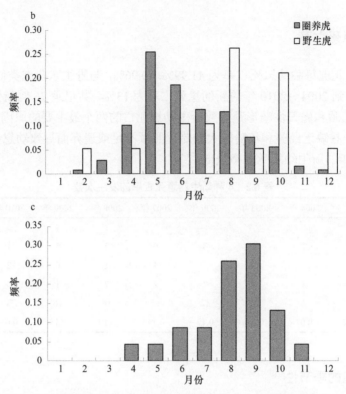

图 6.1 横道河子猫科动物饲养繁育中心圈养虎和俄罗斯野生东北虎繁殖期分布（Kerley et al., 2003）
a. 交配首日分布；b. 正常产仔期分布；c. 当年二胎产仔期分布

在前 6 年（1995～2000 年）的观察中，平均产仔期提早，而 2001～2012 年开始延后。平均产仔期的变化，只在 2001～2012 年有显著延后的趋势（$P=0.01$，$r=0.01$）（图 6.2）。在这同一时期母虎产仔的年龄和之前并没有显著变化（$P=0.68$）（表 6.3）。

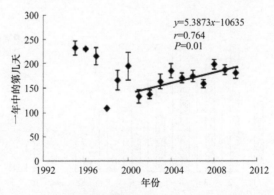

图 6.2 1995～2010 年圈养条件随时间对东北虎产仔期的影响（16 年）
产仔日期已被转换成一年中的第几天进行统计（2001～2010 年产仔期有显著的延迟现象）

表 6.3 1995~2010 年繁殖记录（平均值±标准差）

年份	胎次	产仔期（一年中的第几天）	平均年龄（岁）	最小年龄（岁）	最大年龄（岁）
1995	2	232.50 ± 20.50	4.94 ± 2.55	3.1	6.7
1996	3	230.67 ± 2.51	5.22 ± 2.15	4.0	7.7
1997	4	215.50 ± 35.38	4.77 ± 0.40	4.2	5.0
1998	1	109.00	9.37	9.4	9.4
1999	4	166.00 ± 40.17	7.06 ± 2.50	4.9	10.6
2000	7	195.00 ± 75.74	6.52 ± 3.04	2.9	11.4
2001	7	132.86 ± 35.67	7.09 ± 3.12	3.0	12.3
2002	7	137.14 ± 23.98	4.34 ± 2.54	2.8	9.8
2003	13	163.15 ± 55.03	4.99 ± 2.09	2.7	9.9
2004	20	184.95 ± 64.54	5.21 ± 2.61	2.9	11.8
2005	19	170.79 ± 43.03	4.97 ± 1.85	3.0	10.9
2006	31	174.16 ± 65.35	5.55 ± 1.91	3.0	13.0
2007	32	158.97 ± 46.12	6.33 ± 2.42	3.7	14.0
2008	31	198.71 ± 53.59	5.93 ± 1.86	3.3	10.3
2009	43	187.40 ± 68.25	5.86 ± 1.58	3.3	9.9
2010	28	181.07 ± 57.78	6.84 ± 2.03	3.9	11.3
合计	252	178.12 ± 58.42	5.85 ± 2.16	2.7	14.0

6.3.4 幼崽死亡后再次受孕雌虎的繁殖特征

在研究期间，来自繁育中心的 21 只母虎失去了 26 胎幼崽。幼崽死亡后再次受孕的雌虎妊娠期（108.89 天±2.02 天）和正常的生产周期（108.17 天±3.83 天，$P=0.21$）没有显著变化。平均胎产仔数为 2.73 个±1.15 个，和正常周期平均胎产仔数（2.93 个±0.94 个，$P=0.37$）也没有显著差别。平均幼崽死亡率为 47.35%±42.72%，和正常周期（41.33%±40.69%，$P=0.12$）的死亡率也没有显著差别。但是，产仔期发生较大变化：雌虎通常在 4~11 月产下第二胎，高峰期在夏末（8~10 月）（$x^2=9.17$，df=3，$P=0.01$）（图 6.1c）。

6.4 讨论

圈养东北虎的胎产仔数高于野生东北虎，这些差异可能是由观察胎产仔数的时间不同造成的。我们能够很准确地确定圈养东北虎出生时的产仔数，而 Kerley 等（2003）在野外只能确定近 4 个月大时的平均产仔数，因此无法估计新生幼崽的死亡率，从而明显低估了初始产仔数。与我们的研究结果相似，有研究报道圈养东北虎幼崽在出生前两个月的死亡率接近40%（Christie and Walter，2000）。总的来说，无论在圈养条件下还是在野外环境下，东北虎幼崽死亡率都很高。

我们的研究结果表明，在繁育中心圈养东北虎繁殖季节性很弱，在初夏达到高峰，但全年都在生产。然而，在连续10年的观察中发现，2001~2010年平均产仔期推迟（图6.2）。在此期间，产仔母虎的年龄结构没有显著差异（表6.3），可能与繁育中心管理方式的变化有关，这种现象还需要继续观察和研究。

野生东北虎和圈养东北虎繁殖参数的主要差异是交配/产仔的时间与产仔间隔。我们的研究结果与大部分野生东北虎的繁殖参数结果一致，这些文献报道了1~2月是受孕高峰（Baikov，1925；Dunishenko and Kulikov，1999；Kucherenko，1985），春季（4~5月）是产仔高峰。但是，上述数据均来源于对老虎雪地足迹的追踪，这显然会使结果偏向于那些有雪被覆盖的月份。Seal（1987）也报道了北美洲圈养虎种群在春季出生的一个明显高峰。Kerley等（2003）依赖于一个小样本来估计生育高峰，这可能导致一个偏倚的估计。然而，繁育中心记录的16年来不断变化的产仔日期也表明，繁育中心的管理方式显著地影响了母虎的产仔期（图6.2）。为了解释这些差异，将来还需要进一步对比野生东北虎和圈养东北虎的繁殖参数数据。

在野外正常的产仔间隔是2年，而在繁育中心在6个月龄至1年时将幼虎从母虎身边带走，使得圈养母虎有更短的繁殖间隔（约13个月）。由于很少观察到雌虎在野外过早失去幼崽的情况，这使我们对圈养母虎的观察成为一个宝贵的机会，从而能够更好地理解这一过程的动态，并更好地确定过早失去幼崽后雌虎的产仔间隔。我们的研究表明，失去一窝幼崽后雌虎的繁殖参数与正常的生育周期非常相似，只有产仔期发生了变化。在野外，产仔期的改变会如何影响幼崽的存活率还不得而知。

一般而言，除产仔期和产仔间隔较短外，人工圈养东北虎和野生东北虎的繁殖参数基本一致。我们的研究结果显示，没有证据表明在未来将圈养雌虎用于放归野外的繁殖存在重大问题，但是仍然需要更多的信息来科学评估它们的适应性。我们建议在以下方面进行进一步研究。首先，需要更多关于幼崽生存和死亡原因的信息来评估近交衰退与环境因素对幼崽生存的潜在影响。其次，有必要进一步调查圈养东北虎和野生东北虎大量新生幼崽死亡的原因。最后，需要对圈养东北虎产仔期的变化原因进行深入调查。

6.5 本章小结

如果个体是经过适当的准备且被认为适合野外放归，健康的东北虎圈养种群可能支持中国东北地区野生种群的恢复。我们分析了1995~2010年中国横道河子猫科动物饲养繁育中心 68 只雌性东北虎的繁殖记录，并与野生东北虎的繁殖参数进行了比较。结果表明，圈养东北虎的繁殖参数（初产年龄、交配日期、妊娠时间、产仔存活率）与野生东北虎无显著差异。野生种群和圈养种群在产仔期、产仔数方面的差异可能是由于圈养种群的管理规程或野生种群的野外数据不足。除产仔期外，失胎后生育的雌虎繁殖参数与未失胎的雌虎繁殖参数相似。这些结果并没有表明将圈养母虎用于放归野外的繁殖存在重大问题，但是仍然需要更多的信息来评估它们的适应性。

参 考 文 献

Baikov N. 1925. The Manchurian Tiger. Harbin: Obshchestvolzuchenia Manchurskogo Kraya. (in Russian)

Christie S, Walter O. 2000. European and Australasian Studbook for Tigers (*Panthera tigris*). London: Zoological Society of London.

Dunishenko Y, Kulikov A. 1999. Amur tiger. Khabarovsk, Russia: Khabarovski Izadatilstva. (in Russian)

Kaplanov L G. 1948. Tigers in Sikhote-Alin. Tiger, red deer, and moose. Moscow: Obschestva Ispytateley Prirody. (in Russian)

Kerley L L, Goodrich J M, Miquelle D G, *et al.* 2003. Reproductive parameter of wild female Amur tigers (*Panthera tigris altaica*). Journal of Mammalogy, 84(1): 288-298.

Kitchener A. 1991. The Natural History of the Wild Cats. New York: Comstock Publishing Associates.

Kucherenko S. 1985. Tiger. Moscow, Russia: Agropromizdatilstva. (in Russian)

Lilliefors H W. 1969. On the Kolmogorov-Smirnov test for the exponential distribution with mean unknown. Journal of the American Statistical Association, 64(325): 387-389.

Miquelle D G, Smirnov E N, Zaumyslova O Y, *et al.* 2015. Population dynamics of Amur tigers (*Panthera tigris altaica*) in Sikhote-Alin Biosphere Zapovednik: 1966-2012. Integrative Zoology, 10(4): 315-328.

Miquelle D G, Pikunov D G, Dunishenko Y M. 2007. 2005. Amur tiger census. CAT News, (46): 14-16.

Reed D H, Frankham R. 2003. Correlation between fitness and genetic diversity. Conservation Biology, 17(1): 230-237.

Roelke M E, Martenson J S, Obrien S J. 1993. The consequences of demographic reduction and genetic depletion in the endangered Florida panther. Current Biology, 3(6): 340-350.

Russello M A, Gladyshev E, Miquelle D, *et al.* 2004. Potential genetic consequences of a recent bottleneck in the Amur tiger of the Russian Far East. Conservation Genetics, 5(5): 707-713.

Sankhala K. 1978. Tiger!: The Story of the Indian Tiger. London: HarperCollins.

Seal U S. 1987. Behavioral indicators and endocrine correlates of estrus and anestrus in Siberian tiger. Tigers of the World: 244-254.

Shaer M. 2015. Cinderella story. Smithsonian, (46): 30-41.

Siberia Tiger Park. 2003. Siberia tiger park timeline. http://www.dongbeihu.net.cn/news/view.asp?id=585[2003-9-29].

State Forestry Administration. 2013. Technical Code of Feeding and Management for Wild Animals-Siberian Tiger. http://www.forestry.gov.cn/uploadfile/mama/2013-10/file/2013-10-24-2283c308-

377341588ba7649f025c232e. pdf [2013-10-24].
Yu T, Jianguo W, Xiaojun K, *et al.* 2009. Spatiotemporal pattern and major causes of the Amur tiger population dynamics. Biodiversity Science, 17(3): 211-225.
Zhang C, Zhang M, Stott P. 2013. Does prey density limit Amur tiger *Panthera tigris altaica* recovery in northeastern China. Wildlife Biology, 19(4): 452-461.
Zheng D, Liu X, Ma J. 2005. Patterns of genetic variation within a captive population of Amur tiger (*Panthera tigris altaica*). Acta Theriologica, 50(1): 23-30.

第 7 章 东北虎对自然猎物的偏好本能及其与个性关系的探测研究

7.1 前　言

识别猎物是捕食行为的首要步骤，这对食肉动物的存活至关重要，因为捕食行为是耗能且冒险的（Benoit-Bird, 2004；Liznarova and Pekar, 2013）。为了成功捕食，食肉动物需要权衡捕食行为的代价和收益。有效地识别猎物可以增加捕食效率和能量净收益（McNamara and Houston, 1997；Van Gils et al., 2005）。通常认为食肉动物通过本能和/或学习识别猎物，这被称为"本能假说"和"学习假说"。先前研究有力地佐证了"学习假说"（Darmaillacq et al., 2004；Fabregas et al., 2015；Lillywhite et al., 2015），然而，"本能假说"的证据有限（Dolev and Nelson, 2014）。因此，"本能假说"的有效性和普适性需要更多的检验。

东北虎（*Panthera tigris altaica*）是最濒危的猫科物种之一（Miquelle et al., 2015）。目前，野生东北虎的数量约为 500 只（Miquelle et al., 2006）。大部分野生东北虎栖息于俄罗斯远东地区（Carroll and Miquelle, 2006）。另外，中国东北地区在 20 世纪末时仅有约 20 只野生东北虎（Jiang et al., 2014；Dou et al., 2016）。在野外，东北虎主要捕食本地的中型和大型有蹄类动物（Miller et al., 2013；Kerley et al., 2015），如野猪（wild boar, *Sus scrofa*）、狍（roe deer, *Capreolus pygargus*）、梅花鹿（sika deer, *Cervus nippon*）和马鹿（sika deer, *Cervus elaphus*）。然而，东北虎是否本能地识别自然猎物，人们对此并不清楚。阐明这一问题不仅有助于我们了解东北虎的生物特性，还为检验猎物识别本能假说的有效性提供了良机。

野化圈养动物是保护濒危物种的有效方式（Fabregas et al., 2015）。为了保护东北虎，中国黑龙江东北虎林园成功繁育了约 1000 只东北虎，并计划训练圈养东北虎用于野外放归，这将促进野生东北虎种群的复壮。为了提高放归东北虎的生存率，筛选、训练具有优良行为与认知特性的候选个体十分重要。

研究表明，动物具有不同的个性，这表现为个体之间一贯的行为特异性（Briffa and Weiss, 2010；Stamps and Groothuis, 2010）。鉴于行为、生理和认知之间的关系（Dolan, 2002；Killen et al., 2013），濒危物种的个性研究被用于评估健康状况、提升保护管理（Kenneth and Terry, 1994；Nadja, 1999；Gartner and Weiss, 2013a）。例如，通过圈养猎豹的"紧张–害怕"特性可预测其繁殖能力，这为解决繁育问题提供了新的视角（Nadja，1999）。对于东北虎，如果它本能地识别自然猎物，那么我们可以推测猎物识别本能和个性之间具有相关性，这将为筛选猎物识别本能敏锐的野化候选个体、促进其存活提供重要思路。

在本研究中，圈养东北虎自出生后几乎与其他所有动物隔离，我们假设圈养东北虎对自然猎物刺激物的反应比对非自然猎物刺激物的反应更强烈。这将表明东北虎具有识别猎物的本能。另外，我们还探究了东北虎的猎物识别本能是否与其个性具有相关性。

7.2　材料与方法

7.2.1　实验动物

本研究所用的东北虎繁殖、饲养于东北虎林园（哈尔滨，黑龙江，中国），该园建立于 1996 年，目的为保护东北虎种群。该园的所有东北虎都为园内繁育，几乎与自然猎物和其他所有动物形成地理隔离。共 45 只虎用于我们的实验，年龄为 2～15 岁。每只虎都单独地关在一个笼舍，上午 9 点至下午 3 点开展实验，实验期间虎可自由饮水，实验后喂食。所有实验遵循《中华人民共和国野生动物保护法》和黑龙江东北虎林园的规章制度。

7.2.2　实验设备

东北虎行为由数字视频监控系统（海康威视，中国；视频帧率为 30 帧/s）记录，该监控系统安装在距离虎笼 3m 远的脚手架上，扬声器也固定在脚手架上。脚手架和附属设备在实验前 3 天安装好，以便使东北虎适应。

7.2.3　刺激物

4 种自然猎物、3 种非自然猎物、1 种非动物对照的图像、声音、粪便（或模拟物）分别作为视觉刺激、听觉刺激和嗅觉刺激。我们从哈尔滨北方森林动物园采集 3 种自然猎物（野猪、梅花鹿、马鹿）、3 种非自然猎物（角马、羊驼、河马）的图像、声音、粪便。狍（1 种自然猎物）的图像、声音、粪便样本采集于鸡西市的私人养殖场（黑龙江）。另外，不规则图、白噪声和干草被作为对照组的刺激物。

拍摄的动物图像姿势一致（体侧图像，头部面向观察者）。将裁自 7 种动物的图像随机排布，生成非动物对照组的不规则图。每种图像的面积调整一致（0.66m²，类似于各个动物的实际体侧面积），然后将图像打印在白底防水布（1.5m×1.5m）上。

动物声音通过电子录音笔（爱国者，中国）采集，包括野猪的嚎叫声、狍的吠叫声、梅花鹿的鸣叫声、马鹿的吼叫声、角马的吼叫声、羊驼的咩叫声、河马的咆哮声。白噪声由音频软件（Adobe Audition，美国）生成。所有声音的强度调整一致（距扬声器1m处的强度为65dB）。

采集新鲜的粪便，冻存于–20℃，使用时解冻。

7.2.4 猎物识别测试

为了检测圈养东北虎是否能本能地识别自然猎物刺激物，我们对比了东北虎对自然猎物、非自然猎物和非动物对照的刺激物的反应。

对于每只东北虎，实验中依次被呈递1种视觉刺激、1种听觉刺激和1种嗅觉刺激，刺激间隔为30min。3种刺激物分别独立地按随机顺序呈递，每种刺激物只呈递1次。

7.2.4.1 图像呈递实验

图像悬挂在脚手架上，并用布覆盖。10min 基线期结束时，实验者在远处通过长绳将盖布扯下，然后图像呈递10min（刺激期）。为了避免拉扯盖布惊吓东北虎，在实验开始前，每只东北虎都适应20次拉扯盖布的动作，此时布覆盖在空白的防水布上。我们凭经验发现拉扯盖布不到10次，东北虎即可适应。

7.2.4.2 声音播放实验

基线期结束后，扬声器播放声音10min（刺激期），声音为脉冲式，每分钟播放10s。

7.2.4.3 粪便呈递实验

解冻的粪便或干草装入铁盒内（8cm×8cm×2cm，盒盖上有多个小孔），然后将铁盒呈递给东北虎10min（刺激期）。在实验开始前，每只东北虎适应空铁盒24h。另外，下一次使用前，用水冲洗铁盒，除味、晾干。

7.2.5 个性评估

类似于其他研究，我们通过"个性评分"、"行为观察"和"行为测试"3种方法评估东北虎的个性（Bergvall *et al.*，2011；Gartner and Powell，2012）。

首先，"个性评分"用于评估每只东北虎的个性结构。基于先前的个性研究，我们设计了1份包含25个个性特征的调查表（Gartner and Weiss，2013b；Gartner *et al.*，2014），要求3位饲养员独立地评估每只东北虎的个性，并按0~7尺度对25个个性特征评分（"0"表示最弱，"7"表示最强）。个性评分的可靠性通过组内相关系数（intraclass correlation coefficient，ICC）分析初步判断，由此产生16项可靠的个性特征（F 检验：$n=45$，$P<0.05$；表7.1）。

表7.1 用组内相关系数（ICC）评估个性评分的可靠性

个性特征	ICC（3，1）	ICC（3，k）	P 值
活泼性	0.308	0.572	**0.000**
侵犯人	0.245	0.494	**0.002**
焦虑性	0.017	0.049	0.406
沉着性	0.161	0.365	**0.036**
合作性	0.247	0.496	**0.004**
好奇性	0.423	0.688	**0.000**
沮丧性	0.101	0.252	0.113
分神性	0.175	0.389	**0.007**
古怪性	−0.028	−0.090	0.677
兴奋性	0.373	0.641	**0.000**
害怕人	0.148	0.343	**0.044**
亲人性	0.207	0.439	**0.006**
冲动性	0.049	0.134	0.259
不安全性	0.055	0.148	0.253
易怒性	0.156	0.357	**0.020**
玩耍性	0.403	0.670	**0.000**
鲁莽性	0.144	0.335	**0.049**
自信性	0.098	0.246	0.133
聪明性	0.063	0.168	0.212
怀疑性	0.236	0.480	**0.005**
紧张性	0.122	0.295	0.058
羞怯性	0.308	0.572	**0.000**
信任性	0.295	0.557	**0.001**
警惕性	0.080	0.208	0.153
易叫性	0.429	0.692	**0.000**

注：ICC（3，1）指单个评分的可靠性；ICC（3，k）指平均评分的可靠性。显著 P 值用粗体表示（F 检验，双尾；$n=45$，$P<0.05$）。$P<0.05$ 的个性特征是显著可靠的，用于进一步分析

然后，为了确认个性评分的可靠性，我们进行"行为观察"。所有图像/声音呈递实验的基线期用于观察两种行为，即"活动"和"嚎叫"。"活动频率"：每10min基线期内的平均活动时间，反映"活泼性"（active）；"嚎叫频率"：每10min基线期内的平均嚎叫次数，反映"易叫性"（vocal）。

最后，"行为测试"评估一些个性评分的可靠性，如"好奇性"（curious）、"兴奋性"（excitable）、"玩耍性"（playful）、"鲁莽性"（reckless）、"怀疑

性"（suspicious）、"羞怯性"（timid）和"信任性"（trusting）。在这个实验中，1个新颖物体放置于虎笼内，然后让虎进入笼舍，监测 20min（刺激期）。共使用两种新颖物体，一种是铁质椅子腿，另一种是铁质折叠钢管，每种只使用 1 次，按随机顺序使用。实验间隔为 1h。用水冲洗物体，除味、晾干后，用于下一次实验。"接触潜伏期"，即东北虎从步入虎笼至第一次嗅闻、舔舐或触碰新颖物体的平均时间。"接触时间"，即东北虎嗅闻、舔舐或触碰新颖物体的平均时间。

7.2.6 数据分析

由一名不了解本实验的人员通过视频回放的方式记录行为反应。除特别说明外，数据分析都通过 SPSS 软件进行（SPSS Inc.）。文中数据和图表用"平均值±标准误差"（SEM，standard error of mean）表示。显著性水平设置为 0.05（*$P<0.05$；**$P<0.01$；***$P<0.001$；NS =不显著）。

对于猎物识别测试的数据分析，因为行为反应是非正态分布（KS 检验：$P<0.05$），所以我们用 Friedman 检验分析不同刺激物引起的行为反应差异，如果差异显著（Friedman 检验：$P<0.05$），那么 Wilcoxon 检验用于任意两种刺激物之间的多重比较。

另外，我们用"偏好指数"量化对自然猎物的偏好强度（Dix and Aggleton，1999）。偏好指数（preference index，PI）通过下列公式计算：

$$\mathrm{PI} = \frac{(T_{野猪} + T_{狍} + T_{梅花鹿} + T_{马鹿}) - (T_{角马} + T_{羊驼} + T_{河马} + T_{对照})}{(T_{野猪} + T_{狍} + T_{梅花鹿} + T_{马鹿}) + (T_{角马} + T_{羊驼} + T_{河马} + T_{对照})}$$

式中，"T"表示行为反应时间（注视或嗅闻）；"T"的下标表示东北虎对哪种刺激物反应。理论上，偏好指数的值为+1.0（完全偏好自然猎物）至–1.0（完全偏好非自然猎物），"0"表示对自然猎物和非自然猎物无偏好。

因为偏好指数的分布与正态分布无显著差异（KS 检验：$P > 0.05$），所以我们用线性回归分析虎的年龄与偏好指数的关系，用 t 检验（双尾）分析任意两组偏好指数的差异。将偏好指数公式中的"T"随机更换位置，生成"模拟偏好指数"（simulated preference index），如此操作 100 次。

对于个性评估实验，我们先用组内相关系数（ICC）分析检验"个性评分"的可靠性。将三位评分者评估的显著可靠的个性特征（F 检验，双尾：$P<0.05$）求平均值，用于之后的分析。另外，为了进一步确定个性评分的可靠性，我们分析了个性评分与行为观察/测试中反应参数的相关性（Spearman 相关分析：$n = 45$，P 值经 Bonferroni 校正；表 7.2）。

表 7.2　个性评分与行为观察/测试之间的相关性

个性特征	行为观察		行为测试	
	活动频率	嚎叫频率	接触潜伏期	接触时间
活泼性	**0.811**	0.197	−0.031	−0.022
侵犯人	−0.011	−0.205	0.423	−0.279
沉着性	−0.114	0.141	−0.292	0.266
合作性	−0.020	0.327	**−0.455**	0.269
好奇性	−0.007	−0.037	**−0.696**	**0.806**
分神性	−0.009	−0.211	0.028	0.193
兴奋性	−0.015	0.195	**−0.514**	**0.597**
害怕人	−0.002	0.014	0.166	0.002
亲人性	0.240	0.331	−0.399	0.254
易怒性	−0.078	−0.405	**0.443**	−0.307
玩耍性	0.044	−0.017	−0.424	**0.626**
鲁莽性	0.133	−0.016	**−0.630**	**0.499**
怀疑性	0.000	**−0.442**	0.548	−0.357
羞怯性	0.077	0.025	**0.698**	**−0.677**
信任性	−0.013	0.345	**−0.517**	0.248
易叫性	0.133	**0.753**	−0.126	−0.118

注：显著相关系数用粗体表示（Spearman 相关性检验，经 Bonferroni 校正）

另外，因为个性特征之间存在显著的多重共线性（Bartlett 球形检验：χ^2_{120}=458.951，$P<0.05$），所以我们用主成分分析对个性结构降维（R software version 3.4.1，princomp function；R Development Core Team，2017）。主成分个数通过平行分析和碎石图确定（R software，fa.parallel function；图 7.1）。与先前研究（Gartner et al.，2014）类似，我们定义主成分分析中因子载荷绝对值≥0.4 为显著项（表 7.3）。最后，我们用线性回归分析检验偏好指数与主成分的关系。

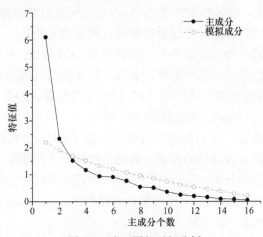

图 7.1　碎石图与平行分析

表 7.3　个性结构的主成分

个性特征	主成分 1	主成分 2
活泼性	−0.08	0.26
侵犯人	−0.38	**−0.72**
沉着性	0.30	0.16
合作性	0.36	**0.65**
好奇性	**0.91**	0.01
分神性	0.14	−0.48
兴奋性	**0.82**	0.18
害怕人	−0.28	**−0.62**
亲人性	0.33	**0.78**
易怒性	**−0.47**	**−0.69**
玩耍性	**0.87**	0.10
鲁莽性	**0.63**	0.03
怀疑性	**−0.40**	**−0.64**
羞怯性	**−0.84**	−0.11
信任性	0.29	**0.64**
易叫性	−0.27	**0.59**

注：粗体为显著载荷

7.3　结　　果

7.3.1　对自然猎物视觉刺激的偏好反应

如图 7.2a 所示，在刺激呈递期，东北虎明显注视自然猎物（野猪、狍、梅花鹿、马鹿）、非自然猎物（角马、羊驼、河马）、非动物对照（不规则图）图像。在呈递刺激的第 1 分钟，相比于其他图像，东北虎注视野猪图像的时间更长（Wilcoxon 检验：$n = 45$，$P < 0.05$；图 7.2a）。东北虎注视野猪图像的总时间也显著多于注视非自然猎物图像或非动物对照图像的总时间（Wilcoxon 检验：$n = 45$，$P < 0.05$）。相比于其他自然猎物图像，东北虎注视野猪图像的时间更长（图 7.2b）。另外，东北虎注视自然猎物图像的平均总时间显著多于注视非自然猎物图像和非动物对照图像的平均总时间（Wilcoxon 检验：$n = 45$，两组比较 $P < 0.001$；图 7.2c）。

因为衰老是损害认知的潜在因素（Burke and Barnes，2006；Bishop et al.，2010），所以我们探究衰老是否影响东北虎对自然猎物的视觉识别本能。我们发现虎的年龄与图像偏好指数之间没有显著的线性关系，其中，图像偏好指数用于量化对自然猎物图像的偏好强度（线性回归分析：$R^2 = 0.011$，$n = 45$，$P = 0.493$；图 7.2d）。另外，亚成年虎（2～3 岁）和成年虎（4～15 岁）的图像偏好指数都显

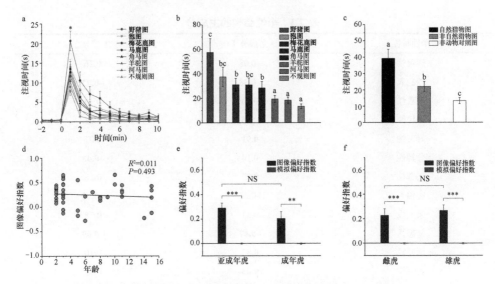

图 7.2 对自然猎物图像的偏好（彩图请扫封底二维码）

a. 注视视觉刺激物的时程。时刻"0"表示给虎呈递图像的起始点。*在刺激期的第 1 分钟, 东北虎注视野猪图像的时间比注视其他图像的时间更长（Wilcoxon 检验: $n=45, P<0.05$)。图 a 和 b 中, 粗体为自然猎物。b. 在 10min 的刺激期内, 注视视觉刺激物的总时间。字母 a、b 或 c 表明任意两组刺激物间具有显著差异（Wilcoxon 检验: $n=45, P<0.05$)。c. 注视自然猎物图像、非自然猎物图像和非动物对照图像的平均总时间。字母 a、b 或 c 表明任意两组间具有显著差异（Wilcoxon 检验: $n=45, P<0.05$)。d. 虎年龄与图像偏好指数的线性回归分析。e、f. 用非配对 t 检验对比亚成年虎和成年虎（$n=24$ 和 21, 分别地）或者雌虎与雄虎（$n=19$ 和 26, 分别地）的图像偏好指数。用单样本 t 检验对比这 4 组的图像偏好指数与相应的模拟偏好指数均值的差异显著性。数据表示为平均值±标准误差; *$P<0.05$; **$P<0.01$; ***$P<0.001$; NS 表示不显著

著大于模拟偏好指数（单样本 t 检验: $t_{23}=7.218, P<0.001$; $t_{20}=3.727, P<0.01$, 分别地; 图 7.2e), 同时这两者的图像偏好指数之间无显著差异（非配对 t 检验: $t_{43}=1.280, P=0.208$; 图 7.2e)。

另外, 雌虎和雄虎的图像偏好指数都显著大于模拟偏好指数（单样本 t 检验: $t_{18}=4.319, P<0.001$; $t_{25}=6.002, P<0.001$, 分别地; 图 7.2f), 这两者的图像偏好指数之间无显著差异（非配对 t 检验: $t_{43}=-0.570, P=0.572$; 图 7.2f)。

7.3.2 对自然猎物听觉刺激的偏好反应

类似于对视觉刺激的反应, 在呈递刺激的前 5min, 相比于其他声音, 东北虎注视野猪声音方向的时间更长（Wilcoxon 检验: $n=45, P<0.05$; 图 7.3a)。东北虎注视野猪声音方向的总时间也显著多于注视其他声音的总时间（Wilcoxon 检验: $n=45, P<0.05$; 图 7.3b)。特别地, 东北虎注视自然猎物声音的平均总时间显著多于注视非自然猎物声音和非动物对照声音的平均总时间（Wilcoxon 检验: $n=45$, 两组 $P<0.001$; 图 7.3c)。

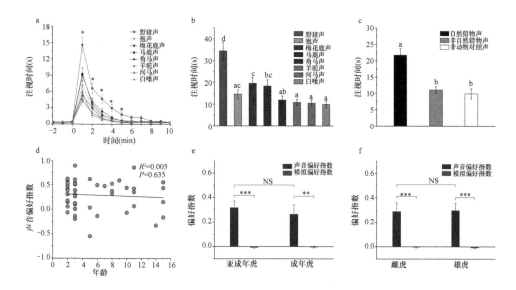

图 7.3 对自然猎物声音的偏好（彩图请扫封底二维码）

a. 注视听觉刺激物的时程。*在刺激期的前 5min，东北虎注视野猪声音方向的时间比注视其他声音方向的时间更长（Wilcoxon 检验：$n=45$，$P<0.05$）。图 a 和 b 中，粗体为自然猎物。b. 在 10min 的刺激期内，对听觉刺激物的注视总时间（Wilcoxon 检验：$n=45$，$P<0.05$）。c. 注视自然猎物声音、非自然猎物声音和非动物对照声音的平均总时间。字母 a、b 或 c 表明任意两组间具有显著差异（Wilcoxon 检验：$n=45$，$P<0.05$）。d. 虎年龄与声音偏好指数的线性回归分析。e、f. 对比亚成年虎和成年虎（$n=24$ 和 21，分别地），或者雌虎与雄虎（$n=19$ 和 26，分别地）声音偏好指数。用单样本 t 检验对比这 4 组的声音偏好指数与相应的模拟偏好指数均值的差异显著性。

数据表示为平均值±标准误差；*$P<0.05$；**$P<0.01$；***$P<0.001$；NS 表示不显著

我们发现年龄与声音偏好指数之间没有显著的线性关系（线性回归分析：$R^2=0.005$，$n=45$，$P=0.635$；图 7.3d）。亚成年虎与成年虎，以及雌虎与雄虎均显著偏好自然猎物声音（图 7.3e、f）。

7.3.3 对自然猎物嗅觉刺激的偏好反应

虽然东北虎对各种粪便的嗅闻时间之间无明显规律（图 7.4a、b），但是东北虎嗅闻自然猎物粪便的平均总时间显著多于嗅闻非自然猎物粪便和非动物对照的平均总时间（Wilcoxon 检验：$n=31$，分别地 $P<0.05$、$P<0.001$；图 7.4c）。

粪便偏好指数与年龄显著负相关（线性回归分析：$R^2=0.146$，$n=31$，$P=0.034$；图 7.4d）。并且亚成年虎（而非成年虎）的粪便偏好指数显著大于模拟偏好指数（单样本 t 检验：$t_{15}=3.632$，$P<0.01$；$t_{14}=0.783$，$P=0.447$，分别地；图 7.4e）。亚成年虎的粪便偏好指数相对地大于成年虎的粪便偏好指数，尽管并不显著（非配对 t 检验：$t_{29}=1.854$，$P=0.074$；图 7.4e）。

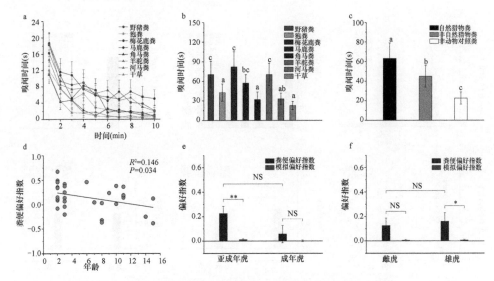

图 7.4 对自然猎物粪便气味的偏好（彩图请扫封底二维码）

a. 嗅闻嗅觉刺激物的时程。图 a 和 b 中，粗体为自然猎物。b. 在 10min 的刺激期内，对嗅觉刺激物的嗅闻总时间（Wilcoxon 检验：$n=31$，$P<0.05$）。c. 嗅闻自然猎物、非自然猎物和非动物对照的嗅觉刺激物的平均总时间。字母 a、b 或 c 表明任意两组间具有显著差异（Wilcoxon 检验：$n=31$，$P<0.05$）。d. 虎年龄与粪便偏好指数的线性回归分析。e、f. 对比亚成年虎和成年虎（$n=16$ 和 15，分别地），或者雌虎与雄虎（$n=14$ 和 17，分别地）的粪便偏好指数。用单样本 t 检验对比这 4 组的粪便偏好指数与相应的模拟偏好指数均值的差异显著性。数据表示为平均值±标准误差；*$P<0.05$；**$P<0.01$；NS 表示不显著

另外，雌虎与雄虎的粪便偏好指数之间无显著差异（非配对 t 检验：$t_{29}=-0.357$，$P=0.724$；图 7.4f），尽管雌虎的粪便偏好指数相对（而非显著）地大于模拟偏好指数（单样本 t 检验：$t_{13}=1.911$，$P=0.078$；图 7.4f）。

7.3.4 对自然猎物的本能偏好及其与个性特征的关系

为了分析偏好指数（量化本能的猎物识别能力）与个性特征的关系，我们首先用主成分分析对个性结构降维，得到个性结构的两个主成分（图 7.1，表 7.3）。第一个主成分的最高载荷为"好奇性"、"玩耍性"、"兴奋性"。第二个主成分的最高载荷为"亲人性"、"合作性"、"信任性"（表 7.3）。图像偏好指数和声音偏好指数都与主成分 1 负相关（线性回归分析：$R^2=0.113$，$n=45$，$P<0.05$；$R^2=0.205$，$n=45$，$P<0.01$，分别地；图 7.5a、c），但是与主成分 2 无显著相关性（线性回归分析：$R^2=0.009$，$P=0.527$；$R^2=0.049$，$P=0.144$，分别地；图 7.5b、d）。然而，粪便偏好指数与两个主成分均无显著相关性（线性回归分析：$R^2=0.020$，$n=31$，$P=0.452$；$R^2=0.014$，$n=31$，$P=0.528$，分别地；图 7.5e、f）。

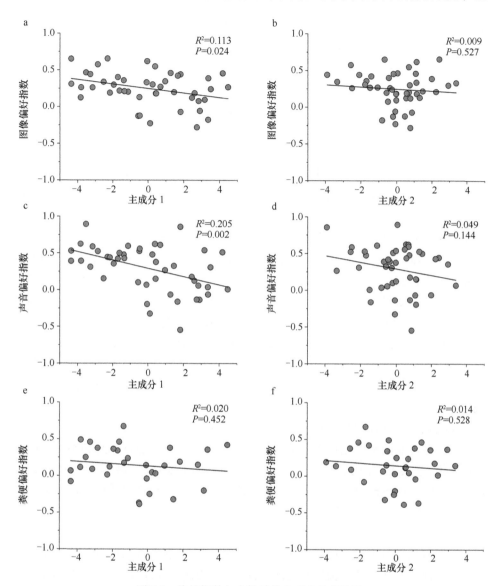

图 7.5　偏好指数与个性结构主成分的相关性

a~d. 图像/声音偏好指数与主成分 1 显著负相关（线性回归分析：$P<0.05$，$P<0.01$，分别地），但是与主成分 2 无显著相关性。e、f. 粪便偏好指数与两个主成分均无显著相关性

7.4　讨　论

在此，我们证明，相比于非自然猎物和非动物对照的刺激物，圈养东北虎对自然猎物刺激物具有偏好反应。因为这些圈养东北虎自出生后就与自然猎物和非自然猎物隔离，东北虎对自然猎物刺激物与生俱来的偏好支持了东北虎本能地识

别自然猎物的假设。鉴于捕食行为不但耗费体能，而且面临危险（Benoit-Bird，2004；Liznarova and Pekar，2013），动物受益于区分易食猎物与非易食猎物的能力。如果猎物识别能力与生俱来，并且只捕易食猎物，那么可避免捕食非易食猎物的风险，因此，可能是选择压力驱动猎物的识别能力进化为保守本能，据此，食肉动物（可能含无脊椎动物和脊椎动物）的适合度增加（Dolev and Nelson，2014；Holding et al.，2016）。尽管一些探索性研究报道了一些本能行为的神经或遗传基础（Musso et al.，2003；Liedtke et al.，2011；Young et al.，2016），但是还没有研究探索猎物识别本能的生物基础，进一步关于猎物识别的研究可聚焦于此。

本研究表明，圈养东北虎最偏好野猪的图像和声音。因为先前研究有力证明了野外环境中东北虎偏好捕食野猪（Kerley et al.，2015；Sugimoto et al.，2016），所以圈养东北虎对野猪图像和声音的本能偏好反应与野生东北虎偏好捕食野猪吻合，这很有趣。事实上，其他研究也发现了类似的现象（Darmaillacq et al.，2006；Portela et al.，2014；Pekar and Cardenas，2015）。例如，相比于非喜食猎物（虾），章鱼本能地偏好捕食喜食猎物（螃蟹），这提示章鱼本能地识别喜食猎物（Darmaillacq et al.，2006）。考虑到单位消化量的喜食猎物能提供更多能量，或者喜食猎物更易捕食（Bittar et al.，2012；Kerley et al.，2015），本能地识别喜食猎物会使食肉动物受益更多。因此，圈养东北虎对野猪图像和声音的本能偏好可能是一种认知适应，类似于虎的生理适应（Meachen-Samuels and Van Valkenburgh，2009a，2009b）。

仅嗅觉（而非视觉和听觉）的本能猎物识别能力与年龄显著负相关，这也很有趣。一种可能的解释是：嗅觉的猎物识别能力对东北虎也许不是必不可少的，更易退化。事实上，相比于昼行性动物，嗅觉更有助于夜行性动物探寻猎物（Bicca-Marques and Garber，2004；Piep et al.，2008），而虎主要在晨昏觅食，所以虎不是夜行性动物（Naha et al.，2016）。

尽管我们主要依靠主观的评分评估东北虎个性，但是个性评分的可靠性可由进一步的客观分析佐证，所用的客观分析有行为观察和新颖物体测试。本研究使用的个性评估方法与其他类似研究的方法一致（Bergvall et al.，2011；Gartner and Powell，2012）。并且本研究中东北虎个性的两个主成分与其他研究中大型猫科动物的个性主成分一致（Gartner et al.，2014）。然而，由于本研究所用的东北虎为独笼饲养，社交类个性难于评估，因此本研究不考虑这类个性。

本研究为圈养东北虎和野生东北虎的保护管理提供了有价值的参考。鉴于本能的可遗传性（Ricker and Hirsch，1985），我们建议筛选并繁育猎物识别本能敏锐的个体。然而，检测本能的常规实验过程很烦琐，因此，通过个性评估初步地估测本能也许具有可行性。另外，个性是影响动物应答外界刺激的重要因子（Stamps and Groothuis，2010）。具有某些个性的个体可能在野外更易存活（Dingemanse *et al.*，2004；Boon *et al.*，2007）。这些结果提示，具有某些个性的虎识别猎物的本能更强，应该优先用于野化训练和野外放归。

7.5 本章小结

识别猎物对野外放归东北虎的捕食和存活至关重要。因此，筛选猎物识别能力优良的适宜候选者具有重要意义。我们的研究表明，东北虎本能地识别自然猎物。这种本能的猎物识别能力与东北虎的个性特征显著相关。这些结果支持猎物识别的本能假说，并且提供一种潜在方法——通过个性评估初步筛选猎物识别本能敏锐的东北虎，用于繁育和野化训练。

参 考 文 献

Benoit-Bird K J. 2004. Prey caloric value and predator energy needs: foraging predictions for wild spinner dolphins. Marine Biology, 145(3): 435-444.

Bergvall U A, Schapers A, Kjellander P, *et al*. 2011. Personality and foraging decisions in fallow deer, *Dama dama*. Animal Behaviour, 81(1): 101-112.

Bicca-Marques J C, Garber P A. 2004. Use of spatial, visual, and olfactory information during foraging in wild nocturnal and diurnal anthropoids: a field experiment comparing *Aotus*, *Callicebus*, and *Saguinus*. American Journal of Primatology, 62(3): 171-187.

Bishop N A, Lu T, Yankner B A. 2010. Neural mechanisms of ageing and cognitive decline. Nature, 464(7288): 529-535.

Bittar V T, Awabdi D R, Tonini W C T, *et al*. 2012. Feeding preference of adult females of ribbonfish *Trichiurus lepturus* through prey proximate-composition and caloric values. Neotropical Ichthyology, 10(1): 197-203.

Boon A K, Reale D, Boutin S. 2007. The interaction between personality, offspring fitness and food abundance in North American red squirrels. Ecology Letters, 10(11): 1094-1104.

Briffa M, Weiss A. 2010. Animal personality. Current Biology, 20(21): 912-914.

Burke S N, Barnes C A. 2006. Neural plasticity in the ageing brain. Nature Reviews Neuroscience, 7(1): 30-40.

Carroll C, Miquelle D G. 2006. Spatial viability analysis of Amur tiger *Panthera tigris altaica* in the Russian Far East: the role of protected areas and landscape matrix in population persistence. Journal of Applied Ecology, 43(6): 1056-1068.

Darmaillacq A S, Chichery R, Poirier R, *et al*. 2004. Effect of early feeding experience on subsequent prey preference by cuttlefish, *Sepia officinalis*. Developmental Psychobiology, 45(4): 239-244.

Darmaillacq A S, Chichery R, Shashar N, *et al*. 2006. Early familiarization overrides innate prey preference in newly hatched *Sepia officinalis* cuttlefish. Animal Behaviour, 71: 511-514.

Dingemanse N J, Both C, Drent P J, *et al*. 2004. Fitness consequences of avian personalities in a fluctuating environment. Proceedings of the Royal Society B-Biological Sciences, 271(1541): 847-852.

Dix S L, Aggleton J P. 1999. Extending the spontaneous preference test of recognition: evidence of object-location and object-context recognition. Behav Brain Res, 99(2): 191-200.

Dolan R J. 2002. Emotion, cognition, and behavior. Science, 298(5596): 1191-1194.

Dolev Y, Nelson X J. 2014. Innate pattern recognition and categorization in a jumping spider. PLoS ONE, 9(6): e97819.

Dou H L, Yang H T, Feng L M, et al. 2016. Estimating the population size and genetic diversity of Amur tigers in northeast China. PLoS ONE, 11(4): e0154254.

Fabregas M C, Fosgate G T, Koehler G M. 2015. Hunting performance of captive-born South China tigers (*Panthera tigris amoyensis*) on free-ranging prey and implications for their reintroduction. Biological Conservation, 192: 57-64.

Gartner M C, Powell D M, Weiss A. 2014. Personality structure in the domestic cat (*Felis silvestris catus*), Scottish wildcat (*Felis silvestris grampia*), clouded Leopard (*Neofelis nebulosa*), snow leopard (*Panthera uncia*), and African lion (*Panthera leo*): a comparative study. Journal of Comparative Psychology, 128(4): 414-426.

Gartner M C, Powell D. 2012. Personality assessment in snow leopards (*Uncia uncia*). Zoo Biology, 31(2): 151-165.

Gartner M C, Weiss A. 2013a. Scottish wildcat (*Felis silvestris grampia*) personality and subjective well-being: implications for captive management. Applied Animal Behaviour Science, 147(3-4): 261-267.

Gartner M C, Weiss A. 2013b. Personality in felids: a review. Applied Animal Behaviour Science, 144(1-2): 1-13.

Holding M L, Kern E H, Denton R D, et al. 2016. Fixed prey cue preferences among dusky pigmy rattlesnakes (*Sistrurus miliarius barbouri*) raised on different long-term diets. Evolutionary Ecology, 30(1): 1-7.

Jiang G S, Sun H Y, Lang J M, et al. 2014. Effects of environmental and anthropogenic drivers on Amur tiger distribution in northeastern China. Ecological Research, 29(5): 801-813.

Kenneth C G, Terry L M. 1994. Personality assessment in the gorilla and its utility as a management tool. Zoo Biology, 13: 509-522.

Kerley L L, Mukhacheva A S, Matyukhina D S, et al. 2015. A comparison of food habits and prey preference of Amur tiger (*Panthera tigris altaica*) at three sites in the Russian Far East. Integrative Zoology, 10(4): 354-364.

Killen S S, Marras S, Metcalfe N B, et al. 2013. Environmental stressors alter relationships between physiology and behaviour. Trends in Ecology & Evolution, 28(11): 651-658.

Liedtke W B, Mckinley M J, Walker L L, et al. 2011. Relation of addiction genes to hypothalamic gene changes subserving genesis and gratification of a classic instinct, sodium appetite. Proceedings of the National Academy of Sciences of the United States of America, 108(30): 12509-12514.

Lillywhite H B, Pfaller J B, Sheehy C M. 2015. Feeding preferences and responses to prey in insular neonatal Florida cottonmouth snakes. Journal of Zoology, 297(2): 156-163.

Liznarova E, Pekar S. 2013. Dangerous prey is associated with a type 4 functional response in spiders. Animal Behaviour, 85(6): 1183-1190.

Mcnamara J M, Houston A I. 1997. Currencies for foraging based on energetic gain. American Naturalist, 150(5): 603-617.

Meachen-Samuels J, Van Valkenburgh B. 2009a. Craniodental indicators of prey size preference in the Felidae. Biological Journal of the Linnean Society, 96(4): 784-799.

Meachen-Samuels J, Van Valkenburgh B. 2009b. Forelimb indicators of prey-size preference in the Felidae. Journal of Morphology, 270(6): 729-744.

Miller C S, Hebblewhite M, Petrunenko Y K, et al. 2013. Estimating Amur tiger (*Panthera tigris altaica*) kill rates and potential consumption rates using global positioning system collars. Journal of Mammalogy, 94(4): 845-855.

Miquelle D G, Pikunov D G, Dunishenko Y M, et al. 2006. A survey of Amur (Siberian) tigers in the Russian Far East, 2004-2005. Final Report to Save the Tiger Fund.

Miquelle D G, Smirnov E N, Zaumyslova O Y, et al. 2015. Population dynamics of Amur tigers (*Panthera tigris altaica*) in Sikhote-Alin Biosphere Zapovednik: 1966-2012. Integrative Zoology, 10(4): 315-328.

Musso M, Moro A, Glauche V, et al. 2003. Broca's area and the language instinct. Nature Neuroscience, 6(7): 774-781.

Nadja C W. 1999. Behavioral differences as predictors of breeding status in captive cheetahs. Zoo Biology, 18: 335-349.

Naha D, Jhala Y V, Qureshi Q, et al. 2016. Ranging, activity and habitat use by tigers in the mangrove forests of the Sundarban. PLoS ONE, 11(4): e0152119.

Pekar S, Cardenas M. 2015. Innate prey preference overridden by familiarisation with detrimental prey in a specialised myrmecophagous predator. Science of Nature, 102(1-2): 8.

Piep M, Radespiel U, Zimmermann E, et al. 2008. The sensory basis of prey detection in captive-born grey mouse lemurs, *Microcebus murinus*. Animal Behaviour, 75: 871-878.

Portela E, Simoes N, Rosas C, et al. 2014. Can preference for crabs in juvenile *Octopus maya* be modified through early experience with alternative prey? Behaviour, 151(11): 1597-1616.

R Development Core Team. 2017. R: a language and environment for statistical computing. Vienna, Austria: R Foundation for Statistical Computing.

Ricker J P, Hirsch J. 1985. Evolution of an instinct under long-term divergent selection for geotaxis in domesticated populations of *Drosophila melanogaster*. J Comp Psychol, 99(4): 380-390.

Stamps J, Groothuis T G G. 2010. The development of animal personality: relevance, concepts and perspectives. Biological Reviews, 85(2): 301-325.

Sugimoto T, Aramilev V V, Nagata J, et al. 2016. Winter food habits of sympatric carnivores, Amur tigers and Far Eastern leopards, in the Russian Far East. Mammalian Biology, 81(2): 214-218.

Van Gils J A, De Rooij S R, Van Belle J, et al. 2005. Digestive bottleneck affects foraging decisions in red knots *Calidris canutus*. I. Prey choice. Journal of Animal Ecology, 74(1): 105-119.

Young K S, Parsons C E, Elmholdt E M J, et al. 2016. Evidence for a caregiving instinct: rapid differentiation of infant from adult vocalizations using magnetoencephalography. Cerebral Cortex, 26(3): 1309-1321.

第8章 利用东北虎粪便DNA进行基因分型的风险评估

8.1 前　　言

长期以来，由于技术手段的限制，很难从粪便和毛发样本中获得有效的遗传信息，但分子遗传标记技术的出现为研究者从粪便或毛发样本中获取遗传信息提供了可能，这种以粪便或毛发为检材的非损伤性遗传分析技术在野生动物检测工作中逐渐被推广（Taberlet and Luikart，1999），尤其是在濒危物种的保护管理实践中，这种方法应用更为广泛（Woodruff，1993）。从粪便样本中提取到DNA的数量和质量在一定程度上受食物组成的影响（Stenglein et al.，2010），食物组成的复杂性与等位基因PCR扩增过程中的出错率之间存在密切联系，由于食肉动物粪便组分比食草动物更为复杂，因此在食肉动物粪便中提取到DNA的数量和质量均较食草动物低很多（Panasci et al.，2011）。

野生东北虎是珍稀濒危的大型猫科动物，数量稀少且野外行踪不定，在野外采集到东北虎的血液或组织样本是极其困难的，但是获取粪便样本相对较易。粪便样本的常规分析流程包括：分离基因组DNA—微卫星位点扩增—生物信息分析。但是，粪便样本被降解或污染会导致DNA模板数量少且质量偏差（Bradley et al.，2000；Regnaut et al.，2006），微卫星位点的扩增和基因分型会在很大程度上影响下游的生物信息分析，错误的信息可能影响野生动物保护管理的决策。近年来，尽管有研究使用东北虎粪便样本作为DNA来源获得种群遗传信息（Rozhnov et al.，2009），但他们并没有通过科学的预实验来减少由较差DNA模板产生的基因分型错误（Arandjelovic et al.，2009）。

影响东北虎微卫星基因分析的主要因素包括DNA提取效率和PCR扩增效率。DNA提取效率在很大程度上取决于样本保存方法（Nsubuga et al.，2004）、样本收集的季节（Hájková et al.，2006）及样本采集后存放的时间（Murphy et al.，2007；Santini et al.，2007）。而PCR扩增效率主要取决于引物序列、扩增片段的长度、二级结构、退火温度及核苷酸组成等因素（Sambrook et al.，2001）。使用软件进行基因分型错误估计，是以统计学模型对于样本大小敏感性为基础从而获得基因分型错误的估计，然而野生东北虎种群很小，且不同个体采样机会存在偏差，从而导致软件对野生东北虎种群的估算效率偏低。本研究不仅改进了DNA分离效率，还研究了基于微卫星遗传标记的PCR方案对东北虎遗传分析的影响。我们选取了 10 只圈养东北虎的 12 对微卫星位点（表 8.1），比较了血液样本和粪便样本中的基因型效率、个体及个体间基因关系的一致性（Bhagavatula and Singh, 2006；Rozhnov et al., 2009）。确切地说，我们通过重复扩增 12 个微卫星位点来评估利用粪便 DNA 估算个体及遗传关系时的潜在风险，为将来以粪便 DNA 分析技术为基础的野生东北虎保护与管理工作奠定基础。

表 8.1 东北虎 12 个微卫星位点聚合酶链反应的特征与条件

微卫星位点	引物序列（5′→3′）	重复单元	退火温度（℃）	PCR 产物片段范围（bp）
E6	CCTGGGGATAATAAAACTAGTA CATGAATGAATCTTTACACTGA	$(TAA)_{11}$	58	147~162
E21B	GCGATAAAGGCTGGCAGAGG CTTTGAGGGTCTGTTCTACTGTGA	$(CA)_{21}$	62	154~168
D10	CCCTCTCTGTCCCTCCCTTG GCCGTTTCCCTCATGCTACA	$(GT)_{14}$	62	134~150
E7	GCCCCAAAGCCCTAAAATAA GCATGTCGGACAGTAAAGCA	$(CA)_{11}CG(CA)_4$	58	136~156
FCA304	TCATTGGCTACCACAAAGTAGG CTGCATGCCATTGGGTAAC	$(GT)_{17}(GG)_1(GT)_6$	58	120~134
FCA043	GAGCCACCCTAGCACATATACC AGACGGGATTGCATGAAAAG		58	116~130
FCA391	GCCTTCTAACTTCCTTGCAGA TTTAGGTAGCCCATTTTCATCA	$(ATGG)_{10}(GATA)_{11}(TAGA)_2$ $TGA(TAGA)_1$	55	190~230
FCA152	TTTAGTCAGCTTAGGCTTCCA CTTCCCAGCTTCCAGAATTG	$(AC)_{21}$	58	129~147
Pti007	ATCAGGAGTTCTATCACC CATGATTAGGGAGTTGAG	$(AC)_{16}$	52	139~193
FCA441	ATCGGTAGGTAGGTAGATATAG GCTTGCTTCAAAATTTTCAC	$(ATAG)_9(GTAG)_1(ATAG)_2$ $AG(ATAG)_1$	58	130~168
FCA094	TCAAGCCCCATTTTACCTTC CACCTGAGCCAAAGGCTATC	$(GT)_{19}(AG)_{22}$	58	193~215
Pti010	GGGACAACTGAGAGAAGA CAAGATATGTTCTCAGACTG	$(AC)_8$	58	118~134

8.2 材料与方法

8.2.1 取样

于 2013 年 11 月在黑龙江省东北虎林园收集 10 只成年东北虎的粪便和血液样本，血液样本是兽医于虎出生三个月后建立遗传谱系时采集的。血液样本中放入 3.8%的柠檬酸钠抗凝血剂并在 DNA 提取前 4℃保存。粪便样本在老虎排泄的 12h 内用塑封袋采集并放至-20℃冰箱保存。

8.2.2 DNA 提取

使用标准苯酚氯仿方法进行血液样本 DNA 的提取（Sambrook et al., 1989），使用改进的 QIAamp DNA Stool Mini Kit 进行粪便样本的提取，具体操作如下：剥取约 5g 重的粪便表面，浸泡在无水乙醇中并室温放置 12h 后 2200r/min 震荡 3min；随后用灭菌后的薄纱对混合物进行过滤，对保留下来的滤液进行 3500r/min 离心 15min，将底部的沉淀物转移到新的 PE 管中进行 DNA 提取；使用 1% 的琼脂糖凝胶进行电泳检测粪便和血液样本中提取到的 DNA 质量和数量。

8.2.3 微卫星数据分析

基于前人的研究，选择 12 个微卫星位点应用于本研究，其中从野生虎研究中选择 D10、FCA043、FCA304、E21B、E6 和 E7 这 6 个位点（Bhagavatula and Singh, 2006; Rozhnov et al., 2009），从已发表的圈养东北虎文章中选择 FCA391、FCA441、FCA094、FCA152、Pti007 和 Pti010 这 6 个位点（Menotti-Raymond et al., 1999; Zhang et al., 2009; Xu et al., 2005）。引物的序列、重复的碱基数、退火温度及期望的等位基因大小的详情见表 8.1。每个正向引物的 5'端用荧光染料进行标记（如 5-FAM、TAMRA 和 HEX）。

PCR 在 20μL 系统中进行，具体包括约 80ng 的总 DNA、1×PCR buffer [50mmol/L Tris-HCl（pH 8.0）、25mmol/L KCl、0.1mmol/L EDTA、1mmol/L DTT]、0.4mmol/L 的 dNTP（TOYOBO 公司）、0.2μmol/L 的正反引物、0.4U 单位的 KOD FX Neo DNA 酶。使用 Model 9700 Thermocycler（Perkin-Elmer）进行 PCR 扩增，具体反应条件：94℃，预变性 2min；98℃，变性 20s；退火 52~62℃，详情见表 8.1，30s；68℃，延伸 20s，运行 35 个循环；最后 68℃持续 20min。每组扩增都包括一个阳性对照和两个空白对照，用来检测目的条带与污染。

ABI 3100（Applied Biosystems）用于分析 PCR 产物，使用 GeneScan3.1 和 Geno-Typer 获取基因型数据。对于粪便样本进行 7 次重复扩增，当血液样本的纯合基因峰图不明确时将进行 2 次重复扩增以确定基因型（刘丹等，2013）。

8.2.4 基因分型的正确性评估

每个微卫星位点上每个老虎血液样本的基因分型被默认为 100% 正确，并作为评价粪便 DNA 的基因分型结果正确性的标准。当一个位点上同一个体的粪便样本的基因分型与血液 DNA 的基因分型匹配，就被认为是"正确的基因分型"，如果不匹配就被认为是"错误的基因分型"。对于所有的位点，计算每个位点上血液样本和粪便样本之间基因分型的累计匹配率（R_m），即正确分型的虎数量除以全部进行分型实验的虎数量，具体公式如下：

$$R_m = \frac{\sum_{i=1}^{m} N_i}{nm}$$

式中，R_m 为 PCR 累计重复 m 次后血液样本和粪便样本之间基因分型的累计匹配率，其中 m 为 PCR 重复的次数；n 为虎的数量（$n = 10$）；N_i 是在第 i 次 PCR 时正确分型的老虎数量。每个位点上的粪便样本基因分型风险表示为 $1-R_m$。

8.2.5 使用粪便样本评估种群遗传学参数

使用 POPGENE v1.32 计算血液样本和粪便样本的种群遗传学参数，包括观测杂合度（observed heterozygosity，Ho）、期望杂合度（expected heterozygosity，He）、等位基因数（the number of alleles，A）、有效等位基因数（effective number of alleles，Ne）、等位基因频率（allelic frequency）和多态信息含量（polymorphism information content，PIC）。

使用 CERVUS v3.0 计算个体排除率（discrimination power，DP）和非父排除率（exclusion probability of paternity，EPP）（Marshall et al.，1998）。应用 SPSS 13.0（SPSS Inc.，Chicago，USA）的两变量间的 Pearson 相关性二元分析，计算血液样本、粪便样本的基因分型风险（$1-R_m$）与使用血液样本、粪便样本计算获得种群遗传学参数间差值绝对值的关系。使用 Coancestry v1.0.12 分别计算血液样本和粪便样本个体间的两两关系（r_R），计算血液样本的 r_R 和粪便样本的 r_R 之间的线性回归关系。

8.3 结　　果

8.3.1 微卫星位点分型正确性评估

每份粪便样本的 12 个微卫星位点 7 次重复 PCR 的基因分型匹配率为 30%～100%，平均匹配率为 71%。根据累计的关联基因分型匹配率（R_m）和 PCR 重复次数，发现 3 次 PCR 后 8 个位点的 R_m 值趋于稳定，5 次 PCR 后 11 个位点的 R_m 值趋于稳定（图 8.1）。位点 Pti010 呈现波动的 R_m 值，表明每次重复扩增获得的 PCR 效率不稳定。

多次 PCR 后不同的微卫星位点达到平稳的 R_m 值（如在第 5 次 PCR 之后）存在显著变化（图 8.1，表 8.2）。位点 E6 的 R_m 最高值为 0.871，而 Pti007 平稳最低值为 0.357，12 个微卫星位点的平均平稳 R_m 值为 0.710±0.139。

8.3.2 基因分型误差对估算种群遗传学参数的影响

每个粪便样本和微卫星位点均进行 7 次重复 PCR 扩增。当粪便样本扩增 4 次或超过 4 次时获得的基因型被认为是正确的基因分型（与同一只老虎的血液样本

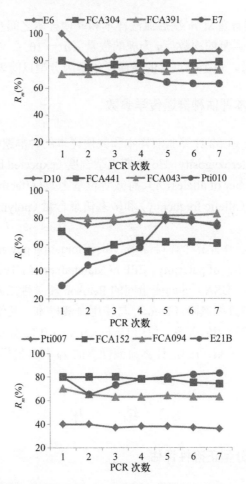

图 8.1 基因分型累计匹配率（R_m）与每个微卫星位点上重复 PCR 次数之间的关联

基因型一致）。12 个微卫星位点的特征及粪便样本和血液样本之间的比较见表 8.2。血液样本每个基因座获得的等位基因数（A）为 3~6 个（$\bar{X}$=4.25），而粪便样本每个基因座获得的等位基因数为 3~5 个（$\bar{X}$=4.08）。除 E6 外所有位点的等位基因频率均不同，基于等位基因频率而计算的种群基因参数也会受到影响。基于粪便样本和血液样本获得的遗传学参数值，如平均有效等位基因数（血液样本平均值 Ne=2.629；粪便样本平均值 Ne=2.439）、平均期望杂合度（血液样本平均值 He=0.616；粪便样本平均值 He=0.597）、多态信息含量（血液样本平均值 PIC=0.543；粪便样本平均值 PIC=0.518）、DP（血液样本平均值 DP=0.228；粪便样本平均值 DP=0.245）和 EPP（血液样本平均值 EPP=0.470；粪便样本平均值 EPP=0.502）均未达到显著水平的不同。Pearson 相关性分析结果显示，基因分型风险（$1-R_m$）与粪便样本和血液样本间种群遗传学参数之差的绝对值之间无显著的相关性

表 8.2　12 个微卫星位点的特征及圈养东北虎种群中粪便样本和血液样本遗传学参数的比较（$n = 10$）

位点		等位基因频率						Ho	He	A	Ne	PIC	DP	EPP	R_m (%)	$1-R_m$ (%)
E6	等位基因	150	153	159											87.1	12.9
	血液	0.650	0.250	0.100				0.700	0.532	3	2.020	0.443	0.308	0.602		
	粪便	0.650	0.250	0.100				0.700	0.532	3	2.020	0.443	0.308	0.602		
E21B	等位基因	156	158	166											82.9	17.1
	血液	0.750	0.150	0.100				0.500	0.426	3	1.681	0.368	0.391	0.655		
	粪便	0.650	0.250	0.100				0.500	0.532	3	2.020	0.443	0.308	0.602		
D10	等位基因	136	138	144	146	148									75.7	24.3
	血液	0.150	0.050	0.050	0.100	0.650		0.600	0.568	5	2.174	0.508	0.244	0.484		
	粪便	0.200	0.000	0.050	0.050	0.700		0.500	0.490	4	1.869	0.421	0.331	0.600		
E7	等位基因	138	142	148	152										62.9	37.1
	血液	0.100	0.250	0.550	0.100			0.900	0.647	4	2.597	0.562	0.201	0.458		
	粪便	0.100	0.100	0.700	0.100			0.600	0.505	4	1.923	0.450	0.300	0.548		
FCA304	等位基因	124	126	130	132	136									78.6	21.4
	血液	0.200	0.200	0.400	0.150	0.050		0.700	0.774	5	3.774	0.694	0.111	0.310		
	粪便	0.150	0.300	0.400	0.100	0.050		0.600	0.753	5	3.509	0.668	0.128	0.342		
FCA043	等位基因	119	121	125	127	129									82.9	17.1
	血液	0.050	0.100	0.200	0.050	0.600		0.700	0.616	5	2.410	0.544	0.213	0.456		
	粪便	0.050	0.100	0.200	0.050	0.550		0.800	0.668	5	2.740	0.595	0.173	0.401		
FCA391	等位基因	198	202	206	210	218	222								72.9	27.1
	血液	0.050	0.450	0.100	0.250	0.050	0.100	0.900	0.747	6	3.448	0.671	0.123	0.319		
	粪便	0.000	0.400	0.050	0.350	0.100	0.100	0.800	0.732	5	3.279	0.643	0.145	0.371		

续表

位点		等位基因频率						Ho	He	A	Ne	PIC	DP	EPP	R_m (%)	$1-R_m$ (%)
FCA152	等位基因	131	137	139	141	143									74.3	25.7
	血液	0.100	0.000	0.150	0.100	0.650		0.400	0.563	4	2.151	0.498	0.253	0.505		
	粪便	0.050	0.050	0.200	0.000	0.700		0.500	0.490	4	1.869	0.421	0.331	0.600		
Pti007	等位基因	141	175	177	191	193									35.7	64.3
	血液	0.150	0.700	0.150	0.000	0.000		0.600	0.490	3	1.869	0.420	0.331	0.610		
	粪便	0.000	0.600	0.100	0.050	0.250		0.800	0.595	4	2.299	0.509	0.245	0.517		
FCA441	等位基因	136	148	152	156	160									61.4	38.6
	血液	0.000	0.400	0.250	0.250	0.100		0.700	0.742	4	3.390	0.652	0.141	0.379		
	粪便	0.050	0.450	0.200	0.250	0.050		0.700	0.726	5	3.226	0.640	0.146	0.376		
FCA094	等位基因	197	201	203	205	207	211								62.9	37.1
	血液	0.150	0.150	0.100	0.150	0.400	0.050	0.700	0.800	6	4.167	0.730	0.088	0.250		
	粪便	0.200	0.100	0.000	0.150	0.550	0.000	0.600	0.658	4	2.667	0.578	0.188	0.436		
Pti010	等位基因	124	126	128											74.3	25.7
	血液	0.150	0.150	0.700				0.400	0.490	3	1.869	0.420	0.331	0.610		
	粪便	0.200	0.100	0.700				0.400	0.484	3	1.852	0.410	0.341	0.624		

(He：$a = 0.431$，$P = 0.161$；Ne：$a = 0.354$，$P = 0.259$；PIC：$a = 0.411$，$P = 0.184$；DP：$a = 0.385$，$P = 0.217$；EPP：$a = 0.365$，$P = 0.244$)。

从血液样本中计算得到的个体间亲缘值为 $-0.36\sim0.22$，而从粪便样本中计算得到的个体间亲缘值为 $-0.41\sim0.23$。如果一个个体的血液样本和粪便样本的 r_R 值是相等的，那么回归系数应为 1，表明粪便样本基因型错误对成对的亲缘关系计算没有产生影响，我们的结果显示其线性关系显著但 R^2 值仅为 0.4911（$P < 0.001$，图 8.2）。该结果与预测的结果相反，表明从粪便样本中获得的基因型误差会影响个体间亲缘关系的计算。

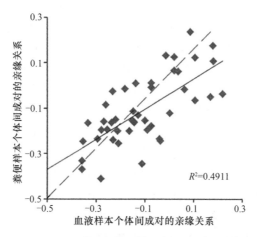

图 8.2　使用 Lynch 和 Ritland（r_R）方法进行粪便样本个体间与血液样本个体间亲缘关系的比较
虚线表示方程 $y = x$

8.4　讨　论

使用粪便样本进行非损伤性遗传分析时，往往由于微卫星错误的基因分型得到不正确的种群遗传学参数估计，从而影响保护与管理决策的科学性（Garshelis et al.，2008）。一些研究者认为提高基因分型的准确性在很大程度取决于保存样本的有效性（Roon et al.，2003）。由于野外样本在采集前接触到各种降解因子，样本质量无法保证（Nsubuga et al.，2004；Brinkman et al.，2010）。因此，做好分析样本保存和 DNA 提取技术优化对质量较差的粪便样本进行遗传分析具有一定的作用（Huber et al.，2003）。虽然多管重复 PCR 可以减少微卫星基因分型错误（Taberlet et al.，1996），但本研究还是使用了常规 PCR 以便我们能有效地控制每个位点的影响因素（如反应温度）。因此，当粪便样本质量较差时会选择放弃对该样品的分析，并通过优化实验流程以改进用于分析总体基因分型的准确度，应优先评估基因分型风险及对后续分析产生的不利影响。

我们发现实验室程序中导致基因型分型错误的来源有两点：PCR 扩增和微卫星位点的性质，我们的基因分型累计匹配率（R_m）指标可包含这两种错误来源的信息。对于 PCR 扩增，12 个微卫星位点的 R_m 在 3~5 次 PCR 后趋于稳定，这表明等位基因的扩增效率波动是由于模板质量差，而且所有 PCR 都有一定的分型风险。用于分型风险比较研究的粪便样本都是在冬天及排泄后 12h 内采集并立即保存于–20℃的。因此，粪便 DNA 质量是较高的。质量更差的样本要达到稳定的分型还需要额外的 PCR 扩增重复。同时，R_m 结果表明，即使使用同一批样本，不同的微卫星位点也有不同的分型准确性（表 8.2，图 8.1）。这说明，每个位点对质量差的模板 DNA 可能有不同的容忍度。也许 DNA 分子不一致的退化模式使得部分基因组 DNA 对降解因素（如细菌核酸酶）更加敏感（Deagle et al., 2006）。因此，从相同样本中提取出的 DNA 模板可能对某些微卫星位点表现出更低的分型成功率。或者，引物二级结构的不同影响了退火效率。尽管可以通过优化退火温度或优化 PCR 体系来提高退火效率，但是这些方法并不能使退火、分型错误完全移除。不同于以前要求对全部微卫星位点提高重复次数的研究（何刚等，2011），我们的结果表明，应对每个微卫星位点进行分型风险评估。

我们的研究结果显示，分型错误会改变 Ne、He、PIC、DP 和 EPP 等种群遗传学参数（表 8.2），但不是显著影响参数，而且与分型风险（$1-R_m$）无显著关系。然而，这并不代表基因分型错误不影响种群遗传学参数。例如，在 12 个微卫星位点的 R_m 中，最低的 R_m 只有 0.357。我们对 10 只虎执行了 7 次 PCR 重复，根据 R_m 方程只有 24.99 次正确分型，与全部 70 次分型相差甚远，这意味着执行 7 次 PCR 只能获得 3 次正确分型。对于其他高 R_m 的位点，分型结果比较准确，但是粪便 DNA 质量不同，3 次重复可能不能应用于所有虎样本中，这意味着不是所有的虎都能被正确地分型，高的 R_m 要求高质量的粪便样品，反之亦然。

总之，PCR 和微卫星的性质增加了粪便 DNA 的微卫星基因分型风险。本研究是东北虎人工圈养种群遗传检测工作的一部分，其结果将应用于未来基于粪便 DNA 进行野生东北虎的基因分型中。

8.5 本章小结

在现代的野生动物生态学研究中，粪便样本是野外最常见的非损伤性 DNA 的来源，以聚合酶链反应（PCR）技术为基础使用微卫星标记挖掘其内在的遗传信息，该遗传标记方法对于濒危物种的研究尤为重要。本研究开发了一个指数——阳性 DNA（血液样本）和粪便 DNA 之间基因型（R_m）的累计匹配率，以探索某一微卫星位点的正确基因分型概率。通过本研究，我们发现了不同的微卫星基因座具有不同的基因分型风险，需要采用不同 PCR 扩增方案。错误的基因分型会改变群体遗传学参数且潜在地影响后续分析。基于以上研究，我们给出如下建议：①12 个微卫星位点中 4 个位点（E7、FCA094、Pti007 和 Pti010）由于 R_m 数值偏低和基因分型数据很难达到平稳状态，不建议应用到东北虎的遗传研究中；②12 个微卫星位点的 R_m 平稳值不同，当使用粪便样品时需要考虑有限的预算及一些位点的扩增时间会延长；③未来野生东北虎的遗传研究应使用基因分型风险（$1-R_m$）进行校正。

参 考 文 献

何刚, 黄康, 郭松涛, 等. 2011. 用多项式分布模型评估低质量 DNA 模板微卫星分型的可靠性. 科学通报, 56(22): 1763-1770.

刘丹, 马跃, 李慧一, 等. 2013. 虎(Panthera tigris altaica)多雄交配与多重父权：对猫科动物圈养种群遗传多样性保护的启示. 科学通报, 58(16): 1539-1545.

张于光, 李迪强, 饶力群, 等. 2003. 东北虎微卫星 DNA 遗传标记的筛选及在亲子鉴定中的应用. 动物学报, 49(1): 118-123.

Arandjelovic M, Guschanski K, Schubert G, et al. 2009. Two-step multiplex polymerase chain reaction improves the speed and accuracy of genotyping using DNA from noninvasive and museum samples. Mol Ecol Resour, 9(1): 28-36.

Bhagavatula J, Singh L. 2006. Genotyping faecal samples of Bengal tiger Panthera tigris tigris for population estimation: a pilot study. BMC Genet, 7(1): 48.

Bradley B J, Boesch C, Vigilant L. 2000. Identification and redesign of human microsatellite markers for genotyping wild chim panzee (Pan troglodytes verus) and gorilla (Gorilla gorilla gorilla) DNA from faeces. Conserv Genet, 1(3): 289-292.

Brinkman T J, Schwartz M K, Person D K, et al. 2010. Effects of time and rainfall on PCR success using DNA extracted from deer fecal pellets. Conserv Genet, 11(4): 1547-1552.

Deagle B E, Eveson J P, Jarman S N. 2006. Quantification of damage in DNA recovered from highly degraded samples—a case study on DNA in faeces. Front Zool, 3(1): 11.

Garshelis D L, Wang H, Wang D J, et al. 2008. Do revised giant panda population estimates aid in their conservation. Ursus, 19(2): 168-176.

Hájková P, Zemanová B, Bryja J, et al. 2006. Factors affecting success of PCR amplification of microsatellite loci from otter faeces. Mol Ecol Notes, 6(2): 559-562.

Huber S, Bruns U, Arnold W. 2003. Genotyping herbivore feces facilitating their further analyses. Wildl Soc B, 31(1): 692-697.

Marshall T C, Slate J, Kruuk L E B, et al. 1998. Statistical confidence for likelihood-based paternity inference in natural populations. Mol Ecol, 7(5): 639-655.

Menotti-Raymond M, David V A, Lyons L A, et al. 1999. A genetic linkage map of microsatellites in the domestic cat (Felis catus). Genomics, 57: 9-23.

Murphy M A, Kendall A, Robinson A, et al. 2007. The impact of time and field conditions on brown bear fecal DNA amplification. Conserv Genet, 8: 1219-1224.

Nazarenko I, Pires R, Lowe B, et al. 2002. Effect of primary and secondary structure of oligodeoxyribonucleotides on the fluorescent properties of conjugated dyes. Nucleic Acids Res, 30(9): 2089-2195.

Nsubuga A M, Robbins M M, Roeder A D, et al. 2004. Factors affecting the amount of genomic DNA extracted from ape faeces and the identification of an improved sample storage method. Mol Ecol, 13(7): 2089-2094.

Panasci M, Ballard W B, Breck S, et al. 2011. Evaluation of fecal DNA preserva tion techniques and effects of sample age and diet on genotyping success. J Wildl Manag, 75(7): 1616-1624.

Regnaut S, Lucas F S, Fumagalli L. 2006. DNA degradation in avian faecal samples and feasibility of non-invasive genetic studies of threatened capercaillie populations. Conserv Genet, 7: 449-453.

Roon D A, Waits L P, Kendall K C. 2003. A quantitative evaluation of two methods for preserving hair samples. Mol Ecol Notes, 3: 163-166.

Rozhnov V V, Sorokin P A, Naidenko S V, et al. 2009. Noninvasive individual identification of the Amur tiger (*Panthera tigris altaica*) by molecular-genetic methods. Dokl Biol Sci, 429(1): 518-522.

Sambrook J, Fritsch E F, Maniatis T. 1989. Molecular Cloning: A Laboratory Manual. 2nd ed. New York: Cold Spring Harbor Laboratory Press: 1626.

Sambrook J, Russell D W, Russell D W. 2001. Molecular Cloning: A Laboratory Manual (3-volume set). New York: Cold Spring Harbor Laboratory Press: 2100.

Santini A, Lucchini V, Fabbri E, et al. 2007. Ageing and environmental factors affect PCR success in wolf (*Canis lupus*) excremental DNA samples. Mol Ecol Notes, 7: 955-961.

Stenglein J L, Barba M D, Ausband D E, et al. 2010. Impacts of sampling location within a faeces on DNA quality in two carnivore species. Mol Ecol Resour, 10(1): 109-114.

Taberlet P, Griffin S, Goossens B, et al. 1996. Reliable genotyping of samples with very low DNA quantities using PCR. Nucleic Acids Res, 24: 3189-3194.

Taberlet P, Luikart G. 1999. Noninvasive genetic sampling and individual identification. Biol J Linn Soc, 68: 41-55.

Woodruff D S. 1993. Non-invasive genotyping of primates. Primates, 34: 333-346.

Xu Y C, Li B, Li W S, et al. 2005. Individualization of tiger by using microsatellites. Forensic Sci Int, 151(1): 45-51.

Yeh F C, Yang R C, Timothy B J, et al. 2001. POPGENE, the user-friendly shareware for population genetic analysis. Edmonton: Molecular Biology and Biotechnology Centre, University of Alberta: 10.

Zhang W P, Zhang Z H, Xu X, et al. 2009. A new method for DNA extraction from feces and hair shafts of the South China tiger (*Panthera tigris amoyensis*). Zoo Biol, 28(1): 49-58.

第9章 应用雪地足迹影像进行东北虎的性别鉴定

9.1 引 言

东北虎（*Panthera tigris altaica*）是世界上体型最大的虎亚种，主要分布于俄罗斯东南部和中国东北部。20 世纪 60 年代，这个虎亚种几乎灭绝，但如今已经恢复到 500 只左右。在中国，东北虎曾经广布于中国东北地区，由于盗猎、栖息地丧失、猎物密度低等，20 世纪末仅 20 只左右（Sun，2011）。值得庆幸的是，在中国境内还尚存较大面积质量较好的栖息地（Kang et al.，2010），中国政府正努力采取各种措施使更多栖息地与俄罗斯东北虎分布区连接，从而使俄罗斯境内的虎扩散至中国并定居。

在中国，常规的东北虎监测技术正面临着诸多挑战（Zhang et al.，2012），现在主要有三种技术方法应用于东北虎的种群监测：①在一些自然保护区利用自动相机获取老虎的影像，用标志重捕法评估种群数量（Kawanishi and Sunquist，2004）；②在俄罗斯远东利用冬季雪地样线调查法，通过样线上东北虎足迹的遇见率来评估虎密度（Miquelle et al.，1996；Hayward et al.，2002）；③分子生物学方法。采集粪便或毛发样本，通过提取样本中宿主的 DNA 进行个体识别（Russello et al.，2004），分子生物学方法通常需要结合上述两种野外调查方法使用。在中国，由于东北虎种群密度太低，样线调查法并不适用，如有一次对一块 1735km^2 的区域进行样线调查，样线总长度达到 609km，调查样线上没有发现任何东北虎的足迹和其他痕迹[WWF-China report（2010～2011 年）未发表数据]。同样，野生东北虎的 DNA 样本在野外也是很难被发现的，2009～2010 年珲春东北虎国家级自然保护区仅采集到 7 份东北虎的粪便样本，但是东北虎的雪地足迹相对较为常见。尽管中国的研究团队已经开始使用自动相机监测东北虎，然而应用范围较小，主要集中在东北虎经常出现的区域。Zhang 等（2012）认为，信息网络搜集结合足迹识别可能是监测东北虎的有效方法，因为雪地足迹与足迹链更容易被发现和采集，足迹也可以提供丰富的东北虎个体信息。

许多研究已经表明，利用沙地或泥地上的足迹进行物种、个体和性别的识别是可行的（Riordan，1998；Jewell et al.，2001；Sharma et al.，2003，2005；Alibhai et al.，2008；Russell et al.，2009；Law et al.，2013）。通过足迹进行物种性别鉴定大致经历 3 个阶段：①基于足迹形状的描述（McDougal，1977；Sankhala，1978；Panwar，1979a，1979b）；②基于简单的测量比较（Gogate et al.，1989；Sagar and

Singh，1991）；③基于使用一个或几个测量值的统计分析（Bhattacharya，1967；Gore *et al.*，1993；McDougal，1999；Sharma *et al.*，2003；Sharma and Wright，2005）。在过去的 20 年，俄罗斯采用雪地足迹和其他痕迹（挂爪、标记等）结合分析的方法进行东北虎种群数量的调查，同时也通过足迹进行性别的判定，主要依据前掌掌垫宽大小来判断，辅助以其他行为痕迹特征的发生时间、距离等（Abramov，1961；Matyushkin *et al.*，1996；Smirnov and Miquelle，1999）。然而，这一技术受限于由人工测量的几个特征值，并没有采集到足够的特征信息，无法区分成年雌性和亚成年雄性东北虎，因为两者的前掌掌垫宽大小是重叠的（Miquelle *et al.*，2006）。

本研究报道了一种可靠的通过东北虎雪地足迹进行性别识别的方法，并成功地将这种方法运用到野生东北虎性别鉴定中，我们认为，这是一种简便、经济、操作性强的非损伤性的野生动物监测方法。

9.2　材料与方法

足迹影像采集来自中国东北地区的圈养东北虎和野生东北虎（图 9.1）。圈养东北虎足迹采集于中国横道河子猫科动物饲养繁育中心，该中心饲养超过 300 只东北虎，每只都有详细的性别等谱系档案。野生东北虎的足迹采自中国黑龙江省完达山地区东方红林业局，该区域长期存在一个数量相对稳定的野生东北虎小种群。

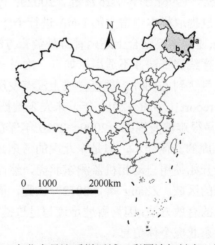

图 9.1　东北虎足迹采样区域（彩图请扫封底二维码）

蓝色表示黑龙江省；位置 a 表示完达山东方红林业局；位置 b 表示牡丹江市附近的横道河子猫科动物饲养繁育中心

9.2.1　名词定义

足迹：足迹链上单一的脚印。

足迹链：动物留下的一连串足迹。

足迹识别技术：the footprint identification technique（FIT）。

9.2.2 足迹照片数据采集和分析方法

用普通的数码相机、带有标签纸的木工折尺来拍摄足迹照片，标签纸上记录照片信息和老虎个体信息。我们用 JMP®（SAS 公司）软件里的 FIT 模块（Alibhai et al., 2008；Jewell et al., 2001）对足迹数据进行分析，我们采集了东北虎的四个足迹以便后续研究，本次研究只采用左后足的照片进行分析。

我们发现了一些采集雪地足迹照片的技巧。例如，我们发现纸质的标签纸不适合用在雪地足迹上，因为它们很容易被雪弄湿；我们采用透明的塑料卡槽，粘在尺子背面，然后用白板笔将具体信息直接写在塑料卡槽表面，留下的笔迹可以很容易被擦去，反复使用，这个小技巧对野外监测很有帮助，既经济又方便。

我们只对脚趾和掌垫轮廓清晰的足迹照片进行分析。我们将足迹照片导入 JMP 软件，然后用 FIT 脚本提取每一张足迹照片上的 128 个变量（包括距离、角度和面积类型变量），为数据分析提供全面的足迹几何特征（Alibhai et al., 2008）。

9.2.3 圈养东北虎的足迹影像采集

由于需要用已知性别东北虎的足迹来建立有效的性别判别的方法，因此个体信息清楚的圈养动物是最理想的选择，我们采集了 3～13 岁的 40 只圈养东北虎的左后足的雪地足迹。每只虎采集了 3～21 张左后足的足迹照片，平均每只虎采集了 13.1 张。然后我们选用至少有 6 张清晰左后足足迹链的照片来建立和检测判别模型。

采集足迹照片时，由于研究区域自然降雪不多，我们在东北虎的笼舍内人工铺雪，最大程度地模仿自然降雪的条件：采集新鲜的雪，过筛，使细小的雪粒铺满东北虎的室外笼舍地面。我们比较了不同的雪深，发现 3～5cm 的雪深（1～2cm 厚比较实的雪加上 2～3cm 松散的雪）可以比较容易地使虎留下清晰的足迹。饲养员用食物或呼唤吸引虎在铺好的雪地上正常行走，然后进行拍照。

根据 Alibhai 等（2008）的 FIT 足迹照片拍摄规程，选择好清晰足迹后，用白板笔在折尺上填写好拍摄日期、东北虎的 ID 号及足迹编号（对于每只虎的每个足迹都有唯一的编号）。然后将折尺拉开，使标尺的一边靠近掌垫底部作为横轴，另一边靠近足迹最左边作为纵轴，如图 9.2 所示，这样折尺方向就与行走方向垂直，确保折尺没有挡住足迹的轮廓。

图 9.2　使用便携式数码相机拍摄足迹图像

我们根据掌垫和四趾的轮廓清晰度建立了评估照片质量的方法，在整个研究期间严格遵守该方法进行照片挑选，图 9.3 显示出照片质量的具体要求。

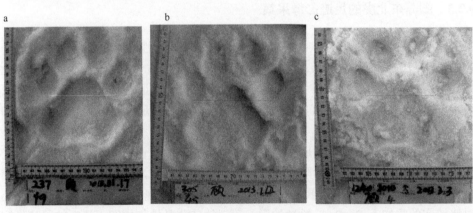

图 9.3　不同质量的足迹图像
a. 图像质量好，轮廓清晰；b. 图像质量较差，但仍有可能确定的轮廓；c. 不能用于分析

由于雪基质的高反射率，缺乏光的对比，因此轮廓难以看清。实践证明，在早上或傍晚，阳光斜照能造成部分阴影，或通过人为遮光制造阴影，此时拍摄的足迹轮廓清晰，在阴天及能见度较低时需要闪光灯加以辅助。拍照片时，要将相机置于足迹的垂直上方 30~50cm 的位置，构图时，尽量让相机画面中只有足迹和带标签的折尺。

9.2.4 野生东北虎的足迹影像采集

于 2011 年年底至 2012 年年初的冬季在东方红林业局搜集野生东北虎出现的信息。一旦发现野生东北虎的足迹，就进行足迹跟踪，搜集每条足迹链上尽可能多的清晰左后足足迹照片。

尽管野外环境复杂，但找到清晰的足迹并不难。根据经验，东北虎喜欢行走在平坦区域，如结冰的河面、道路、山脊、田地等，这些区域都比较容易留下清晰的足迹（图9.4）。在雪较深的区域，沿着足迹链继续跟踪，通常在倒木、大岩石周围重新获取清晰的足迹。遇到雪深但清晰的足迹时，需要将足迹周围的雪移开，让标尺和足迹轮廓在同一水平面。有诸多影响足迹形态变化的因素，如降雪、风、结冰、融化，1~2 月新的降雪是导致足迹变化最主要的因素（82%），3 月由于气温回暖和风的作用，足迹变化会更快，通常只能保持 2.1 天（Hayward et al., 2002）。

图 9.4 野生东北虎留下的清晰足迹

野生东北虎的足迹影像采集方法与圈养东北虎一致，但是每只虎的个体信息是未知的。我们按照以下方法记录足迹链上的足迹：第一条采集的足迹链命名为足迹链 1，这条足迹链上的足迹分别记为 1a、1b、1c 等。第二条采集的足迹链命名为足迹链 2，这条足迹链上的足迹分别命名为 2a、2b、2c 等。如果一条足迹链因为被覆盖、破坏而跟踪中断，之后找到下一组足迹时需要命名成不同的足迹链编号。对每一条足迹链同时采集其 GPS 点坐标。

9.2.5 足迹几何轮廓的提取

东北虎足迹几何轮廓的提取与 Alibhai 等（2008）对沙地面上白犀牛足迹的提取方法类似，但此次是 FIT 首次应用到东北虎雪地足迹的识别。足迹照片经一系列处理后提取足迹几何特征的测量值。每一张照片调整方向使脚趾朝上方；裁剪照片，使得画面内只有足迹和折尺；旋转，以第 1 趾和第 4 趾最低点的连线为基准线，旋转照片使基准线水平，这样可以保证足迹的四趾垂直朝前；调光，使得轮廓更清晰；固定照片的长宽比，缩小照片以适合 JMP 的使用。将处理好的足迹照片导入 JMP 的图形窗口，填写好照片相关信息，根据计算程序脚本提示按顺序在足迹照片上加上 25 个原始点，根据标尺和这些点线，计算程序脚本会对每一个脚印提取出 128 个形态测量变量，自动生成到 JMP 的数据表格中（图 9.5）。

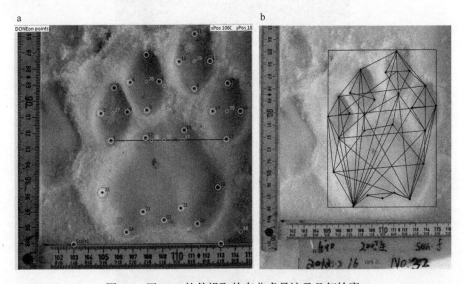

图 9.5 用 JMP 软件提取的东北虎足迹及几何轮廓

a. 用 JMP 软件标记了 25 个特征点的左后足脚印图像；b. 标完特征点后，在同一张照片衍生出的点和线，通过这些点线画出 128 个形态测量变量，这些构成了足迹的几何轮廓

9.2.6 数据分析

我们取 40 只圈养东北虎（19 只雌虎、21 只雄虎）的 523 个足迹照片进行性别判别。雌雄东北虎年龄为 3~13 岁，并且两个性别组年龄没有显著性差异（雌虎平均年龄 = 8.07 岁±0.18 岁，雄虎平均年龄 = 8.36 岁± 0.19 岁，$F = 1.18$，$P > 0.05$）。用 JMP 中的线性判别分析（linear discriminant analysis，LDA）来进行性

别识别。LDA 是基于线性的多变量识别进行分类的,在本研究中就是对不同性别的足迹链进行分类。使用 JMP 中的逐步变量选择功能来进行变量选择,通过 F 值找到能提供最好判别效果的变量组合。逐步变量选择功能同样能帮助剔除自相关高的变量。通过变化变量个数,以及不同变量个数对应的判别率,从而找到最合适的变量个数,以保证预测模型的有效性(图 9.6)。

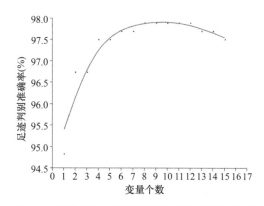

图 9.6 变量个数与足迹判别准确率的渐近关系

为了检验 LDA 建立性别判别模型的准确率,我们采用两种方法:刀切法与训练组和检验组数据各一半的方法。刀切法检测数据的顺序子集即一条足迹链上的全部足迹,而不是单一的足迹,而训练组和检验组数据各一半的方法可以在足迹水平拆分数据来对比不同算法的检测结果。最后,以圈养东北虎足迹数据建立的包含 10 个变量的算法作为最佳预测模型来判别野生东北虎足迹的个体性别。

9.3 结　果

9.3.1 圈养东北虎的性别判别模型

实验证明,利用 JMP 中的逐步变量选择功能选择变量得出的判别率最高。为了避免过多的变量和过度拟合的问题,使用 LDA 找出判别率最高的变量个数。如图 9.6 所示,渐近线显示当变量数为 8~10 个时,渐近线开始变得平滑。因此,我们用逐步选择的方式选取了 10 个变量建模。图 9.7 为使用 10 个变量的 LDA 生成的前两个典型变量的示意图。在共 523 个足迹中,有 11 个被错误分类(7 个雌性脚印被划分为雄性,4 个雄性脚印被划分为雌性)。

利用逐步选取的 10 个变量对圈养东北虎的数据集进行了刀切法验证,共产生了 13 个足迹误分类,性别判定的准确率为 97.5%。这几个错误的分类可能是由于微小的足迹形状扭曲,以及虎的步态或雪基质的变化。为了进一步验证精度水平,

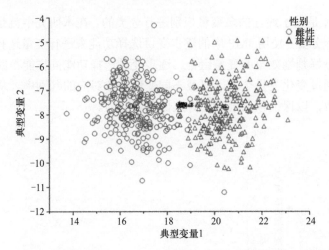

图 9.7 使用 10 个变量进行判别的雄性（三角形）和雌性（圆形）脚印的分布（彩图请扫封底二维码）
分类错误数=11；误分类百分比=2.103%；–2 log likelihood = 63.699

我们使用 FIT 中的一个判别验证平台，将圈养东北虎的数据集的 50%随机分配给训练集、50%分配给测试集。这个过程重复了 5 次，每次随机分配。训练集的准确率为 96.9%~98.4%，测试集的准确率为 97.4%~98.5%。

9.3.2 基于足迹识别技术对野生东北虎进行性别鉴定

利用圈养东北虎数据生成的性别识别算法来判别野生东北虎的性别。我们从 5 个林场（HYS、QY、XNC、WLD、QS）采集到 8 条野生东北虎的足迹链，共采集到 83 个符合模型判别的野生虎的足迹影像。结果表明：8 条足迹链中雌性东北虎为 5 条，雄性东北虎为 3 条（表 9.1）。在每一条足迹链中，92%的足迹被划分为同一性别。

表 9.1 8 只野生东北虎雪地足迹链性别判别结果

足迹链名	足迹个数	判定为雌性的足迹个数	判定为雄性的足迹个数	在采集的足迹链中被判断为雌性/雄性的（也就是判别成统一性别）百分比 [a]	预测性别
HYS	13	11	02	82	F
QY	12	11	01	91	F
WLD-B	14	13	01	92	F
XNC-A	13	13	0	100	F
WLD-A	07	05	02	71	F
XNC-B	05	0	05	100	M
QS-B	13	0	13	100	M
QS-A	06	01	05	80	M

注：HYS. 海音山，QY. 奇源，WLD. 五林洞，XNC. 西南岔，QS. 青山。F.雌性；M.雄性。a 指一条足迹链上，和最后性别预测结果一样的脚印个数占整条足迹链的脚印总数的比例

9.4 讨 论

9.4.1 采集和测量足迹

本研究仅采集了后足的足迹，因为在正常行走时，后足的脚印通常会覆盖在前足的脚印上，导致完整的前脚脚印更少（Sharma et al., 2005）。为了方便野外足迹采集和最小化采集量，我们开发了只使用左后足脚印的识别系统。Riordan（1998）发现，右后足的错判率略高于左后足。印度专家在采集足迹样本时，至少采集 5 个清晰的后足足迹（Sharma et al., 2003）。俄罗斯专家只测量前掌掌垫宽，作为东北虎个体数量的识别标准（Abramov, 1961; Matyushkin et al., 1996; Smirnov and Miquelle, 1999）。

我们认为应尽可能对足迹测量不同的特征值（128 个），以使判别能力最大化，Sharma 等（2003，2005）、Sharma 和 Wright（2005）对足迹提取了 93 个测量值。我们采用了许多前人研究认为有用的变量（Das and Sanyal, 1995; Gogate et al., 1989; Gore et al., 1993）。我们使用 10 个变量来构建和测试性别判别模型，但是这 10 个变量（128 个变量中的 10 个）不是不可改变的，这取决于输入模型的数据特征。

9.4.2 雪地足迹数据的客观性

研究人员很容易在短期训练后熟练地使用数码相机记录足迹，从而去除了临摹和复制图像带来的误差（Rishi, 1997）。因此，有雪被覆盖的虎分布区的当地居民、专家等都可以参与到数据采集工作中来。

根据我们的经验，当足迹在深雪（>5cm）时，表面的雪会掩盖足迹的真实轮廓和形状细节。此外，在深雪中获取图像时，由于尺子与足迹轮廓的距离增加，因此在深雪中所获得的图像比浅雪中的要小。同理，如果足迹印模在非常浅的雪中（<3cm），雪被就没有足够的深度来反映足迹的全部细节。因此，在圈养或野外环境中，最好在 3~5cm 的积雪深度采集足迹，并在拍摄足迹时将尺子和足迹轮廓保持在同一平面上。在本研究中我们发现雪基板是相对稳定的，因此可以在野外收集到高质量的雪地足迹数据（Hayward et al., 2002）。

9.4.3 样本量和建模方法对性别判别的影响

本研究从以下几方面使得通过足迹进行性别判别的方法更严格。首先，在数据收集方面，对每只动物采集的足迹数量显著增加，比以往的研究都要多（Riordan,

1998；Sharma et al., 2003, 2005)。我们排除了年龄的影响，因为我们使用的年龄范围更广（3~13 岁），雄性组和雌性组之间年龄没有显著差异。其次，我们建立了严格的照片质量控制过程和筛选体系来决定足迹照片是否能使用。我们从足迹图像中提取数据，而不是临摹图片。再次，在变量较多的情况下，利用训练集和测试集对模型进行了仔细的验证，并建立了以适合度判别的验证平台。最后，我们开发了一款用户界面友好的软件（FIT），可以在一个软件平台上完成图像处理和数据分析。

9.4.4 管理启示

我们认为在中国东北地区通过雪地足迹来确定东北虎信息是最为理想的方法之一。几十年以来，俄罗斯和中国的科学家通过简单的前掌掌垫宽的测量来对东北虎进行性别与个体的识别。这一方法的主要限制是难以区分成年雌性和亚成年雄性。尤其在中国，由于东北虎种群密度极低，传统的监测方法面临各种挑战，新的东北虎监测网络（Zhang et al., 2012）需要更准确的方法，我们相信 FIT 是理想的方法，这项技术也可以帮助我们评估雌性定居东北虎的数量。雌性东北虎的扩散范围有限，且日活动距离短于雄性（Goodrich et al., 2010），雌性定居虎的监测数量是当地老虎种群恢复的一个重要指标。

使用非损伤性足迹识别技术对野生动物进行监测具有积极的管理意义。足迹收集不会干扰或影响动物，非损伤性的监测方法在保护工作中会带来更好的科研结果（Jewell, 2013）。

综上所述，FIT 的优点是易于用于东北虎大区域景观尺度的监测，是中国政府正在进行的评估该物种恢复的一个潜在有应用价值的工具。目前，我们正在对东北虎的个体和年龄鉴定进行进一步研究，并与俄罗斯同行沟通，将这种非损伤性方法作为中俄跨境种群调查的技术方法，以便对这类濒危种群进行更加科学的评估和保护。

9.5 本章小结

在中国曾经分布很广的东北虎，如今渐少到约 20 只。中国政府十分关注恢复东北虎种群。然而，恢复工作面临着有效监测技术的挑战。我们建立了一种稳健、非损伤、经济的技术来通过雪地足迹进行性别的鉴定。2011 年 12 月至 2012 年 12 月，我们共拍摄了 523 张圈养东北虎的雪地足迹照片，它们来自 3~13 岁 40 只已知个体信息的东北虎的左后足，其中雌虎 19 只，雄虎 21 只（雄虎平均年龄为 8.36 岁±0.19 岁，雌虎平均年龄为 8.07 岁±0.18 岁，$P>0.05$)。我们用便携式数码相机按照统一标准的拍照流程拍摄每一张足迹照片，然后使用 SAS 公司的 JMP 软件对每一张足迹照片进行 128 个变量的测量，该方法由 Alibhai 等（2008）建立。我们对这 128 个变量进行逐步选择，发现当采用其中 10 个变量时，用刀切法与 50%留出法进行检测，最终性别判别的算法可以达到 98%的准确率。这套来自已知性别圈养虎的算法被运用到 2011 年年底到 2012 年年初采集到的未知个体信息的野生东北虎的雪地足迹上，这套算法对来自野生东北虎的 8 条足迹链上 83 个足迹的判别结果为有 5 条足迹来自雌虎，3 条足迹来自雄虎。这套算法可以作为中国东北地区东北虎种群恢复的有效监测工具。

参 考 文 献

Abramov K G. 1961. On the procedure for tiger census. Moscow: Obschestvo Ispytaltely Prirody.

Alibhai S K, Jewell Z C, Law P R. 2008. A footprint technique to identify white rhino *Ceratotherium simum* at individual and species levels. Endangered Species Research, 4: 205-218.

Bhattacharya C G. 1967. A simple method of resolution of a distribution into gaussian components. Biometrics, 23(1): 115.

Das P K, Sanyal P. 1995. Assessment of stable pug measurement variables for identification of tiger. Tiger Paper, (22): 20-26.

Gogate M, Joshi P, Gore A, et al. 1989. Tiger pugmark studies: a statistical perspective. Directorate, Project Tiger Technical Bulletin No. 2.

Goodrich J M, Miquelle D G, Smirnov E N, et al. 2010. Spatial structure of Amur (Siberian) tigers (*Panthera tigris altaica*) on Sikhote-Alin Biosphere Zapovednik, Russia. Journal of Mammalogy, 91(3): 737-748.

Gore A P, Paranjape S A, Rajgopalan G, et al. 1993. Tiger census: role of quantification. Current Science, 64(10): 711-714.

Hayward G D, Miquelle D G, Smirnov E N, et al. 2002. Monitoring Amur tiger populations: characteristics of track surveys in snow. Wildlife Society Bulletin, 30(4): 1150-1159.

Jewell Z C. 2013. Effect of monitoring technique on quality of conservation science. Conservation Biology, 27(3): 501-508.

Jewell Z C, Alibhai S K, Law P R. 2001. Censusing and monitoring black rhino (*Diceros bicornis*) using an objective spoor (footprint) identification technique. Journal of Zoology, 254(1): 1-16.

Kang A, Xie Y, Tang J, et al. 2010. Historic distribution and recent loss of tigers in China. Integrative Zoology, 5(4): 335-341.

Kawanishi K, Sunquist M E. 2004. Conservation status of tigers in a primary rainforest of Peninsular Malaysia. Biological Conservation, 120(3): 329-344.

Law P R, Jewell Z C, Alibhai S K. 2013. Using shape and size to quantify variation in footprints for individual identification: case study with white rhinoceros (*Ceratotherium simum*). Wildlife Society Bulletin, 37(2): 433-438.

Matyushkin E, Pikunov D, Dunishenko Y, et al. 1996. Numbers, distribution, and habitat status of the Amur tiger in the Russian Far East. Moscow: Final Report to the USAID Russian Far East Environmental Policy and Technology Project.

McDougal C. 1977. The face of the tiger. New York: Rivington & Deutsch.

McDougal C. 1999. You can tell some tigers by their footprints with confidence. Riding the Tiger:

Tiger Conservation in Human-dominated Landscapes, 383: 190-191.
Miquelle D, Pikunov D, Dunishenko Y, et al. 2006. A survey of Amur (Siberian) tigers in the Russian Far East, 2004-2005. Wildlife Conservation Society, World Wildlife Fund.
Miquelle D G, Smirnov E N, Quigley H G, et al. 1996. Food habits of Amur tigers in Sikhote-Alin Zapovednik and the Russian Far East, and implications for conservation. Journal of Wildlife Research, 1(2): 138.
Panwar H. 1979a. A note on tiger census technique based on pugmark tracings. Tigerpaper FAO, 6: 16-18.
Panwar H. 1979b. Population dynamics and land tenure of tigers in Kanha National Park. Indian Forester, (Special Issue): 18-36.
Riordan P. 1998. Unsupervised recognition of individual tigers and snow leopards from their footprints. Animal Conservation, 1(4): 253-262.
Rishi V. 1997. Monitoring tiger populations by impression-pad method. The Indian Forester, 123(7): 583-600.
Russell J C, Hasler N, Klette R, et al. 2009. Automatic track recognition of footprints for identifying cryptic species. Ecology, 90(7): 2007-2013.
Russello M A, Gladyshev E, Miquelle D, et al. 2004. Potential genetic consequences of a recent bottleneck in the Amur tiger of the Russian Far East. Conservation Genetics, 5(5): 707-713.
Sagar S R, Singh L A K. 1991. Technique to distinguish sex of tiger (*Panthera tigris*) from pug-marks. The Indian Forester, 117(1): 24-28.
Sankhala K. 1978. Tiger! The Story of the Indian Tiger. London: Harper Collins.
Sharma S, Jhala Y, Sawarkar V B. 2003. Gender discrimination of tigers by using their pugmarks. Wildlife Society Bulletin, 31(1): 258-264.
Sharma S, Jhala Y V, Sawarkar V B. 2005. Identification of individual tigers (*Panthera tigris*) from their pugmarks. Journal of Zoology, 267(1): 9-18.
Sharma S, Wright B. 2005. Monitoring tigers in Ranthambore using the digital pugmark technique. New Delhi: Wildlife Protection Society of India.
Smirnov E, Miquelle D. 1999. Population dynamics of the Amur tiger in Sikhote-Alin State biosphere Reserve//Seidensticker J, Christie S, Jackson P. Riding the Tiger: Tiger Conservation in Human-dominated Landscapes. Cambridge: Cambridge University Press: 61-70.
Sun H Y. 2011. Amur Tiger. Harbin: Northeast Forestry University Publishing House.
Zhang C Z, Zhang M H, Jiang G S. 2012. Assessment of monitoring methods for population abundance of Amur tiger in Northeast China. Acta Ecologica Sinica, 32(19): 5943-5952.

第 10 章　中国东北虎栖息地等级保护优先区域和廊道确定的适宜性模型技术

10.1　引　　言

栖息地保护对拯救濒危大型哺乳动物及其栖息地的生态系统尤为重要（Ripple et al., 2014）。对虎（*Panthera tigris*）这样广域分布的大型哺乳动物而言，栖息地保护十分困难，面临诸多挑战，主要原因是需要为它们有效扩散在大空间上维持可渗透的景观，并且栖息地保护还面临着政治与社会经济、土地利用决策之间竞争的问题（Chapron et al., 2014）。已经存在的保护区面积能否拯救大型哺乳动物成为科学界一个悬而未决的问题（McLaughlin, 2016）。

截止到 2015 年，中国的自然保护区覆盖了中国陆地面积的 14.8%，远超过世界的平均水平（8.8%）（Zhu et al., 2018）。尽管存在如此大面积的自然保护区，但制定的保护措施在濒危物种数量下降和生态退化的趋势上是不成功的（Liu et al., 2001；Tian et al., 2009）。Viña 等（2016）报道了中国天然林保护工程的益处，即在恢复和保护大型猫科动物及其所处的生态系统方面取得了巨大成功，这也正是驱动 2000～2014 年中国东北虎种群增长的根本原因（Jiang et al., 2017）。这些成果和经验也使中国政府意识到当前保护区管理体系上存在的不足，因此，2015 年 12 月中国政府决定在大空间尺度上重新设计类似国家公园形式的由中央直管的自然保护地体系（Li et al., 2016）。

尽管中国政府在国家公园体系建设过程中取得的保护方面的进展非常令人鼓舞（Li et al., 2016），但是仍有几个问题未能解决：这些保护区域是否应该成为某些物种生存的优先区域；这些保护区域是否满足物种的需求；保护行动是否可以同时解决生态需求与社会需求的矛盾（Sala et al., 2002）。针对上述问题，当务之急是要识别出可以实现上述目标的斑块与生态廊道（Fazey et al., 2005）。在划定和实施保护优先区域的过程中，需要平衡在以人为主导的景观中的资源配置、社会成本或冲突（Wikramanayake et al., 2004）。另外，在制定保护目标时，应该确定哪些目标应该最大化或最小化（Game et al., 2013）。本研究要讨论的是：以中国东北虎（*Panthera tigris tigris*）保护为例，在探索制定其保护目标与计划时，使用预先选定的等级优先区是否可以加强对东北虎的保护。

为回答这个问题，首先，运用栖息地适宜性模型识别出栖息地优先保护的斑块和廊道；其次，通过建立目标函数最大化栖息地适宜性和适宜斑块面积，同时考虑保护投资成本的权重，并最小化人为干扰强度；最后，选用目标函数的最大值，该值代表了生态与社会因素的最佳组合，最终可得到一个栖息地适宜性的临界值以划定优先保护核心区域。如此，我们将建立起一个自适应模型来等级化优先保护区域，以确保政府在共享土地决策时充分考量动物的栖息地需求（Jiang et al., 2017）。

10.2 研究方法

10.2.1 数据收集与栖息地斑块确定

首先，从公共网站上下载了关于气候、地形、植被及人为干扰因子的数据。在4个冬季采用雪地样线调查方法收集了有蹄类猎物的分布数据（Stephens et al., 2006），通过 2000~2017 年东北虎出现点信息收集了东北虎在中国的分布数据（Jiang et al., 2017）。东北虎个体及其移动数据通过 2012~2017 年东北虎及其猎物自动相机监测网络获得（Wang et al., 2018）。其次，使用 MaxEnt 软件建立了栖息地预测模型，并通过物种出现的临界值概率确立了适宜栖息地斑块（Phillips et al., 2006）。如果有东北虎在某栖息地斑块出现，我们就认为该斑块被虎占有。再次，使用逻辑斯谛回归拟合东北虎是否出现与斑块面积之间的关系（Qing et al., 2016）。最后，我们通过比较拟合的逻辑斯谛曲线与零响应曲线，检验了东北虎是否出现对斑块面积的敏感程度。零响应曲线定义为函数 $1-(1-p_b)^{a/a_b}$，式中，a_b 表示东北虎最小家域面积；a（$a > a_b$）表示斑块面积；p_b 表示斑块面积为 a_b 时的占有概率（Qing et al., 2016）。然后，利用药理学上的量效曲线估计东北虎的面积需求，量效曲线的作用就是测量产生某特定概率时需要的有效剂量（ED）。ED_p 可以看作当东北虎有 $p\%$ 的概率出现在一个斑块内时这个斑块的面积。使用 ED_{90} 作为东北虎生存的最小面积需求（Qing et al., 2016）。

收集和使用气象、地形、植被及人为干扰因子的组合模拟东北虎与有蹄类动物的潜在分布（表 10.1）。雪深数据来自加拿大气象中心（CMC，http://nsidc.org/data/NSIDC-0447）。村庄、道路、河流及铁路的地理矢量数据来自国家青藏高原科学数据中心（http://westdc.westgis.ac.cn/）。数字高程模型（DEM）数据与归一化植被指数（NDVI）数据来自地理空间数据云（http://www.gscloud.cn/）。坡向与坡度通过 ArcMap 10.3 空间分析模块计算得到。植被数据来自国家地理空间信息中心全球 30m 空间分辨率数据，植被类型主要分为森林、灌丛、草地、耕地和湿地。全部生境因子都经重采样至 3km 空间分辨率来预测猎物分布和东北虎分布。

第 10 章　中国东北虎栖息地等级保护优先区域和廊道确定的适宜性模型技术

表 10.1　与东北虎和有蹄类动物分布相关的栖息地变量

栖息地因子	栖息地因子描述	数据类型	单位
雪深	2005~2015 年每年 12 月至翌年 3 月的平均雪深（源数据分辨率为 24km）	连续	cm
海拔	数字高程模型（30m 分辨率）	连续	m
坡向	由数字高程模型计算而来的坡向（30m 分辨率）	连续	
坡度	由数字高程模型计算而来的坡度（30m 分辨率）	连续	(°)
到铁路的距离	栅格中心点距最近铁路的距离	连续	km
到村庄的距离	栅格中心点距最近村庄的距离	连续	km
道路密度	栅格中心点 10km 半径内道路密度	连续	km/km^2
距河流距离	栅格中心点距最近河流的距离	连续	km
归一化植被指数	2012~2017 年 9 月 NDVI 平均值（源数据分辨率 500m）	连续	%
森林比例	栅格内森林面积比例	连续	%
耕地比例	栅格内耕地面积比例	连续	%

我们通过雪地样线调查方法收集了 4 个冬季有蹄类猎物的雪地足迹数据（2014~2017 年 11 月至翌年 3 月）。在已知东北虎分布区内按照 36km/100km^2 的抽样强度进行样线调查。每条样线最小长度为 5km，两条样线之间的最小间隔为 3km。科研人员与林业工人在野外工作前会接受物种足迹识别和数据记录及收集的技术培训。样线 24h 以内的新鲜足迹才被记录（动物种类与数量），如果调查人员能确认之后的足迹与该足迹来自同一个个体，那么之后的足迹可以忽略不计（Qi et al., 2015; Stephens et al., 2006）。我们共调查了 345 条样线，样线总长度 1876km，覆盖 11 928km^2（调查抽样强度为 16.4%）的典型东北虎生境（图 10.1）。考虑到东北马鹿（Cervus elaphus）与梅花鹿（Cervus nippon）仅在局部分布且数量稀少，我们仅将野猪（Sus scrofa）与狍（Capreolus pygargus）作为东北虎的潜在猎物并将猎物的分布作为变量来模拟东北虎的分布。

我们收集了 2000~2017 年的东北虎出现点信息，这些信息主要来自东北虎捕食家畜补偿记录、样线调查发现的老虎足迹、捕食现场、东北虎粪便或毛发（DNA 鉴定）、经专家鉴定的东北虎出现点（通常为疑似足迹）。2012~2017 年，在东北虎分布区内我们进行了自动相机监测。首先将当前东北虎分布区划分为 3km×3km 的栅格，在栅格内相机位置通常选在林间道路、兽道或其他可以以最大概率捕捉到虎影像的地点。在东北虎分布区内共 1052 个相机点被选中（即相机监测覆盖了 9468km^2）。相机点分布图见图 10.2。为了尽量拍到东北虎两侧的体侧花纹，每个相机位点对装两台自动相机。每 4 个月更换一次内存卡和电池。当捕捉到老虎影像时，在两侧花纹均确认的情况下，该个体才被计入总数（即仅有单侧花纹的影像不计入虎总数）。此外，新生个体的扩散和每只个体的迁移模式都通过 2012~2017 年相机捕捉到的影像来反映。数据表明：雌性东北虎首次繁殖的平均年龄为（4±0.4）岁，两胎之间的平均间隔为（21.4±4.4）月（Kerley et al., 2003）。因此，我们认为至少需要 4 年才能探测到一只雌性东北虎是否成功扩散并繁殖，或者至

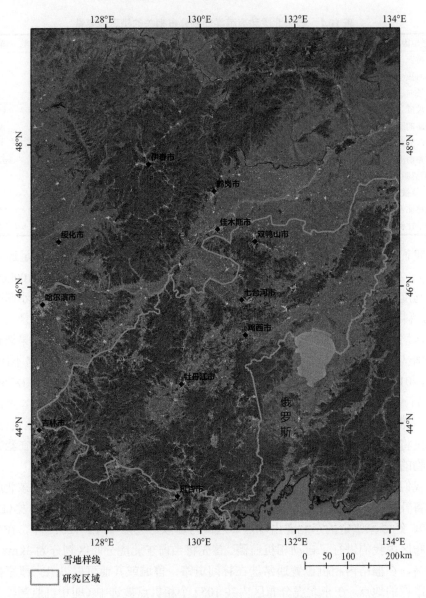

图 10.1 研究区内东北虎猎物调查样线的分布图（彩图请扫封底二维码）
地图底图为谷歌卫星影像

少 4 年才能监测到新的雌性个体扩散地点、方向和距离等信息，以检验生态廊道对于扩散的潜在有效性。

猎物的分布是大型食肉动物生存和扩散的重要限制因子（Karanth and Chellam, 2009）。首先，我们利用猎物出现点数据和栖息地变量模拟猎物的分布。然后，我们结合猎物分布与栖息地变量模拟东北虎的分布。我们以 3km×3km 栅格内有蹄类动物

第 10 章　中国东北虎栖息地等级保护优先区域和廊道确定的适宜性模型技术

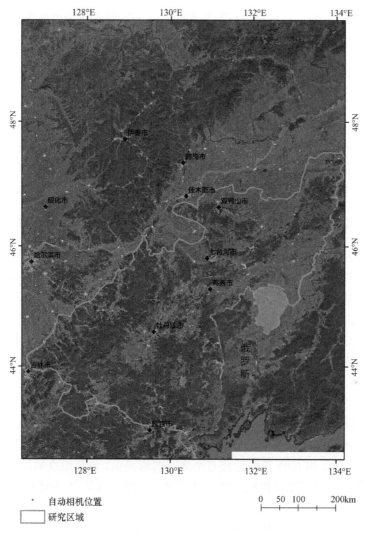

图 10.2　研究区自动相机位置分布图（彩图请扫封底二维码）
地图底图为谷歌卫星影像

或东北虎出现与否作为因变量。使用 R 语言 car 包的方差膨胀因子和 Spearman 秩相关系数检验自变量间的共线性问题。不相关（$r<0.6$）的栖息地变量被选中输入模型（Ramsay et al., 2003）。最大熵模型常被用来模拟物种分布（Phillips et al., 2006），我们使用 MaxEnt 软件（版本 3.4.1）来模拟有蹄类猎物和东北虎分布。东北虎出现点数据集被随机地分为两个子集：训练集（75%）与测试集（25%）。训练集用来建立模型，测试集用来验证模型。我们使用测试集 AUC 值来评估模型的预测能力，模型 AUC 值在 0.7~0.9 可认为模型有效，模型 AUC 值大于 0.9 可认为模型预测能力非常高。每个栖息地变量的重要性通过刀切法检验获得。最后，我们以当测试集模型敏

感性与特异性总和最大时的概率作为临界点将 MaxEnt 模型适宜性地图重分类为二值图（即适宜栖息地与非适宜栖息地）（GuÉnette and Villard，2005）。

10.2.2 栖息地廊道识别

以面积超过东北虎最小需求面积（MAR）的斑块作为源斑块（Walston et al.，2010），使用由物种分布模型（MaxEnt）得到的出现概率并通过 Linkage Mapper 软件进行连通性分析（McRae and Kavanagh，2011）。Linkage Mapper 采用最小代价理论识别东北虎适宜栖息地斑块间的潜在廊道。动物移动的栖息地阻力值与 MaxEnt 模拟的东北虎出现概率值呈倒数相关（Chetkiewicz and Boyce，2009）。

10.2.3 优化算法模型、适宜栖息地斑块与廊道内优先核心栖息地识别

一般认为，越大的栖息地面积对物种保护越好，但保护区域的面积往往受到社会和经济因素的限制（Ferson and Burgman，2006）。我们的目标是识别出适合东北虎生存的最大的栖息地范围，同时最小化其中的人为干扰因子。Sala 等（2002）创建了一个模型，通过权衡保护成效与冲突从而得到决定核心栖息地斑块的概率的临界值。

首先，在 ArcMap 空间分析工具中使用多尺度移动窗口（从 0 栅格到 6 栅格，即物种空间感知范围半径为 0~18km）处理 MaxEnt 模型得到适宜概率图（Ribeiro et al.，2017）。物种空间感知范围是动物对景观属性有反应的空间区域，相关专家建议可以用动物的移动距离或家域大小估计物种的空间感知范围（Ribeiro et al.，2017）。雌性东北虎的平均日活动距离为 7km（Carroll and Miquelle，2006），该距离也在 ArcMap 多尺度移动窗口范围内。在每一个移动窗口尺度情景下，将出现概率从 0 至 1 间，每 0.05 作为适宜与不适宜的临界值，计算适宜栖息地斑块面积、平均阻力值、道路密度、耕地比例及村庄密度。优化算法模型-适应性模型如下：

$$Z = \max_i Z_i = \max_i \max_{p \in [0,1]} \lambda_p S_p p \exp\left(-\left(\sum \omega_{pj} \frac{L_{pj}}{\max(L_{pj})}\right)\right) \quad (10.1)$$

式中，Z 为目标函数得分；i 为移动窗口尺度，范围 0~6；p 为 MaxEnt 模型预测得到的东北虎出现概率；λ_p 为惩罚因子，为适宜栖息地内所有栅格的平均阻力值，代表了保护成本；S_p 为当概率临界值为 p 时，适宜栖息地的面积；L_{pj} 为人为干扰类型 j 的面积比例或密度；ω_{pj} 为任何一种人为干扰类型的选择因子，当 $L_{pj}=0$ 时，$\omega_{pj}=0$，否则 $\omega_{pj}=1$。该方程使得适宜栖息地面积最大同时其中的人为干扰强度最小，通过使 7 种移动窗口空间尺度下或 10 种廊道缓冲区宽度（1~10km）下的 Z 值最大，从而获得最佳的 p 值。

最后，用概率临界值 p 获得的核心栖息地斑块或廊道的位置和面积作为在 MaxEnt 模型中识别出的适宜栖息地内的优先保护区域。7km 的缓冲区宽度为廊道

模拟了某空间范围内优先保护区域的等级模型。适应性等级模型的所有统计分析都使用 SPSS 10.0 完成。

10.3 结　　果

10.3.1 中国东北虎种群和栖息地分布

基于 2000～2017 年的 4028 个东北虎出现点信息，目前中国东北虎分布区主要集中在三个区域：老爷岭、张广才岭和完达山，覆盖了 137 085km² 的区域（图 10.3a）。此外，2012 年 1 月至 2017 年 12 月，自动相机监测记录到 27 只东北虎的影像具有两侧花纹，另外还有 9 只仅记录到右侧花纹，3 只仅记录到左侧花纹。考虑到仅有一侧花纹可能会造成重复识别，因此本研究中至少有 36 只（36～39 只）东北虎个体被确认，其中一些个体的母体也被识别出来（表 10.2）。

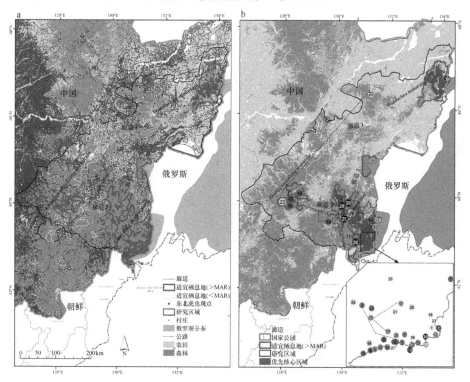

图 10.3　生物学模型与适应性模型的比较（彩图请扫封底二维码）

a. 传统生物学模型产生的适宜栖息地斑块与廊道，红色点为 2000～2017 年的东北虎出现点。b. 在 1 栅格大小情景下，使用适应性模型，最大化栖息地斑块适宜性，最小化栖息地斑块人为干扰，在目标函数值最大时求得适宜性概率切断点，从而得到优先核心区域。紫色线表示国家公园；传统生物学模型适宜斑块通过概率切断点将栖息地分为适宜栖息地与非适宜栖息地，当模型敏感性和特异性总和最大时得到该切断点，为 0.324。2012～2017年，相机监测得到的东北虎个体分布由图中编号的小圆圈表示。个体的移动线路用黑色虚线表示。地图底图为 World Ocean Reference，http://goto.arcgisonline.com/maps/Ocean_Basemap

表10.2　2012年1月至2017年12月自动相机记录到的中国东北虎个体（包括被拍摄到的体侧花纹）

个体编号	性别	捕获的体侧花纹	捕获历史						母本
			2012年	2013年	2014年	2015年	2016年	2017年	
CT1	F	LR	Y	Y	Y	Y	Y	Y	
CT2	M	LR		Y	Y	Y			
CT3	F	LR		Y	Y	Y	Y		
CT4	M	LR	Y	Y	Y				
CT5	F	LR		Y	Y	Y			
CT6	U	R				Y			CT18
CT7	M	LR		Y	Y	Y			
CT8	F	LR		Y	Y				
CT9	M	LR			Y	Y			
CT10	M	LR			Y	Y	Y	Y	
CT11	M	LR			Y	Y	Y		CT1
CT12	F	LR			Y				CT1
CT13	M	LR		Y	Y	Y			
CT14	U	R			Y				CT18
CT15	M	LR			Y	Y			
CT16	F	LR			Y	Y	Y		CT1
CT17	F	LR		Y	Y	Y	Y		
CT18	F	LR		Y	Y	Y	Y		
CT19	U	R			Y				CT3
CT20	M	R	Y						
CT21	U	L			Y				CT5
CT22	M	LR			Y	Y			
CT23	M	LR				Y	Y	Y	
CT24	U	L			Y				CT5
CT25	U	R	Y						
CT26	M	LR			Y	Y	Y	Y	
CT27	M	LR				Y	Y		
CT28	M	LR					Y	Y	
CT29	M	LR		Y	Y	Y	Y	Y	
CT30	U	R	Y						
CT31	F	R					Y	Y	
CT32	F	L						Y	

续表

个体编号	性别	捕获的体侧花纹	捕获历史						母本
			2012 年	2013 年	2014 年	2015 年	2016 年	2017 年	
CT33	M	LR					Y		
CT34	M	LR						Y	
CT35	M	LR						Y	
CT36	U	R					Y		CT3
CT37	U	R					Y		CT3
CT38	U	LR			Y			Y	CT8
CT39	M	LR						Y	

注：F. 雌性；M. 雄性；U. 性别未知；LR. 左侧和右侧花纹；L. 左侧花纹；R. 右侧花纹。Y. 是

10.3.2 东北虎栖息地模拟及最小需求面积

运用东北虎出现记录、环境变量及预测出的猎物分布结果，模拟了栖息地适宜性。测试集的 AUC 值为 0.834，表明模型具有较好的辨识能力。检验变量重要性的刀切法结果显示，猎物分布是最有价值的信息（图 10.4）。以 0.324 为概率临界值，我们将适宜性图分为适宜栖息地与非适宜栖息地二值图（图 10.5）。栖息地适宜性模型识别出 4 个大于最小需求面积的斑块（图 10.3a）。

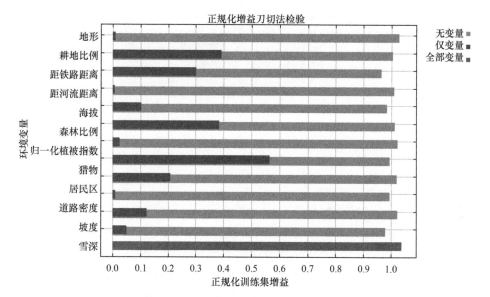

图 10.4 刀切法检验模型变量重要性的贡献（彩图请扫封底二维码）

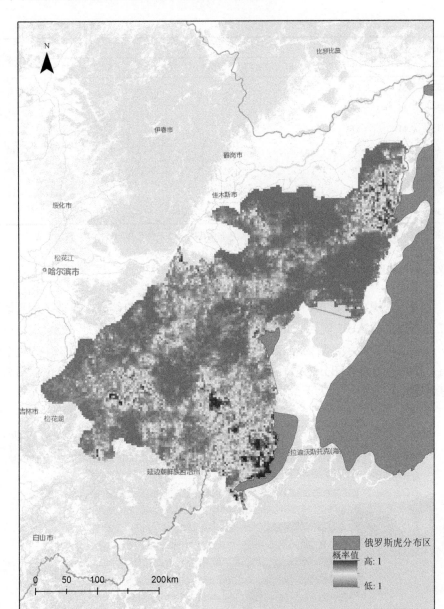

图 10.5 MaxEnt 模型预测结果显示的中国东北虎出现概率分布（彩图请扫封底二维码）
红色栅格表示出现概率高，蓝色表示出现概率低

利用逻辑斯谛回归模型对出现概率与斑块面积进行拟合，模型结果显著（$P<0.001$）（表 10.3，图 10.6），因此，我们认为东北虎对斑块面积敏感。我们计算了 ED_{90} 的值（能使东北虎有 90% 的概率出现的斑块面积）并将其作为东北虎最小需求面积，ED_{90} 的估计值为 992 km^2。

第 10 章　中国东北虎栖息地等级保护优先区域和廊道确定的适宜性模型技术 | 139

表 10.3　东北虎出现概率与斑块面积的逻辑斯谛回归参数

变量	估计值	标准差	t	P>|t|
常数	−3.5238	0.3211	−10.975	<0.001
斑块面积	0.0058	0.0017	3.402	<0.001

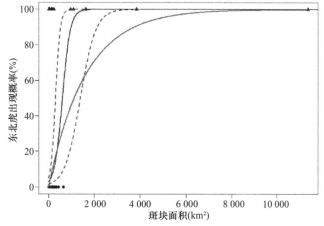

图 10.6　东北虎出现概率与斑块面积大小的关系（彩图请扫封底二维码）
实心圆圈表示没有东北虎出现的斑块，三角形表示有东北虎出现的斑块。黑色实线为逻辑斯谛拟合曲线，蓝色虚线表示 95% 置信区间，红色线代表零响应曲线

10.3.3　等级优先保护区域确定

我们计算了适宜斑块面积、斑块内平均阻力值、村庄密度、道路密度、耕地面积比例，得出了多种人为干扰因子与目标函数得分在每种情景下随适宜性变化的情况（图 10.7）。当移动窗口大小为 1 栅格、p=0.57 时，目标函数得分最高（图 10.8）。生境适宜性模型识别出 4 个大于东北虎最小需求面积的适宜栖息地斑块，以及它们之间的模拟廊道（图 10.3a）。考虑社会成本后，适应性等级模型识别出 3 个大于最小需求面积的优先核心斑块（图 10.3b：基于自动相机监测获得的东北虎个体分布图；图 10.9：等级优先保护区域模型的判别过程）。

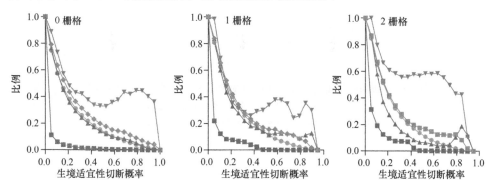

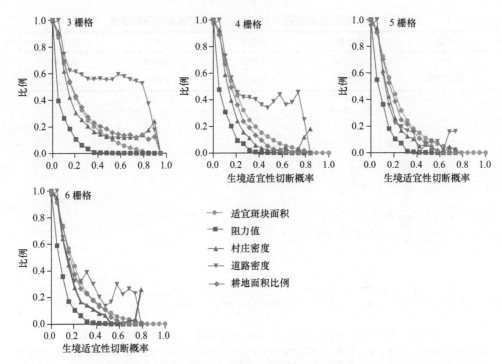

图 10.7　窗口多空间尺度下，随着 MaxEnt 模型出现概率不同切断点的变化，适宜栖息地面积及其内部的人为干扰强度的动态情况（彩图请扫封底二维码）

0~6 栅格代表了 0~18km 的移动窗口半径。所有变量都进行了归一化处理

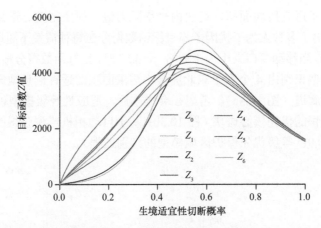

图 10.8　目标函数 Z 值随着生境适宜性切断概率变化而变化的情况（用来确定优先核心栖息地斑块的最优处理尺度）（彩图请扫封底二维码）

Z 值最大时的 p 值作为优先核心区域的切断点，在每一个情景尺度下代表生物和社会因子的最优组合。所有变量都经归一化处理。Z_0~Z_6 表示 0~6 栅格

第 10 章　中国东北虎栖息地等级保护优先区域和廊道确定的适宜性模型技术

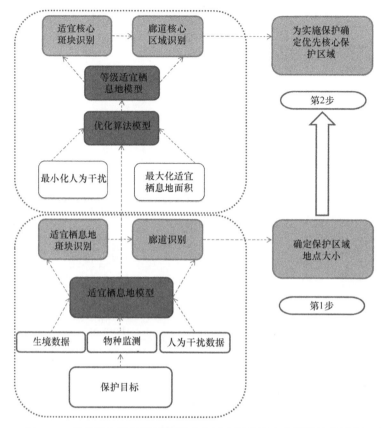

图 10.9　等级优先保护区域模型确定适宜栖息地和廊道区域的过程

对于 10 个不同缓冲宽度的廊道区域,采用同样的算法(式 10.1)得到了缓冲区内最优的廊道长度与适宜生境面积,通过 MaxEnt 得到了最大化适宜性值和适宜生境面积,同时得到了最小化的人为干扰强度。使得式(10.1)中函数值最大时,廊道缓冲宽度越大,生境适宜性临界值越小(图 10.10,图 10.11),这说明大廊道会包含大面积的低质量生境,需要更多的投入来改善生境质量(图 10.12)。

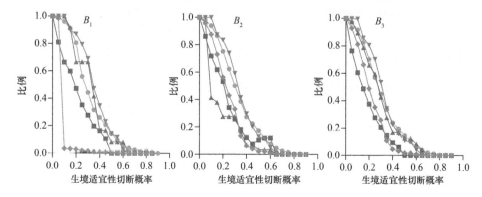

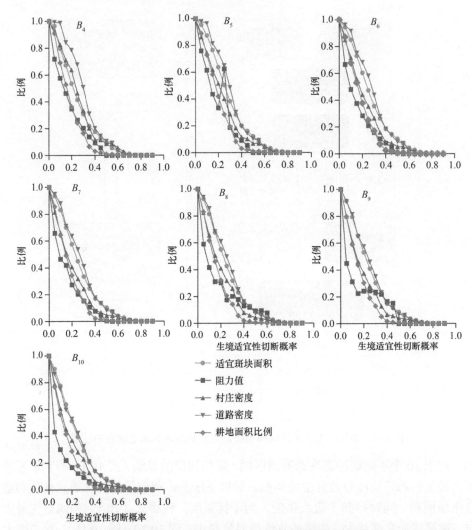

图 10.10 廊道不同缓冲宽度下，不同的 MaxEnt 模型预测概率切断点与适宜栖息地面积及人为干扰形式之间的变化关系（彩图请扫封底二维码）

村庄密度、道路密度和耕地面积比例分别指不同缓冲宽度下村庄、道路和耕地在适宜栖息地的数量、长度和面积。$B_1 \sim B_{10}$ 代表不同缓冲宽度，下角 $1 \sim 10$ 代表千米数

10.3.4 潜在保护功能的比较

基于东北虎自动相机监测数据库，我们统计了中国境内的东北虎总数量，将每个 3km 栅格中有东北虎出现看作被东北虎占据，得到了 2012~2017 年被东北虎占据的总面积为 2025km²。GIS 空间统计得到的研究区域内村庄总数量为 8186 个，耕地总面积为 49 205km²，道路总长度 23 271km。我们统计并比较了老虎个体和生境面积的比例，以及优先核心生境斑块内人为干扰因子的数量和比例（表 10.4）。

第 10 章 中国东北虎栖息地等级保护优先区域和廊道确定的适宜性模型技术 | 143

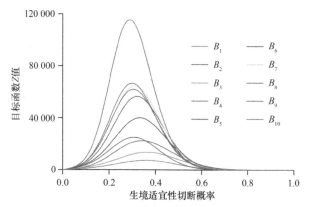

图 10.11 目标函数 Z 值随着生境适宜性切断概率变化而变化（用来确定不同缓冲区域内的优先核心栖息地）（彩图请扫封底二维码）

Z 值最大时的 p 值作为优先核心区域的切断点，在每一个情景尺度下代表生物和社会因子的最优组合。$B_1 \sim B_{10}$ 代表不同缓冲宽度，下角 $1 \sim 10$ 代表千米数

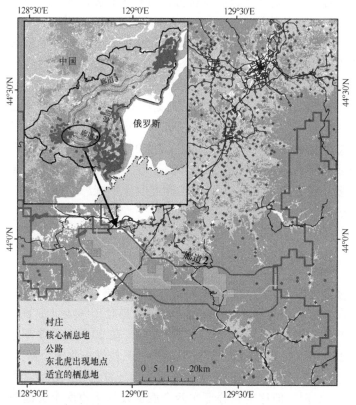

图 10.12 中国东北虎廊道线 $1 \sim 10$km 缓冲区内条块栖息地和人为干扰因子的空间分布（彩图请扫封底二维码）

图中廊道 2 为 7km 缓冲区内核心栖息地。地图底图为 World Light Gray Base，http://goto.arcgisonline.com/maps/World_Light_Gray_Base

表10.4 中国东北虎分布区生物学栖息地模型和适应性模型分别产生的适宜栖息地斑块与优先核心斑块潜在保护功能的比较

	中国东北虎分布区	适宜栖息地斑块（>MAR）	优先核心斑块（>MAR）	国家公园
村庄数量	8 186	246	80	249
耕地面积（km²）	49 205	1 105	320	949
道路长度（km）	23 271	2 078	809	1 281
2012～2017年，不同类型栖息地内被东北虎所占据的比例（%）	1.48	7.41	12.06	7.31
栖息地内东北虎种群数量比例（%）	100（36/36）	100（36/36）	88.89（32/36）	94.44（34/36）

生境适宜性模型表明：4个大于最小需求面积的斑块占目前被东北虎占据区域的7.41%，覆盖了所有记录到老虎个体信息的出现点，且可以潜在满足现有老虎在斑块间的移动；考虑到社会成本，生境适宜性模型识别出3个大于最小需求面积的优先核心斑块，这3个核心斑块占目前被东北虎占据区域的12.06%，覆盖现有记录东北虎出现总数的88.89%（表10.4，图10.3b）。此外，在优先核心区域内，人为干扰强度（村庄数量80个、耕地面积320km²、道路长度809km）较生境适宜性模型识别出的斑块（村庄数量246个、耕地面积1105km²、道路长度2078km）明显减少，而且与东北虎豹国家公园区域（村庄数量249个、耕地面积949km²、道路长度1281km）相比也显著减少（表10.4）。尽管现已监测到的东北虎总数的94.44%都在东北虎豹国家公园范围内，但是它仅覆盖了东北虎占区的7.31%，低于优先核心区域的12.06%（表10.4）。国家公园范围内高密度的东北虎种群预示着老虎可能向国家公园外围扩散，因此，在公园外部广阔区域的保护工作需要引起足够的重视，如建立廊道或保护张广才岭区域的潜在核心栖息地（图10.3a）。

10.4 讨 论

生物学模型揭示了大范围的生态廊道在东北虎渗透景观中是非常有必要的，以满足老虎在各斑块间的移动（Theobald et al., 2012）。实际上，我们发现了一只出生在俄罗斯的雄性东北虎个体（编号为CT10）在2014～2017年进入中国境内后沿着廊道2移动超过300km（图10.3b）。另外一只雄性东北虎个体（编号为CT22）自2013年3月至2015年1月沿廊道3从完达山迁移到张广才岭（图10.3b）。尽管参考案例仅2例，但这也充分证实两个廊道发挥了将两个关键斑块连接起来的重要作用。但是，在模型分析过程中有些阻碍也不应该被忽视，如即使随着生境适宜性的提高也很难将优先核心斑块内的道路清除（图10.7）。监测数据显示，2017～2018年，高速公路、农田和村庄阻碍了两只东北虎个体由中俄边境向中国腹地的扩散（未发表数据）。优先核心斑块和廊道及栖息地恢复不仅应该考虑采取退耕还林、移民、关闭道路等措施，还应该考虑建设隧道或高架桥等动物通道以连接核心生境斑块（图 10.3）。

东北虎豹国家公园是中国野生东北虎种群恢复的重要种源（Walston *et al.*，2010）。我们目前已经在国家公园范围内发现了 5 个东北虎繁殖家族及 11 个亚成体，国家公园内东北虎种群密度已是周边区域的 3 倍（Wang *et al.*，2018），该种群急迫向周边区域扩散，尤其是向中国长白山腹地扩散。通过中俄联合开展的东北虎跨境监测，在该区域共监测到至少 45 只成年东北虎个体，尽管该小种群的增长速度令人鼓舞，但我们应清醒地认识到，该小种群仍面临着种群崩溃的风险。俄罗斯 Sikhote-Alin Biosphere Zapovednik 地区一个东北虎小种群 1966～2012 年快速增长，一度达到约 40 只的种群规模，但由于暴发疾病，种群崩溃（Miquelle *et al.*，2015）。另外，这个中俄边境的东北虎小种群已经和俄罗斯东北虎北部大种群（野生东北虎数量的 95%）产生了生殖隔离（Henry *et al.*，2009），因此，需要尽快与俄罗斯大种群连接从而降低种群退化的风险。本研究中，我们提供了一种该小种群与俄罗斯锡霍特-阿林大种群连接的可能性：通过老爷岭斑块—张广才岭斑块—完达山斑块与俄罗斯大种群连接。遗憾的是，东北虎豹国家公园的设计尚未认识到通过生态廊道以满足该公园内小种群与其他大斑块连接的迫切需要（图 10.3b），目前东北虎豹国家公园的区域范围未必能满足中国东北虎种群恢复的需要。

我们的研究表明：通过栖息地适宜性模型能识别出大尺度优先保护区域，为政府制定适宜栖息地评估和生态廊道建设等管理措施提供参考，而且，由于兼顾考量了生态学需求和社会经济因素，我们的等级模型提供了一个更加科学严谨的前瞻性设计。此外，栖息地适应性模型可以精确评估每个变量的作用（Castilla，1999）（尽管这些变量间是相关的），而且可以判断斑块面积是否满足动物长期生存的需求（Qing *et al.*，2016）。我们的研究为保护管理者在濒危动物栖息地景观规划和斑块管理及在物种需求与社会经济因素限制情况下如何划定优先保护区方面提供了一个范例。

10.5 本章小结

保护濒危大型动物急需识别出其生境的优先保护区，尤其要同时评估其生态需求和当地的社会经济因素。通过平衡最大化适宜栖息地和最小化人为干扰的优化算法，我们分等级识别出了中国东北虎的优先保护区域。相较于普通生境适宜性模型确定的适宜生境和东北虎豹国家公园，生境适宜性模型显示面积大于最小需求面积（MAR）的等级优先核心区域可以保护更多被老虎占据的区域，并且有更少的人为干扰。因此，东北虎豹国家公园的设计与管理应重点放在连接虎生境斑块上。该等级量化方法展示了如何使用广泛适用的透明工具为保护大型陆生哺乳动物在区域尺度上选址建新保护区，并为新保护区域设计核心区域。

参 考 文 献

Carroll C, Miquelle D G. 2006. Spatial viability analysis of Amur tiger *Panthera tigris altaica* in the Russian Far East: the role of protected areas and landscape matrix in population persistence. Journal of Applied Ecology, 43(6): 1056-1068.

Castilla J C. 1999. Coastal marine communities: trends and perspectives from human-exclusion experiments. Trends in Ecology & Evolution, 14(7): 280-283.

Chapron G, Kaczensky P, Linnell J D, et al. 2014. Recovery of large carnivores in Europe's modern human-dominated landscapes. Science, 346(6216): 1517-1519.

Chetkiewicz C L B, Boyce M S. 2009. Use of resource selection functions to identify conservation corridors. Journal of Applied Ecology, 46(5): 1036-1047.

Fazey I, Fischer J, Lindenmayer D B. 2005. What do conservation biologists publish? Biological Conservation, 124(1): 63-73.

Ferson S, Burgman M. 2006. Quantitative methods for conservation biology. Berlin: Springer Science & Business Media.

Game E T, Kareiva P, Possingham H P. 2013. Six common mistakes in conservation priority setting. Conservation Biology, 27(3): 480-485.

GuÉnette J S, Villard M A. 2005. Thresholds in forest bird response to habitat alteration as quantitative targets for conservation. Conservation Biology, 19(4): 1168-1180.

Henry P, Miquelle D, Sugimoto T, et al. 2009. *In situ* population structure and *ex situ* representation of the endangered Amur tiger. Molecular Ecology, 18(15): 3173-3184.

Jiang G, Wang G, Holyoak M, et al. 2017. Land sharing and land sparing reveal social and ecological synergy in big cat conservation. Biological Conservation, 211: 142-149.

Karanth K U, Chellam R. 2009. Carnivore conservation at the crossroads. Oryx, 43(1): 1-2.

Kerley L L, Goodrich J M, Miquelle D G, et al. 2003. Reproductive parameters of wild female Amur (Siberian) tigers (*Panthera tigris altaica*). Journal of Mammalogy, 84(1): 288-298.

Li J, Wang W, Axmacher J C, et al. 2016. Streamlining China's protected areas. Science, 351(6278): 1160.

Liu J, Linderman M, Ouyang Z, et al. 2001. Ecological degradation in protected areas: the case of Wolong Nature Reserve for giant pandas. Science, 292: 98-101.

McLaughlin K. 2016. Can a new park save China's big cats? https://www.sciencemag.org/news/2016/08/can-new-park-save-chinas-big-cats[2017-1-10].

Mcrae B, Kavanagh D. 2011. Linkage mapper connectivity analysis software. Seattle WA: The Nature Conservancy.

Miquelle D G, Rozhnov V V, Ermoshin V, et al. 2015. Identifying ecological corridors for Amur tigers (*Panthera tigris altaica*) and Amur leopards (*Panthera pardus orientalis*). Integrative Zoology, 10(4): 389-402.

Phillips S J, Anderson R P, Schapire R E. 2006. Maximum entropy modeling of species geographic distributions. Ecological Modelling, 190(3-4): 231-259.

Qi J Z, Shi Q H, Wang G M, et al. 2015. Spatial distribution drivers of Amur leopard density in northeast China. Biological Conservation, 191: 258-265.

Qing J, Yang Z, He K, et al. 2016. The minimum area requirements (MAR) for giant panda: an empirical study. Scientific Reports, 6: 37715.

Ramsay T, Burnett R, Krewski D. 2003. Exploring bias in a generalized additive model for spatial air pollution data. Environmental Health Perspectives, 111(10): 1283.

Ribeiro J W, Silveira Dos Santos J, Dodonov P, et al. 2017. LandScape Corridors (LSCORRIDORS): a new software package for modelling ecological corridors based on landscape patterns and species requirements. Methods in Ecology and Evolution, 8(11): 1425-1432.

Ripple W J, Estes J A, Beschta R L, et al. 2014. Status and ecological effects of the world's largest carnivores. Science, 343(6167): 151-162.

Sala E, Aburto-Oropeza O, Paredes G, et al. 2002. A general model for designing networks of marine reserves. Science, 298(5600): 1991-1993.

Stephens P A, Zaumyslova O Y, Miquelle D G, et al. 2006. Estimating population density from indirect sign: track counts and the Formozov-Malyshev-Pereleshin formula. Animal Conservation, 9(3): 339-348.

Theobald D M, Reed S E, Fields K, et al. 2012. Connecting natural landscapes using a landscape permeability model to prioritize conservation activities in the United States. Conservation Letters, 5(2): 123-133.

Tian Y, Wu J, Kou X, et al. 2009. Spatiotemporal pattern and major causes of the Amur tiger population dynamics. Biodiversity Science, 17: 211-225.

Viña A, Mcconnell W J, Yang H, et al. 2016. Effects of conservation policy on China's forest recovery. Science Advances, 2(3): e1500965.

Walston J, Robinson J G, Bennett E L, et al. 2010. Bringing the tiger back from the brink—the six percent solution. PLoS Biology, 8(9): e1000485.

Wang T, Royle J A, Smith J L, et al. 2018. Living on the edge: opportunities for Amur tiger recovery in China. Biological Conservation, 217: 269-279.

Wikramanayake E, Mcknight M, Dinerstein E, et al. 2004. Designing a conservation landscape for tigers in human-dominated environments. Conservation Biology, 18(3): 839-844.

Zhu P, Huang L, Xiao T, et al. 2018. Dynamic changes of habitats in China's typical national nature reserves on spatial and temporal scales. Journal of Geographical Sciences, 28(6): 778-790.

第 11 章　东北虎种群和栖息地精准管理的生态阈值

11.1　引　言

大型食肉动物作为顶级捕食者在构建和调节生态系统、影响生态系统功能强度方面发挥着重要作用（Ripple et al., 2014）。但是，生活史较长、种群密度低、繁殖率低、需要广阔的家域以获取足够的猎物资源等因素都会导致大型食肉动物更易受到威胁，种群恢复更加困难（Ripple et al., 2014）。此外，大型食肉动物和人类、牲畜之间的冲突也是威胁其生存的重要原因之一（Madden, 2008）。截止到 2014 年，全球大型食肉物种（成年个体平均体重≥15kg）中有 61%被世界自然保护联盟（IUCN）列为受胁物种（易危、濒危、极危），77%的物种种群数量持续降低（Ripple et al., 2014）。大型食肉动物的保护一直是全球性的难题。

东北虎（*Panthera tigris altaica*）是分布于中国东北和俄罗斯远东地区的大型食肉动物，处于濒危状态，是中国和俄罗斯政府的重点保护野生动物。从 2015 年 12 月开始，中国政府开始设计一个横跨黑龙江、吉林两省并直接由中央管理的国家公园，其目的是组建一个统一的管理系统，以保护东北虎和东北豹（*Panthera pardus orientalis*）。另外，为了加强对中国东北虎豹的保护及东北虎豹国家公园的建设，中国政府将在东北虎豹分布区全面禁止自然资源的开采。中国作为人口众多的发展中国家，人类与野生动物冲突的问题也不容忽视。在一些地区因野生动物引起的牲畜损失甚至伤人事件，造成经济损失，进而导致了人类与野生动物冲突矛盾的加剧（Liu et al., 2011）。

动物种群和生态系统都有一定的抵抗环境因子波动的能力，以维持种群或者生态系统的稳定，这种能力被称为环境可塑性或适应性，但是它只能在一定范围内起作用（Vogt et al., 2015）。环境因子的变化若超出生态系统或动物种群调节承受能力的范围，就会导致其结构和功能的改变。生态系统或种群结构和功能的改变通常会非常迅速，并且以非线性的形式发生，难以恢复。生态阈值是指生态系统或动物种群能够忍受的环境因子的最大值和最小值，当达到阈值时，哪怕一个或多个关键因子发生微小变化，也会导致生态系统或种群快速的状态变化（Bennett and Radford, 2003）。有时，一定的环境条件范围内，可能会出现多个稳定状态，该范围内任何一点的轻微波动都可能导致这些状态的相互转换

（Scheffer and Carpenter，2003），此时，这一范围也可以视为一个生态阈值。超过阈值会导致生态系统功能和生物多样性的丧失（Bestelmeyer，2006）。撒哈拉沙漠的形成是表述生态阈值这一概念的很好例证（Demenocal et al.，2000），在其成为沙漠前，撒哈拉地区是一片碧绿的景观，几乎完全被当年生的草本和灌木所覆盖，数千年间，撒哈拉地区的植被覆盖率虽然在降低，但这种下降趋势一直保持平稳，5000~6000年前，当夏季日照超过了470W/m^{2}的阈值时，撒哈拉地区的生态在几十年到几个世纪内就会完全崩溃（Demenocal et al.，2000），成为现如今的沙漠景观。

自从1977年5月阈值的概念被引入生态学科以来，生态学家就持续关注这一领域（Briske et al.，2006；李春贵和袁振，2017）。生态阈值的实际应用在生物保护和可持续生态系统管理中具有非凡的吸引力（Huggett，2005）。一般而言，生态阈值在自然保护中的应用主要包括以下4个方面：①确定物种对干扰的敏感性；②协调生物多样性保护与人类发展；③建立生境保育、修复和恢复的目标；④完成生物多样性保护的景观设计（Huggett，2005）。

阈值反映了生态系统或野生动物种群对不同环境因子或其他驱动因子的最大或最小耐受性，它非常适合为自然资源管理制定标准规则或目标，以便将环境条件保持或恢复到目标物种的可承受范围内（Choquenot and Parkes，2001）。从这个意义上讲，生态阈值的概念与中国政府强调的"生态保护红线"非常一致（李干杰，2014）。"生态保护红线"是指应被严格执行的空间边界和管理上的限制值，包括自然生态服务功能、环境质量和安全，以及自然资源利用（李干杰，2014）。它代表着必须遵守的生态保护的定量指标，一旦超过将造成不可逆转的损害。对于东北虎保护和国家公园建设而言，我们需要了解所有不同的环境因子可能对这种大型食肉动物和整个生态系统的阈值的效应，这些研究成果将对国家公园的设计、政策法规制定与实施起到巨大帮助。例如，东北虎栖息地破碎化阈值，将成为国家公园区域范围划定和生态廊道建设的重要参考；如果管理人员能掌握人为活动频率强度引起东北虎对该区域的回避阈值，就可以通过相应的政策或办法将人为活动限制在合理范围内。另外，如果我们了解雪深影响有蹄类动物觅食的阈值，我们就可以有的放矢地开展冬季补饲，从而节省不必要的人力和物力支出。

本章我们讨论的是不同形式的生态阈值是如何通过上行效应直接或间接地影响东北虎种群及其生存的生态系统的。引起阈值效应的因子很复杂，如气候、极端天气、栖息地、人为干扰、植被、猎物和竞争者等。然而，哪些因子对物种和生态系统的保护起关键作用？为什么？回答上述问题的同时也可为全球其他受胁的大型食肉动物的保护提供有益的参考。我们的研究拟回答以下两个问题：①如何制定出更明确、更实际的东北虎保护和管理措施；②有哪些普遍适用的策略能够促进生态阈值概念在大型食肉动物保护中的实践应用？我们认为对生态阈值的研究应该成为东北虎保护、国家公园建设的重要组成部分，并且广泛应用于其他地区大型食肉动物的精准保护管理工作中。

11.2 基于上行效应确定东北虎保护中的生态阈值

11.2.1 气候和极端天气引起的阈值效应

气候变化已经引起了广泛的社会关注。气候状况的改变或极端天气事件的相对增多都会极大地影响物种的分布和行为（Seddon et al., 2016），以及生态系统的结构和功能（Doughty et al., 2015）。气候变化对植物及其生态系统的影响通常是非线性的。例如，根据预测结果，当全球温度上升3～4℃时，就会导致亚马孙雨林的退化（Cox et al., 2004）。同样，极端低温的频次下降可能会导致红树林因其能够适应的最小温度阈值的改变而发生其地理分布的变化，逐渐向两极方向扩张（Cavanaugh et al., 2014）。气候变化会引起植物群落结构及植被成熟时间的变化，进而影响动物群落的种间互作关系和地理分布。例如，随着气候变暖，喜温植物将在与耐寒植物的竞争中获得优势，生存更多依赖于耐寒植物的消费者将会受到更多限制，与此同时，高度依赖于喜温植物的消费者将获得更多的生存机会，消费者群落的变化会影响食物网中更高等级的物种，作为顶级捕食者的东北虎也会受到影响。虽然还没有直接证据阐明这种影响如何作用于顶级捕食者，但是在食物网底层物种或群落的互相影响方面已经有过大量研究（Visser, 1998；Kaiser-Bunbury et al., 2017）。Seddon 等（2016）使用自回归模型定量分析生态系统对气候变量（气温、降水和云层覆盖）的相对响应率。在全世界范围内共确定了9个已经接近引起状态彻底改变的关键阈值的生态群落，北方森林带的部分地区就在其中，这些地区是东北虎的重要分布区。中国东北虎分布区中的许多森林属于次生林或人工林，这类森林对环境干扰特别敏感（Lugo, 2010），因此，应该进一步开展阈值研究，以了解气候变化在什么程度上会造成森林退化，引起多米诺骨牌效应从而危及东北虎的生存。

对于动物而言，温度或其他环境条件在时间上的快速变化可能迫使动物为适应不同条件下的能量消耗而改变其日常活动模式（Bellard et al., 2012）。生态学家也开始探究气候变化对动物生活史特征的影响，包括繁殖时间、繁殖成功率、秋季迁徙时间（Visser, 1998）。还有一些研究发现了气候变化对动物地理分布、物种互作（Both and Visser, 2001）和种群动态（Pounds et al., 1999）的影响。Root 等（2003）通过荟萃分析（meta-analysis）认为，为了应对气候变化，在过去50年中，动物和植物的关键物候行为平均每10年提前5.1天。气候变化可能也会增加捕食者-猎物系统的不同步性（Parmesan，2006），从而导致捕食者的局部灭绝。气候变化对东北虎的影响可能是直接的，也可能是间接的。一方面，长期的进化使得东北虎适应了寒冷气候，气候变化对其他动物的活动模式、繁殖时间、繁殖成功率、地理分布、物种互作等产生的改变，同样也会发生在东北虎身上。另一方面，猎物直接决定了捕食者的生存和分布，猎物分布的变化、活动模式和生活史特征会明显影响捕食者的生存与活动。因此，气候驱动的阈值效应对东北虎的影响会很复杂。

极端天气在影响生态系统或动物种群发生状态变化中也可能发挥着重要作用（Parker et al.，1984；Visser，1998）。有证据表明，拥有大型捕食者的生态系统可能更易受到极端天气事件的影响，从而发生状态变化。极端天气可能会破坏种子萌发，减缓森林再生（Bradford，2005），极端天气导致植物产量的减少或成熟时间的改变与动物对植物食物的需求时间往往不一致（Visser，1998）。类似暴雪这种极端天气也会导致有蹄类动物直接死亡。据报道，当雪深达到有蹄类动物胸高40%~60%时会大幅度增加其活动的能量消耗，威胁其冬季生存（Parker et al.，1984）。有蹄类动物的死亡使大型食肉动物的猎物数量减少，最终可能导致捕食者的地理分布发生变化（Serrouya et al.，2015）。然而，这些阈值效应对东北虎的潜在影响仍然很不明确。

据推测，中国北部地区将会经历较其他地区更大程度的气候变暖，中国东北的北部地区降水量会增加，而南部部分区域降水量会减少（秦大河等，2005）。研究气候变化和极端天气事件对植被、有蹄类种群及东北虎的影响的阈值将会有助于管理者制定预防措施与应急预案，以提前应对气候变化和极端天气对虎及整个生态系统的影响。

11.2.2　栖息地丧失和人为干扰引起的阈值效应

许多物种及它们的行为都具有高度特异性的生态和环境需求。例如，觅食和交配，都需要在特定的环境条件下进行。特别是顶级捕食者，都需要一个广阔而相对稳定的家域，有充足的猎物种群来保证其生存。GPS 项圈数据显示，成年雌性东北虎稳定的核心家域的平均面积是 401km^2（401km^2±205km^2）（Hernandez-Blanco et al.，2015）。成年雌性东北豹的核心家域面积是 23.3km^2（Rozhnov et al.，2015），雄性个体通常有更大的家域需求。了解这两种大型食肉动物生存的最小栖息地需求，是进行国家公园设计的重要依据，也是开展东北虎豹保护工作的重要参数。

人类活动范围的增加加速了野生动物栖息地的丧失，人为干扰对残留的栖息地也会造成持续影响（Fahrig，2001），从而导致斑块（或栖息地）大小与种群生存之间的关系更为复杂。一方面，人类活动直接导致栖息地破碎化或栖息地丧失。人为干扰是全球物种生存面临的主要威胁，而且是个体在景观内扩散和迁移的重要决定因素（Fahrig，2001）。在高度破碎化的栖息地景观中，残留适宜栖息地内物种活动模式也会发生显著的变化（Andren，1994）。另一方面，野生动物常离开那些具有强人为活动干扰的区域，迁移到其他区域活动。例如，马鹿生性胆小，会与人类景观保持一定距离，它们更喜欢在距离村庄 8.2km 以上的地区活动，而对森林道路和废弃公路的回避距离分别是 1.6km 和 2.2km（Jiang et al., 2007）。Jiang 等（2014）发现，铁路的修建会严重限制沿途 15km 范围内东北虎的出现概率。在黑龙江省完达山区域，在面积为 196km² 的研究栅格区域内农田面积超过 25% 时，东北虎的相对出现概率最低（Jiang et al., 2014）。动物因回避人为干扰而导致的栖息地丧失可能比直接杀死动物本身对其种群生存的影响更大（Jaeger et al., 2005）。因此，开展东北虎栖息地评价研究时必须考虑到人为干扰因子。为了维持东北虎种群的持续生存，在国家公园整体设计中，对栖息地丧失和破碎化阈值的评估至关重要。

不仅是栖息地丧失，人类活动同样正在以前所未有的速度影响着生态系统的其他方面（Leemans and Groot，2003）。对于东北虎而言，人类活动不仅会破坏东北虎的栖息地，还是东北虎直接或间接的致死因素。在中国，人为盗猎是导致东北虎濒危的主要原因之一（周晓禹，2008）。成年雌性东北虎的生存受到道路数量和道路类型的严重威胁（Kerley et al., 2002）。牲畜破坏植被与有蹄类动物的竞争导致东北虎猎物种群数量下降，同样威胁着东北虎的生存（Li et al., 2016）。尽管中国政府已经全面禁止狩猎活动，但为了更好地管理人为活动并减少人类与野生动物冲突，我们依然需要进一步量化动物或生态系统内各种人为干扰的频率和强度的耐受阈值，如放牧密度阈值、机动车出现频率阈值等。

许多研究人员和管理人员已经开始尝试研究能够体现目标物种或生态系统对人为干扰的最大或最小耐受性的阈值，以便于更好地规范人类活动（Spooner and Allcock，2006；Rodrigues et al., 2016）。其中一个好的例子就是利用生态阈值指导牧场和林地管理人为干扰的频率与密度，以保持牧场和林地系统的健康与可持续利用（Spooner and Allcock，2006）。不同物种对人类活动的敏感程度不同，一些适应性强的物种能容忍 90% 的栖息地丧失，而对栖息地特异性要求较高的物种只能在栖息地丧失程度低于 60% 的景观中生存（Rompré et al., 2010）。最近，我们分析了公元 218 年到 2015 年期间的 3428 条关于老虎信息的历史记录，揭示了人口密度影响老虎灭绝概率的阈值效应，结果表明，一旦人口密度超过每平方千米 400 人，该区域老虎在 50 年内局部灭绝的概率将超过 60%（未发表数据）。这些基本信息对中国东北虎的保护非常重要。

11.2.3 植被的生态阈值

植被（或植物）对动物的重要性不仅反映在食物供应上，还影响着动物御寒和隐蔽场所的数量（Reimoser and Gossow，1996）。因此，植被会通过不同机制引起阈值效应。植物作为动物的食物，对食草动物和杂食动物的限制性影响通常通过植物质量与数量来体现。或者说，如果植被的质量或数量低于动物对食物的最低需求，消费者就不会在这片区域觅食，我们把此时的植被密度称为放弃密度（Brown，1988）。植物作为食草动物的食物来源，对食草动物的影响决定了捕食者猎物的丰富度，通过上行效应间接影响大型食肉动物的生存。此外，有学者认为食草动物多样性常与植物多样性密切相关，这种关系很可能是非线性的（Siemann et al.，1998），更多样化的消费者群体更有利于东北虎和东北豹生态位分离与共存（Richman and Lovvorn，2009）。

为动物提供隐蔽所是植被的另一个关键功能，植被可提供的隐蔽所的数量和质量往往与动物的存在呈非线性关系。Smith 和 Long（1987）发现，落基山山脉马鹿（*Cervus elaphus nelsonii*）对隐蔽条件的要求是胸径大于 $3.14cm/m^2$ 或林冠郁闭度大于 $4.75cm/m^2$ 的植被，而其维持自身热量对植被依赖更严格。对于许多其他有蹄类动物而言，隐蔽条件对其的影响也通常是可测量的，通常隐蔽条件对其的影响的阈值比御寒要求要小（Mysterud and Østbye，1999）。这些结果说明了植物在维持动物热量和提供隐蔽所等方面对动物抵抗恶劣天气与捕食者的意义重大。此外，消费者必须在捕食风险和觅食之间进行权衡，这种权衡和选择将改变猎物的活动模式和食物的可获得性（Esparza-Carlos et al.，2016），这种关系将会反过来影响植被管理。

因此，植被对猎物或食草动物的生存至关重要，而猎物或食草动物是虎生存的基础。一旦植被不足以维持有蹄类动物生存，有蹄类动物将会消失或离开这片区域，对东北虎的保护便无从谈起。东北虎栖息地的植被恢复，需要重视对植被影响有蹄类动物丰富度和多样性及寒冷地区个体越冬情况的量化研究与阈值确认，同时需要考虑植被结构和植被作为食物的可获得性。

11.2.4 猎物或食草动物的生态阈值

相对于上文讨论的植被对捕食者的间接影响，捕食者-猎物之间的关系一直受到野生动物保护人员的重点关注，特别是对大型捕食者而言更为突出（Johnson et al.，2007）。据了解，虎会根据猎物丰富度调整其家域范围，而虎的密度也与猎物密度有关（Karanth et al.，2004）。在加拿大 Yukon 西南部地区，当北美野兔（*Lepus americanus*)的密度从 14.7 只/0.01km^2 下降到 0.2 只/0.01km^2 时，加拿大猞猁（*Lynx canadensis*）的家域会从 13.2km^2 增加到 39.2km^2（Ward and Krebs，1985）。在北美野兔的密度降低到 0.5 只/0.01km^2 时，部分猞猁会离开其原来的栖息地（Ward and Krebs，1985）。这种现象与生态阈值-放弃密度的概念是一致的（如 Nolet *et al.*，2006）。

提高猎物种群密度是保护大型肉食动物的重要组成部分。一项关于非洲水牛（*Syncerus caffer*）、疣猪（*Phacochoerus africanus*）与狮子（*Panthera leo*）种间互作关系的阈值效应的研究表明，想要实现三个物种的共存，水牛的种群密度应该维持在环境容纳量的一半以上（Ddumba *et al.*，2012）。Miller 等（2014）认为在猎物的管理中，需要深入了解满足捕食者生存和繁殖所需能量的最低猎物密度。东北虎的平均捕食成功率为 1 次/6.5 天，每只东北虎每天需要消费的猎物生物量为 8.9kg（Miller *et al.*，2013）。当猎物种群密度小于 0.5 只/km^2 且在老虎家域内有蹄类猎物低于 250～500 只时，雌性东北虎个体会停止繁殖（李振新等，2010）。这些信息表明，猎物密度阈值对于东北虎保护的重要性需要深入研究并具有挑战性。

不同于猎物密度决定捕食者个体的生存和繁殖，猎物的多样性在影响不同捕食者共存方面起着重要作用（Tucker and Rogers，2014）。不同体型的捕食者捕猎不同体型的猎物（Dickman，1988）。通常，东北虎更倾向于将野猪作为其主要猎物，而东北豹则更喜欢捕食狍（Yang *et al.*，2018a）。东北虎和东北豹对猎物的偏好差异降低了这两种大型食肉动物在食物资源上的竞争（Yang *et al.*，2018a）。此外，不同体型的猎物种群分布的空间异质性也有利于猎物与捕食者间的共存（Richman and Lovvorn，2009），而这将有助于维持物种多样性和生态系统的稳定性。

另外，除通过上行效应影响捕食者以外，食草动物在很大程度上也会通过下行效应影响植被。食草动物的数量如果超过环境容纳量，就会给森林林下灌木层带来严重的损害，导致其喜欢采食的植物丰富度下降和分布范围缩小（Rooney *et al.*，2004）。为了保持采伐之后阔叶林的生物多样性，De Leo 和 Levin（1997）认为，管理白尾鹿种群，应将其种群密度控制在 4 只/km^2 以内。为了维持处于早期演替阶段的北方香脂冷杉林的稳定，Tremblay 等（2006）认为鹿的适宜密度应该是 7.5～15 只/km^2。一般来说，食草动物种群存在一个会引起植被状态变化的食草动物密度阈值，这种植被状态的变化也会影响食草动物的种群密度，从而影响捕食者的密度（Sasaki *et al.*，2008）。这种间接影响印证了 Qi 等（2015）发现的一旦野猪和狍的总生物量超过 100～150kg/0.25km^2 东北豹的密度就会下降这一结论。因此，猎物或食草动物的管理应避免其密度太高而对植被造成太大压力，从而间接影响捕食者的生存。全面考虑猎物的多样性和丰富度、上行效应与下行效应，会使猎物的阈值研究变得更加复杂。

11.2.5　东北虎和东北豹的共存阈值

对于同域分布的大型食肉动物，某一个物种种群的快速增加会加剧竞争。相关研究发现，在印度，随着老虎密度从 3.31 只/100km^2 增加到 5.81 只/100km^2，豹的食物组成发生了明显的变化，种群密度也急剧下降（Harihar et al., 2011）。东北虎和东北豹是同域分布的两种捕食者，两者之间的竞争会在很大程度上影响保护工作的最终结果。一般来说，关于捕食者共存的研究主要集中在潜在的共存机制方面，如猎物偏好（Andheria et al., 2007）、活动模式（Fedriani et al., 1999）和空间利用（Palomares et al., 1996）等的差异。这些因素同样影响着东北虎和东北豹的共存。例如，东北豹更倾向于利用险峭悬崖地形以回避东北虎（Jiang et al., 2015），而且它们的猎物偏好和时间活动模式也是截然不同的（Yang et al., 2018a, 2018b）。

一项在南非开展的研究体现了竞争性互作的可能性。在不同的竞争压力下，满足 10 只猎豹（*Acinonyx jubatus*）生存的最小区域面积具有很大差异。当没有其他捕食者时，它们需要 48~466km^2 满足 10 只猎豹的生存；当存在同等数量的狮子（*Panthera leo*）时，最小栖息地面积就需要 166~2806km^2；当存在同等数量的金钱豹（*Panthera pardus*）、斑点鬣狗（*Crocuta crocuta*）、非洲野犬（*Lycaon pictus*）和狮子时，这一数值增长为 727~3739km^2（Lindsey et al., 2011）。对于俄罗斯远东地区的东北虎和东北豹，其冬季 Pianka's 猎物资源生态位重叠指数达 0.77 之高（Sugimoto et al., 2016）。基于上述数据及物种竞争优势理论，我们认为某一区域东北虎数量增加到一定程度，可能导致东北豹离开该区域。因此，一旦老虎种群密度增加，管理者应采取措施来创造适于豹生存的微生境，或者促进豹个体向其他适宜栖息地扩散。此外，如果我们掌握了两种顶级捕食者营养生态位重叠指数的阈值，管理部门就可以通过管理猎物的密度和组成，以促进两种捕食者的共存。

11.3　未来生态阈值研究应用和面临的挑战

综上所述，阈值效应更可能是由几个互相作用的因子驱动的结果，而不是由单个特定因子引起的（van der Ree et al., 2004；Huggett, 2005）。例如，气候变化可能会影响食物的供应，从而影响动物的密度和分布（Both et al., 2006）。物种间的相互作用也可能受气候变化的直接影响（Tylianakis et al., 2008）。保证捕食者种群生存的家域阈值可能会受栖息地破碎化、种间竞争、气候变化、猎物可获得性等因素的综合影响出现波动，而这些因子又受到人类活动的影响。多变量互作引起的潜在的混合效应使阈值效应更加复杂，使得阈值的探测更为困难（Bennett and Radford，2003）。

变量之间的互作也使阈值的应用更依赖于具体环境背景。例如，在某一特定区域确定的东北虎生存的最小栖息地阈值不一定适用于其他物种或区域，因为不同人为活动频率、栖息地破碎化程度、猎物丰富度及其他影响因素导致物种和区域间生态阈值估计的巨大时空差异，因此没有哪一个阈值是可以普遍使用的（Huggett，2005）。所以，我们认为对生态阈值的研究应该侧重于特定区域，并从整体角度把握限制因子的研究，以找出不同保护和管理阶段中涉及的相关生态阈值。

东北虎豹国家公园的建设旨在保护这两个濒危物种及整个生态系统，这需要投入巨大的人力、物力和财力，以及合理配置公园内的自然资源。但是人类活动干扰及逐年增加的巨额投入都将是东北虎豹国家公园未来面临的挑战。庆幸的是，生态阈值评估可以成为特定条件下利用有限资源实现有效保护的合适工具。例如，了解满足东北虎和东北豹共存的最小森林栖息地面积，就可以避免将过多的土地规划为保护用地，避免更大面积的移民和资源禁采。保护东北虎豹，不仅需要维持大面积的可渗透栖息地景观以保证这些大型捕食者的有效扩散，还要统筹协调地方经济的发展。所以，我们只需在核心栖息地斑块和廊道区域把人为干扰控制在阈值范围以下，而不必全面禁止该区域的所有人类活动（Chapron et al.，2014）。此外，生态阈值能够为生境管理及栖息地中的植被和猎物管理提供量化指标。东北虎豹国家公园的建设为生态阈值理论和方法应用于大型捕食者保护提供了良好的平台与机遇，而通过阈值研究了解生物群落的恢复机制将助力东北虎豹国家公园的建设。

对于东北虎及其他大型食肉动物保护中的阈值研究和应用，首先，我们需要理清栖息地中基于上行效应的各种环境因子与物种间的逻辑关系，并考虑到特定区域和情况下的所有相关驱动因子。这些驱动因子可能包括气候和极端天气事件、栖息地丧失与其他人为干扰、植被、猎物，以及两种捕食者的种间竞争。所有直接或间接因子都很重要，如植被不仅为捕食者提供了隐蔽所，还为其猎物的生存提供了食物来源，间接地有益于捕食者。其次，我们试图找出可能影响或改变目标物种状态的关键因子。最后，确定这些因子和彼此间的相互作用是否表现出阈值效应，以及阈值是多少（图11.1）。在野生动物研究和保护工作中，阈值理论和观点的应用可以使保护政策的制定与具体实施更加完善、精准。阈值研究也应该成为虎豹国家公园的常规科研工作，在整个东北虎豹保护进程中持续发挥重要作用。

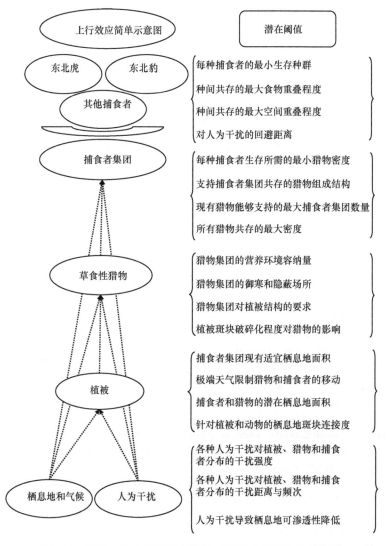

图 11.1 基于上行效应的顶级捕食者保护中潜在的生态阈值

11.4 本章小结

生态阈值是指由一个或多个生态因子的改变而导致整个生态系统状态发生完全改变的现象。生态阈值的概念强调对限制性因子的研究，能够反映生态系统或生物体对限制性因子变化的耐受性。目前，越来越多的保护实践证明了生态阈值确实存在，这为其在野生动物保护中的实际应用奠定了基础。将生态阈值概念用于指导东北虎豹国家公园体制试点建设，进而提高东北虎和东北豹的保护成效，其作用和发展潜力巨大。本研究基于上行效应的生态学理论，阐述了各种形式的生态阈值及其对东北虎保护可能产生的影响，以表明生态阈值在东北虎保护中的巨大应用潜力。需要注意的是，东北虎作为大型食肉动物，位于食物链顶端，生态阈值通常是通过上行效应对其产生影响的，可能存在阈值效应并最终影响东北虎的环境因子，包括气候、极端天气、栖息地、人为干扰、植被、猎物和竞争者。另外，因子间相互作用和生态阈值的环境依赖性是阈值研究与保护实践中必须考虑的两个重要前提。阈值的应用使我们能够更全面地评估保护需求，并指导中国东北虎未来的研究和保护工作。生态阈值在东北虎保护中的应用也将对世界上其他大型食肉动物的保护产生重要的启示作用。

参 考 文 献

李春贵, 袁振. 2017. 生态阈值研究进展及其应用. 河北林业科技, 3: 54-57.

李干杰. 2014. "生态保护红线"——确保国家生态安全的生命线. 求是, 2: 44-46.

李振新, Zimmermann F, Hebblewhite M, et al. 2010. 中国长白山区东北虎潜在栖息地研究. 北京: 中国林业出版社.

秦大河, 丁一汇, 苏纪兰, 等. 2005. 中国气候与环境演变评估(I): 中国气候与环境变化及未来趋势. 气候变化研究进展, 1(1): 4-9.

周晓禹. 2008. 野生东北虎保护现状及保护对策探讨. 野生动物, 29(1): 40-43.

Andheria A P, Karanth K U, Kumar N S. 2007. Diet and prey profiles of three sympatric large carnivores in Bandipur Tiger Reserve, India. Journal of Zoology, 273(2): 169-175.

Andren H. 1994. Effects of habitat fragmentation on birds and mammals in landscapes with different proportions of suitable habitat: a review. Oikos, 71(3): 355-366.

Bellard C, Bertelsmeier C, Leadley P, et al. 2012. Impacts of climate change on the future of biodiversity. Ecology Letters, 15(4): 365-377.

Bennett A F, Radford J Q. 2003. Know your ecological thresholds. Thinking Bush, 2: 1-3.

Bestelmeyer B T. 2006. Threshold concepts and their use in rangeland management and restoration: the good, the bad, and the insidious. Restoration Ecology, 14(3): 325-329.

Both C, Bouwhuis S, Lessells C M, et al. 2006. Climate change and population declines in a long-distance migratory bird. Nature, 441(7089): 81-83.

Both C, Visser M E. 2001. Adjustment to climate change is constrained by arrival date in a long-distance migrant bird. Nature, 411: 296-298.

Bradford K J. 2005. Threshold models applied to seed germination ecology. New Phytologist, 165(2): 338-341.

Briske D D, Fuhlendorf S D, Smeins F E. 2006. A unified framework for assessment and application of ecological thresholds. Rangeland Ecology and Management, 59(3): 225-236.

Brown J S. 1988. Patch use as an indicator of habitat preference, predation risk, and competition. Behavioral Ecology and Sociobiology, 22(1): 37-47.

Cavanaugh K C, Kellner J R, Forde A J, et al. 2014. Poleward expansion of mangroves is a threshold response to decreased frequency of extreme cold events. Proceedings of the National Academy of Sciences of the United States of America, 111(2): 723-727.

Chapron G, Kaczensky P, Linnel J D, et al. 2014. Recovery of large carnivores in Europe's modern human-dominated landscapes. Science, 346: 1517-1519.

Choquenot D, Parkes J. 2001. Setting thresholds for pest control: how does pest density affect resource viability? Biological Conservation, 99(1): 29-46.

Cox P M, Betts R A, Collins M, et al. 2004. Amazonian forest dieback under climate-carbon cycle projections for the 21st century. Theoretical and Applied Climatology, 78(1-3): 137-156.

Ddumba H, Mugisha J Y T, Gonsalves J W, et al. 2012. The role of predator fertility and prey threshold bounds on the global and local dynamics of a predator-prey model with a prey out-flux dilution effect. Applied Mathematics and Computation, 218(18): 9169-9186.

De Leo G A, Levin S. 1997. The multifaceted aspects of ecosystem integrity. Conservation Ecology, 1(1): 3.

Demenocal P, Ortiz J, Guilderson T, et al. 2000. Abrupt onset and termination of the African humid period: rapid climate responses to gradual insolation forcing. Quaternary Science Reviews, 19(1-5): 347-361.

Dickman C R. 1988. Body size, prey size, and community structure in insectivorous mammals. Ecology, 69(3): 569-580.

Doughty C E, Metcalfe D B, Girardin C A J, et al. 2015. Drought impact on forest carbon dynamics and fluxes in Amazonia. Nature, 519(7541): 78-82.

Esparza-Carlos J P, Laundré J W, Hernández L, et al. 2016. Apprehension affecting foraging patterns and landscape use of mule deer in arid environments. Mammalian Biology-Zeitschrift für Säugetierkunde, 81(6): 543-550.

Fahrig L. 2001. How much habitat is enough? Biological Conservation, 100(1): 65-74.

Fedriani J M, Palomares F, Delibes M. 1999. Niche relations among three sympatric Mediterranean carnivores. Oecologia, 121(1): 138-148.

Harihar A, Pandav B, Goyal S P. 2011. Responses of leopard *Panthera pardus* to the recovery of a tiger *Panthera tigris* population. Journal of Applied Ecology, 48(3): 806-814.

Hernandez-Blanco J A, Naidenko S V, Chistopolova M D, et al. 2015. Social structure and space use of amur tigers (*Panthera tigris altaica*) in southern Russian far east based on GPS telemetry data. Integrative Zoology, 10(4): 365-375.

Huggett A J. 2005. The concept and utility of 'ecological thresholds' in biodiversity conservation. Biological Conservation, 124(3): 301-310.

Jaeger J A G, Bowman J, Brennan J, et al. 2005. Predicting when animal populations are at risk from roads: an interactive model of road avoidance behavior. Ecological Modelling, 185(2): 329-348.

Jiang G, Qi J, Wang G, et al. 2015. New hope for the survival of the amur leopard in China. Scientific Reports, 5: 15475.

Jiang G, Sun H, Lang J, et al. 2014. Effects of environmental and anthropogenic drivers on Amur tiger distribution in northeastern China. Ecological Research, 29(5): 801-813.

Jiang G, Zhang M, Ma J. 2007. Effects of human disturbance on movement, foraging and bed selection in red deer *Cervus elaphus xanthopygus* from the Wandashan Mountains, Northeastern China. Acta Theriologica, 52(4): 435-446.

Johnson C N, Isaac J L, Fisher D O. 2007. Rarity of a top predator triggers continent-wide collapse of mammal prey: dingoes and marsupials in Australia. Proceedings of the Royal Society of London B: Biological Sciences, 274(1608): 341-346.

Kaiser-Bunbury C N, Mougal J, Whittington A E, et al. 2017. Ecosystem restoration strengthens pollination network resilience and function. Nature, 542(7640): 223.

Karanth K U, Nichols J D, Kumar N S, et al. 2004. Tigers and their prey: predicting carnivore densities from prey abundance. Proceedings of the National Academy of Sciences of the United

States of America, 101(14): 4854-4858.

Kerley L L, Goodrich J M, Miquelle D G, et al. 2002. Effects of roads and human disturbance on amur tigers. Conservation Biology, 16(1): 97-108.

Leemans H B J, Groot R S D. 2003. Millennium ecosystem assessment: ecosystems and human well-being: a framework for assessment. Physics Teacher, 34(9): 534.

Li Z, Kang A, Gu J, et al. 2016. Effects of human disturbance on vegetation, prey and Amur tigers in Hunchun Nature Reserve, China. Ecological Modelling, 353: 28-36.

Lindsey P, Tambling C J, Brummer R, et al. 2011. Minimum prey and area requirements of the vulnerable cheetah *Acinonyx jubatus*: implications for reintroduction and management of the species in South Africa. Oryx, 45(4): 587-599.

Liu F, Mcshea W J, Garshelis D L, et al. 2011. Human-wildlife conflicts influence attitudes but not necessarily behaviors: factors driving the poaching of bears in China. Biological Conservation, 144(1): 538-547.

Lugo A E. 2010. Visible and invisible effects of hurricanes on forest ecosystems: an international review. Austral Ecology, 33(4): 368-398.

Madden F M. 2008. The growing conflict between humans and wildlife: law and policy as contributing and mitigating factors. Journal of International Wildlife Law and Policy, 11: 189-206.

May R M. 1977. Thresholds and breakpoints in ecosystems with a multiplicity of stable states. Nature, 269(5628): 471-477.

Miller C S, Hebblewhite M, Petrunenko Y K, et al. 2013. Estimating amur tiger (*Panthera tigris altaica*) kill rates and potential consumption rates using global positioning system collars. Journal of Mammalogy, 94(4): 845-855.

Miller C S, Hebblewhite M, Petrunenko Y K, et al. 2014. Amur tiger (*Panthera tigris altaica*) energetic requirements: implications for conserving wild tigers. Biological Conservation, 170: 120-129.

Mysterud A, Østbye E. 1999. Cover as a habitat element for temperate ungulates: effects on habitat selection and demography. Wildlife Society Bulletin, 27(2): 385-394.

Nolet B A, Fuld V N, Rijswijk M E C, et al. 2006. Foraging costs and accessibility as determinants of giving-up densities in a swan-pondweed system. Oikos, 112(2): 353-362.

Palomares F, Ferreras P, Fedriani J M, et al. 1996. Spatial relationships between Iberian lynx and other carnivores in an area of south-western Spain. Journal of Applied Ecology, 33(1): 5-13.

Parker K L, Robbins C T, Hanley T A. 1984. Energy expenditures for locomotion by mule deer and elk. The Journal of Wildlife Management, 48(2): 474-488.

Parmesan C. 2006. Ecological and evolutionary responses to recent climate change. Annual Review of Ecology Evolution and Systematics, 37(1): 637-669.

Pounds J A, Fogden M P L, Campbell J H. 1999. Biological response to climate change on a tropical mountain. Nature, 398(6728): 611-615.

Powell R A. 2000. Animal home ranges and territories and home range estimators. *In*: Research Techniques in Animal Ecology: Controversies and Consequences. Cambridge: Colombia University Press: 64-110.

Qi J, Shi Q, Wang G, et al. 2015. Spatial distribution drivers of Amur leopard density in northeast China. Biological Conservation, 191: 258-265.

Reimoser F, Gossow H. 1996. Impact of ungulates on forest vegetation and its dependence on the silvicultural system. Forest Ecology and Management, 88(1): 107-119.

Richman S E, Lovvorn J R. 2009. Predator size, prey size and threshold food densities of diving ducks: does a common prey base support fewer large animals? Journal of Animal Ecology, 78(5): 1033-1042.

Ripple W J, Estes J A, Beschta R L, et al. 2014. Status and ecological effects of the world's largest carnivores. Science, 343(6167): 1241484.

Rodrigues M E, Roqueb F O, Quintero J M O, et al. 2016. Nonlinear responses in damselfly community along a gradient of habitat loss in a savanna landscape. Biol Conserv, 194: 113-120.

Rompré G, Boucher Y, Bélanger L, et al. 2010. Conserving biodiversity in managed forest landscapes: the use of critical thresholds for habitat. Forestry Chronicle, 86(5): 589-596.

Rooney T P, Wiegmann S M, Rogers D A, et al. 2004. Biotic impoverishment and homogenization in unfragmented forest understory communities. Conservation Biology, 18(3): 787-798.

Root T L, Price J T, Hall K R, et al. 2003. Fingerprints of global warming on wild animals and plants. Nature, 421: 47-60.

Rozhnov V V, Chistopolova M D, Lukarevskii V S, et al. 2015. Home range structure and space use of a female amur leopard, *Panthera pardus orientalis* (Carnivora, Felidae). Biology Bulletin, 42(9): 821-830.

Sasaki T, Okayasu T, Jamsran U, et al. 2008. Threshold changes in vegetation along a grazing gradient in Mongolian rangelands. Journal of Ecology, 96(1): 145-154.

Scheffer M, Carpenter S R. 2003. Catastrophic regime shifts in ecosystems: linking theory to observation. Trends in Ecology and Evolution, 18(12): 648-656.

Seddon A W R, Macias-Fauria M, Long P R, et al. 2016. Sensitivity of global terrestrial ecosystems to climate variability. Nature, 531(7593): 229-232.

Serrouya R, Wittmann M J, McLellan B N, et al. 2015. Using predator-prey theory to predict outcomes of broadscale experiments to reduce apparent competition. The American Naturalist, 185(5): 665-679.

Siemann E, Tilman D, Haarstad J, et al. 1998. Experimental tests of the dependence of arthropod diversity on plant diversity. The American Naturalist, 152: 738-750.

Smith F W, Long J N. 1987. Elk hiding and thermal cover guidelines in the context of lodgepole pine stand density. Western Journal of Applied Forestry, 2(1): 6-10.

Spooner P G, Allcock K G. 2006. Using a state-and-transition approach to manage endangered *Eucalyptus albens* (White Box) woodlands. Environmental Management, 38(5): 771-783.

Sugimoto T, Aramilev V V, Nagata J, et al. 2016. Winter food habits of sympatric carnivores, Amur tigers and Far Eastern leopards, in the Russian Far East. Mammalian Biology-Zeitschrift für Säugetierkunde, 81(2): 214-218.

Tremblay J P, Huot J, Potvin F. 2006. Divergent nonlinear responses of the boreal forest field layer along an experimental gradient of deer densities. Oecologia, 150(1): 78-88.

Tucker M A, Rogers T L. 2014. Examining the prey mass of terrestrial and aquatic carnivorous mammals: minimum, maximum and range. PLoS ONE, 9(8): e106402.

Tylianakis J M, Didham R K, Bascompte J, et al. 2008. Global change and species interactions in terrestrial ecosystems. Ecology Letters, 11(12): 1351-1363.

van der Ree R, Bennett A F, Gilmore D C. 2004. Gap-crossing by gliding marsupials: thresholds for use of isolated woodland patches in an agricultural landscape. Biological Conservation, 115(2): 241-249.

Visser M. 1998. Warmer springs lead to mistimed reproduction in great tits (*Parus major*). Proceedings of the Royal Society of London B: Biological Sciences, 265(1408): 1867-1870.

Vogt N D, Pinedo-Vasquez M, BrondãZio E S, et al. 2015. Forest transitions in mosaic landscapes: smallholder's flexibility in land-resource use decisions and livelihood strategies from World War

II to the present in the amazon estuary. Society and Natural Resources, 28(10): 1043-1058.

Ward R M P, Krebs C J. 1985. Behavioural responses of lynx to declining snowshoe hare abundance. Canadian Journal of Zoology, 63(12): 2817-2824.

Yang H, Dou H, Baniya R K, et al. 2018a. Seasonal food habits and prey selection of Amur tigers and Amur leopards in northeast China. Scientific Reports, 8(1): 6930.

Yang H, Zhao X, Han B, et al. 2018b. Spatiotemporal patterns of Amur leopards in northeast China: influence of tigers, prey, and humans. Mammalian Biology, 92: 120-128.

第12章 东北虎精准保护管理工作中的挑战、机遇与举措

12.1 引 言

中国是世界上虎的发源地,也是世界上虎亚种和豹亚种分布最多的国家。虎豹种群的健康稳定,体现了食物链和生态过程的完整,意味着特定自然生态系统还保持原真性和完整性,标志着整个自然生态系统健康稳定,生态服务功能正常发挥。20 世纪 50 年代,我国仍有近 200 只野生东北虎广泛分布于东北林区,而当时俄罗斯仅分布有 30~40 只。经粗略评估,20 世纪 90 年代末,中国境内分布的虎种群数量,已经减少至不到 20 只(Jiang et al., 2017),俄罗斯虎种群数量通过保护和恢复却增加到 530 只左右(2015 年俄罗斯官方发布数据)。

庆幸的是,由于天然林保护工程的实施,我国东北虎豹的栖息地得到逐步恢复,种群数量也稳步上升。尽管如此,这两个物种依然面临着可能灭绝的危险,其主要威胁因素有:①栖息地破碎化严重,种群分布区隔离,特别是全球仅有的两个东北虎种群在俄罗斯境内相互隔离,种群基因交流困难;②我国境内由于高速公路、铁路、村屯、农田等人为景观的阻隔,完达山、老爷岭和张广才岭种群之间没有有效的连通,而老爷岭南部珲春—汪清—绥阳的中俄边境区域东北虎种群密度较高,向内陆扩散不畅,种群面临近亲繁殖、疾病暴发乃至种群崩溃的风险;③历史上的森林过度采伐留下的影响,当前林下放牧、非木质林产品的采集与开发等人为干扰活动依然强烈;④由于振兴东北经济需要和国防需要,边境围栏、铁路和公路等道路、农田开发、矿业开采等基础设施建设仍发展迅猛;⑤东北虎豹分布区内猎物种类不全、分布不均匀,虎豹偏好猎物密度偏低;⑥局部区域依然存在非法狩猎的威胁。总体上,从大尺度栖息地景观来看,扩散障碍较大,森林景观可渗透性不强,阻碍东北虎和东北豹的自由迁移扩散与恢复,从人兽关系来看,东北虎豹的恢复和当地经济发展需求之间仍存在矛盾(马建章等,2015)。

作为森林生态系统中的顶级捕食者,东北虎活动范围大,它的保护恢复涉及栖息地恢复与连通、猎物种群监测与恢复、盗猎的控制、人虎冲突的缓和等问题,解决这些问题涉及的管理和技术需求复杂,也需要国际合作的帮助和推进,可谓任重道远。下面三方面的工作至关重要。

12.2 顶层设计、统一规划

东北虎保护需要建立保护地斑块连接网络，建设国际和国内廊道，形成可渗透性栖息地生态景观。

中国东北虎目前主要分布于老爷岭、张广才岭和完达山3个区域（图12.1）。我国在东北虎豹分布关键区域的老爷岭区域建立了吉林珲春东北虎国家级自然保护区、黑龙江老爷岭东北虎国家级自然保护区、吉林汪清国家级自然保护区、黑龙江穆棱红豆杉国家级自然保护区、黑龙江鸟青山省级自然保护区、吉林天桥岭东北虎省级自然保护区等，在完达山区域建立了黑龙江东方红湿地国家级自然保护区、黑龙江大佳河省级自然保护区、黑龙江七星砬子东北虎国家级自然保护区等，在张广才岭区域建立了吉林黄泥河自然保护区、黑龙江宁安小北湖国家级自然保护区等，覆盖了3个区域东北虎豹出现最多的核心区域。而这3个大区域之间，正在筹建的黑龙江东京城东北虎自然保护区位于老爷岭和张广才岭的连接处，在进一步巩固了中国内陆虎种群分布栖息地的同时也提供了东北虎种群由张广才岭向北扩散的机会；黑龙江完达山西部的黑龙江七星砬子东北虎国家级自然保护区及周围林业局的栖息地恢复也将有效促进完达山东部虎种群向西部的扩散，进而实现张广才岭和老爷岭东北虎种群的连接。因此，中国东北虎豹分布区内的自然保护区网络框架已基本形成。

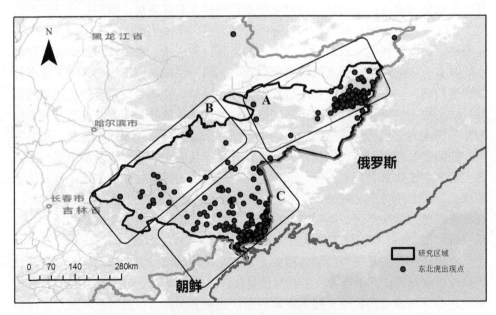

图12.1 研究区地理位置示意图（彩图请扫封底二维码）
A. 完达山区域；B. 张广才岭区域；C. 老爷岭区域

虽然中国东北虎总栖息地范围已达 137 085km²，但近 5 年我国累计记录到的东北虎个体的数量仅接近 40 只，且绝大多数个体分布在老爷岭中俄边境区域。

世界上野生东北虎主要在两个孤立的区域生存：一为俄罗斯锡霍特-阿林（Sikhote-Alin）山脉及与其连接的中国完达山区域，这一地区东北虎数量约占整个种群的 90%（约 500 只）；另一区域则为俄罗斯滨海边疆区（Primorskii Krai）西南地区及与之相连的中国老爷岭，该区域现存的这个孤立的东北虎小种群（40多只）是我国东北虎种群的主要来源（图 12.2）。这两个区域已经由于城市发展和湿地阻隔，在俄罗斯境内被分隔已久（Henry et al., 2009）。相关研究表明，南部这个独立的小种群存在近亲繁殖、已达到环境容纳量、易感疾病等问题，存在灭绝风险（Miquelle et al., 2015）。由于滨海边疆区西南地区东南侧就是日本海，因此这个小种群数量增加时，只能往中国腹地方向扩散，并且也只能通过中国黑龙江省境内的完达山区域与俄罗斯北部约 500 只的大种群连接（图 12.2）。东北虎豹国家公园覆盖了东北虎在中国境内老爷岭分布区的核心区域，有效保护了与俄罗斯相连的小的隔离种群（图 12.2），其建立将有效打开边境廊道，庇护东北虎繁殖种群，尤其对只分布于俄罗斯西南滨海边疆区域和中国老爷岭区域的东北豹种群的保护意义更加重大，但这并不意味着，只要保护好东北虎豹国家公园中的虎豹，中国的东北虎豹就能有效恢复，因为对于东北虎来说，东北虎豹国家公园仅约占总分布区的 11%（图 12.2），并且没有和俄罗斯北部 500 多只的大种群相连；对于

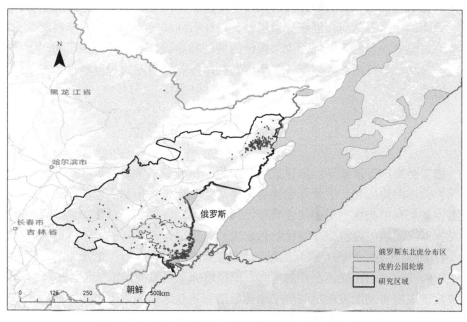

图 12.2　东北虎种群分布图（彩图请扫封底二维码）
红点为 2000 年起东北虎出现点

东北豹来说，国家公园内密度较高的东北虎种群也一定程度上通过种间竞争限制了其数量恢复，东北豹也需要更多的栖息地。国际专家的研究成果表明，保护大型猫科动物最有效的手段是创造能保证其自由移动的有效连通廊道和可渗透的栖息地景观（Wikramanayake et al., 2011）。

当我们把保护注意力放到中国的整个虎豹分布区而非仅仅是东北虎豹国家公园时，就能很自然地意识到，更重要的保护举措应该是考虑如何加强我国东北虎豹种群栖息地斑块（老爷岭、张广才岭、完达山甚至小兴安岭）间的连接，尤其是老爷岭区域的东北虎种群与俄罗斯北部大种群的连通。俄罗斯北部大种群扩散到中国境内后，就要打通完达山东部到完达山西部、完达山西部到张广才岭、张广才岭到老爷岭的通道，或完达山东部到完达山西部再直接到老爷岭的通道（图12.3）。只有这样，中国主要的东北虎种群才能与俄罗斯北部大种群相连接。但是多年的监测发现，由于建鸡高速及道路两侧大片的农田，俄罗斯北部大种群的个体进入完达山东部后，在往完达山西部扩散时就已经遇到严重阻碍。2017年9月，一只雄性成年东北虎在建鸡高速东侧徘徊了半个月才找到公路两侧的一小片林带，从而穿过了高速公路来到完达山西部的桦南林区。2017年11月，另一只从俄罗斯入境完达山东部的成年雄性东北虎试图向完达山西部扩散，却始终没有找到能穿越建鸡高速的区域，沿着高速围栏徘徊几日后无功而返。

目前，东北虎豹国家公园范围以外的区域，保护力度小，并受体制等因素制约，更容易忽视栖息地间的连通问题。要解决这一问题迫切需要国家层面协调交通、林业、环保等部门开展统一顶层设计，在巩固和完善国家公园与自然保护区网络的基础上，大力开展虎豹生态廊道的规划和建设，切实打开国际和国内虎豹廊道，建立一个可渗透的景观，保障虎豹在国际和国内的大尺度栖息地内自由迁移、活动。

12.3　专家咨询、科学规范

建立坚实的专家团队，以保证保护与管理措施的科学、先进和规范，从而实现保护技术的国际化。"不谋全局者，不足谋一域"，东北虎豹的保护是一个系统工程，只有中央统一思想，主管部门科学统筹，各行业多部门协调参与，并充分发挥专家团队的力量，才能推动东北虎豹保护工作有序开展。

自2000年起，我国专家开始建立虎豹监测信息网络，在不断加强与国内外政府组织和非政府组织交流合作的同时，积极培训基层监测技术队伍。2011年开始逐步尝试应用自动触发相机与分子遗传学等先进技术监测中国虎豹种群和栖息地的动态变化。目前，国家林业和草原局猫科动物研究中心已经建立起了信息网络汇报信息数据库、虎豹影像花纹数据库、虎豹DNA数据库、数码足迹信息数据

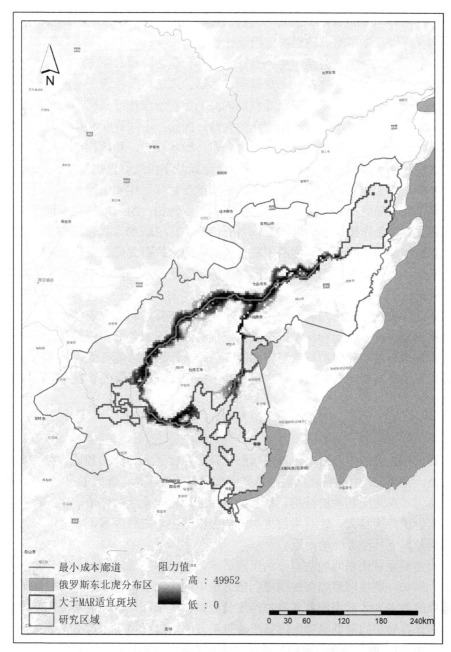

图 12.3 东北虎核心栖息地和廊道分析（彩图请扫封底二维码）

库、猎物种群数量和分布数据库、虎豹疾病与寄生虫数据库。2000~2016 年，国家林业局虎豹监测数据库记录了 4028 次虎捕食、足迹、影像、粪便等信息位点，确定的空间分布和个体扩散信息表明，东北虎分布区正在稳定扩张，东北豹也呈现相似的趋势（Jiang et al., 2017）。这些数据和成果对东北虎豹国家公园的规划、设计意义重大，是国家虎豹保护工程设计的重要依据。

虎豹保护涉及生态公路、生态友好型森林、生态补偿、生态廊道等领域，也需要社会学、经济学、建筑学、交通学、林学、地理信息学等学科的专家的参与（Jiang et al., 2017）。我国目前已经获得的虎豹保护阶段性成果离不开各行业的专家团队，未来更需要跨学科科研团队的支持。因此，不同行业和部门，要打破体制的束缚，统一思想，科学规划，通力合作，稳步推进，积极调动全社会的力量共同关注和广泛参与东北虎豹保护工作，让全民共享由虎豹保护带来的生态红利。

此外，由于虎豹跨中俄分布，虎豹的保护与管理工作要国际化、标准化，特别需要与俄罗斯等虎豹分布国的专家深入合作。例如，由国家林业局批准，在联合国亚太经济合作组织委员会东亚项目的支持下，中俄双方联合开展了东北虎、东北豹跨境移动研究项目，为两国联合开展虎豹保护研究建立了良好的国际专家队伍平台，也为将东北虎豹国家公园的保护与管理提升到国际水平提供了技术支持（Shevtsova et al., 2018）。

总之，虎豹保护研究需要稳定的科研团队，在持续积累生态学数据的同时，还要建设好高校野生动物保护国家级重点学科，为国家各级保护管理部门不断地培养和输送野生动物保护与管理方面的专门人才，这样才能有效保障国家管理部门遵循野生动物管理学方法和野生动物的生态规律进行科学的管理决策。

12.4 创新举措、人虎和谐

探究东北虎的生存需求，开展精细化评估，促进人与虎豹和谐共存，实现生态安全。现阶段，中国东北虎豹种群资源状况已基本摸清，管理工作也初步理顺。我们急需对东北虎豹种群结构、栖息地质量进行科学系统的评估，以促进东北虎豹种群健康持续壮大。针对当前虎豹种群面临的主要威胁因素，我们下一步要重点搞清楚以下几个方面的问题。

一是需要量化东北虎豹真正的生态需求，才能保证采取的保护措施是科学有效的。例如，要搞清楚如何控制虎豹栖息地中的人为干扰强度，首先要知道人为干扰强度到多大才对虎豹及其猎物产生负面影响。又如，研究表明，东北虎能够繁殖和抚育幼崽的最低猎物密度需求是每平方千米的有蹄类猎物不少于0.5头（Li et al., 2010），那么现在，虎豹分布区内还有多大区域不能达标，这些区域都在哪里；保护区或国家公园要控制人为干扰强度及其影响范围到多大，才能保证不影响虎豹和猎物；虎豹最小栖息地斑块需要多大；国内国际廊道在哪里，要多宽才能使东北虎豹国家公园内的虎豹种群由边境向我国内陆纵深扩散；这些廊道要如何建设；等等。解决这些保护与管理工作中迫切需要解决的问题，就需要真正了解东北虎豹的生态需求。

二是在解决上述管理问题和技术标准问题的基础上，还要考虑如何在有效保护虎豹的同时兼顾人的经济发展需求。要对虎豹分布区的保护区内外都进行精细化评估，对于面临的不同威胁有针对性地采取栖息地恢复和猎物恢复措施，找到适合当地的绿色经济发展之路，努力让当地居民在虎豹保护工程中受益，从而调动他们参与保护和从事适当的生态经济活动的积极性。

三是根据管理和评估需要，建立创新性的野生动物和栖息地恢复监测体系。尽管天地空一体化监测的设计结合了多种现代先进技术，局部区域的信息监控设施建设也是必要的，但需注意，一定要将监测措施和监测活动给植被与野生动物带来的负面影响降到最低，监测是手段不是目的，应根据监测目标和需求，尽可能地应用或研发对环境和野生动物无损伤或损伤性弱的监测技术手段，并尽量减少人为活动频次和人工设施带来的噪声、辐射等，以确保虎豹和猎物乃至森林的生态安全（Jewell，2013）。

四是要研究建立以国家公园为主体的自然保护地体系。国家公园是最重要的自然保护地类型，但不是唯一的，还需要有其他自然保护地作为补充，自然保护区、森林公园、天然林保护等保护管理措施只能加强不能削弱。

12.5 本章小结

综上所述，为保护东北虎豹种群资源及其所在的自然生态系统，应遵循东北虎豹种群可持续生存与发展的保护生态学规律，并建立"管理体制与技术创新统一平台"，利用当前我国东北地区天然林禁伐、棚户区改造、国家公园建设等工程开展的良好机遇，创新保护举措，筹建大空间尺度的东北虎豹自然保护区网络，促进我国东北虎豹栖息地质量的全面提升，逐步实现老爷岭、张广才岭与完达山东北虎种群的有效连通，实现人与虎豹的和谐共存，让东北虎豹成为我国重要的生态、经济和文化财富。

参 考 文 献

马建章, 张明海, 姜广顺, 等. 2015. 我国老虎及其栖息地保护面临的挑战与对策. 野生动物学报, 36(2): 129-133.

Henry P, Miquelle D, Sugimoto T, et al. 2009. In situ population structure and ex situ representation of the endangered Amur tiger. Molecular Ecology, 18(15): 3173-3184.

Jewell Z. 2013. Effect of monitoring technique on quality of conservation science. Conservation Biology the Journal of the Socicty for Conservation Biology, 27(3): 501-508.

Jiang G, Wang G, Holyoak M, et al. 2017. Land sharing and land sparing reveal social and ecological synergy in big cat conservation. Biological Conservation, 211: 142-149.

Li Z, Zimmermann F, Hebblewhite M. 2010. Study on the Potential Tiger Habitat in the Changbaishan Area, China. Beijing: China Forestry Publishing House.

Miquelle D G, Smirnov E N, Zaumyslova O Y, et al. 2015. Population dynamics of Amur tigers (*Panthera tigris altaica*) in Sikhote-Alin Biosphere Zapovednik: 1966-2012. Integrative Zoology, 10(4): 315-328.

Shevtsova E, Jiang G, Vitkalova A, et al. 2018. Saving the Amur tiger and Amur leopard. NEASPEC.

Wikramanayake E, Dinerstein E, Seidensticker J, et al. 2011. A landscape-based conservation strategy to double the wild tiger population. Conservation Letters, 4(3): 219-227.

Chapter 1 Environmental factors drive the habitat distribution of Amur tiger

1.1 Introduction

Anthropogenic and environmental factors drive range loss in mammals (Shenko *et al.*, 2012), and the human pressure index is a reasonable proxy for threats to large mammals (Yackulic *et al.*, 2011). Spatial variation of environmental conditions induces variation in the size and demography of local populations, influencing their ability to tolerate anthropogenic threats (Lawton, 1993; Channell and Lomolino, 2000). Edge populations are more vulnerable than core populations (Yackulic *et al.*, 2011) and the distribution of a species often starts shrinking from its edge. Amur tigers (*Panthera tigris altaica*) are the largest feline (Kitchener and Yamaguchi, 2010). They are adaptive animals and important ecological actors in their ecosystems (Seidensticker, 1996). In China, Amur tiger range was relatively large historically, spreading over the forests and grasslands: from the northern Greater Khingan Mountains to the southern Yanshan mountains, and from the eastern Wusuli River to the western Erguna River (Ma, 2005; Tian *et al.*, 2009). Forested regions of Changbaishan, Wandashan, as well as the Greater and Lesser Khingan mountains were considered the core of the population (Kang *et al.*, 2010). In the mid-1950s, there were approximately 200 individuals left in China and probably less than 50 in Russia (Tian *et al.*, 2009). The distribution of the species in northeast China has drastically decreased over the last 60 years and by the end of the 20th century approximately only 20 individuals seem to survive in two core habitats, i.e., south of Laoyeling and east of Wandashan near the Sino-Russia border (Li *et al.*, 2008; Zhou *et al.*, 2008). On the positive side of this drastic population reduction, frequent observations recorded in these areas over the past 10 years seem to indicate that these populations are stable and increasing (Li *et al.*, 2008; Zhou *et al.*, 2008). Furthermore, large patches of forest remain in these two subregions and could certainly sustain healthy populations of tigers, as well as their ungulate species prey. At the same time, the tiger population in Russia has increased and there is now estimated to be just fewer than 500 individuals (Carroll and Miquelle, 2006).

According to Hemmer (1976), China was the center of tiger evolution and this hypothesis has been partially confirmed by a recent genetic study (Luo *et al.*, 2004). If the current forested areas are maintained and managed properly, the tiger population may recover in these regions and play a central role in the long-term conservation of the entire Amur tiger population (Kang et al., 2010). Urgent needs for successful field conservation practices include knowing the distribution of suitable habitat that could be further re-colonized by the tiger and what the factors influence habitat suitability and tiger presence.

Li et al. (2010) and Hebblewhite et al. (2012) combined inferences from (a) an expert based tiger habitat model (Luan et al., 2011), (b) an environmental niche factor analysis (ENFA) on tiger presences and pseudo-absences from an adjacent area of the Russian Far East, and (c) a resource selection function (RSF) on the same data as the ENFA model. However, they did find important potential habitat patches for the Amur tiger in Changbaishan in China by average prediction of the three models. We recognized that they used data based on a robust sampling survey in the adjacent Russian Far East to predict the Chinese habitat suitability. However, they did not use Chinese Amur tiger occurrence information to validate the rusticity of the prediction results and predicted only one of two main Amur tiger habitats in China. Furthermore, the researches mentioned above identified some positive or negative effects of habitat factors, but they did not provide quantitative responses to these factors, especially human disturbance. According to their results, conservation managers cannot control, mitigate or manage these disturbances effectively. Therefore, it is urgent for further inclusive assessment on Chinese Amur tiger habitat suitability and to quantitatively identify effects of threatened factors based on occurrence data of tigers within this area. In addition, wildlife population distribution often showed various clustering patterns, for the species like Amur tiger which always has large home range, the distribution of their habitat patches may pose the impact of spatial pseudo-replication sampling on habitat suitability prediction (José and Mariana, 2002). It is also an urgent technical problem for scientific identification of suitable habitat for species.

In this study, based on the actual situation of the Amur tiger distribution in China, we hypothesized that Amur tiger occurrence in China is associated with diffusion from the source population in Russia, local climatic, topographic, vegetation and anthropogenic factors. Furthermore, we hypothesized that spatial models will make more biologically sound habitat suitability predictions. Then, we expect we can get a robust habitat suitability rank map, with currently occupied tiger habitat as a function of distance to Sino-Russia border and make the quantitative relationships between tiger occurrences and environmental or anthropogenic factors.

1.2 Materials and methods

1.2.1 Study area

We restricted our study area to 158 564km^2 with north latitude of 42.358°–47.605°, east longitude of 126.631°–134.290° based on the Amur tiger occurrence information range of China from the year 2000 to 2012. This includes the Changbaishan areas of 107 604km² with north latitude of 42.358°–46.655°, east longitude of 126.659°–131.708° and the Wandashan areas of 50 764km² with north latitude of 45.027°–47.605°, east longitude of 130.257°–134.277°(Fig. 1.1). The Changbaishan tiger range connects to the southwest Primorye tiger range in Russia, while the Wandashan range connects to the Sikhote-Alin mountain range in Russia.

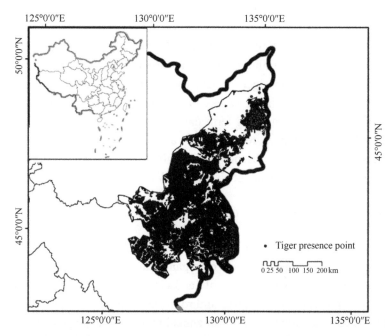

Fig. 1.1 Illustration of the research area in Northeast China. Black areas were covered by natural forest. Please scan the QR code on the back cover to see the color image.

The climate of the research area is temperate continental monsoonal precipitation. Vegetation types include Korean pine *Pinus koraiensis* mixed with deciduous forests, meadows, natural shrub lands, coniferous plantation forests and agricultural areas. The majority of forests have been logged, and many low-elevation forests have been converted to secondary deciduous forests (Hebblewhite et al., 2012). Far Eastern leopards *Panthera pardus orientalis*, wolves *Canis lupus*, Eurasian lynx *Lynx lynx*, red deer *Cervus elaphus*, wild boar *Sus scrofa*, sika deer *Cervus nippon* and roe deer *Capreolus pygargus* and other rare and diverse animal species coexist with the Amur tiger in this area.

1.2.2 Amur tiger population data

Since the year 2000, experts and staff of both forest management departments and nature reserves collected Amur tiger presence information based on field surveys (Li et al., 2001; Zhou et al., 2008), livestock depredation (Li et al., 2009), and information network collection work (Zhang et al., 2012). Other signs of presence were recorded as well, such as the front pad width of tiger footprints in snow or mud, occurrence GPS coordinates and times that tigers preyed on ungulate or livestock.

To reveal the real factors threatening the tiger and to test for subregional differences in habitat associations, we used the mean daily travel distance of female tigers (i.e., 7km) as half width to be the square scale for regional presence modeling of Amur tiger (Goodrich et al., 2005; Carroll and Miquelle, 2006). Because the data were collected from opportunistic findings of forest workers active in forest harvesting, we used one time records from tiger occurrence records repeatedly within one pixel (14km×14km = 196km^2) within 2 weeks to minimize the risk of pseudo-replication (Hurlbert, 1984). Based on the long-term data collection, from 2000 to 2012, 584 of 600 signs of tiger occurrences remained for analysis. We then counted occurrence frequency, defined presence or absence of each pixel, and treated the sample units into two subregional datasets, i.e., Changbaishan and Wandashan datasets.

1.2.3 Environmental and anthropogenic variables

Spatial distribution modeling must account for characteristic scales of habitat factors associated with different levels of organization of the species (Latimer et al., 2006). We considered five categories of landscape-scale habitat variables: climate; Russian tiger population; vegetation; anthropogenic factors; topography and river (Table 1.1). Climatic variables (i.e., mean annual max snow depth for the years 2000–2010) derived from China's deep snow time series datasets (1978–2010) with 25km resolution and re-sampled to a 14km resolution (http://www.datatang.com/data/16192). Distance of the center of each pixel to the Sino-Russia border as source population proxy, meaning where tiger populations are connected to the Russian tiger population through immigration or emigration from China. Because different vegetation types may influence the Amur tiger presence (Carroll and Miquelle, 2006), vegetation was classified mainly as vegetation types typical to northeastern China, i.e., deciduous forest, coniferous forest, and mixed coniferous and deciduous forest; however, mixed type forest is small proportionally and not considered here. Most forest vegetation is beneficial for tiger presence relative to farmlands or other land types, so we considered forest occurrence as one variable. All forest type vector data were converted into a grid with 1km resolution. Furthermore, because of the importance of ungulate prey within carnivore habitat (Karanth et al., 2004), we used the forest area occurrence proportion of each pixel as proxy for the forest ungulate index because of the absence of standardized and reliable prey density data in China across such a large range.

Table 1.1 Habitat variables tested for association with Amur tiger (*Panthera tigris altaica*) presence in northeastern China

Habitat factor	Description of the habitat factor	Data type	Unit
Climate			
Snow depth	Mean annual max snow depth for 2000-2010 year (source grid has 25km resolution)	Continuous	cm
Russia tiger population			
Distance to Sino-Russia border	Distance from the central point of each pixel (14km×14km) to Sino-Russia boarder line segments where Amur tigers often migrate with each other of two countries observed and confirmed during recently 10 years. One line segment located in Wandashan, another one in Changbaishan	Continuous	m
Vegetation			
Forest	Forest area including all types of forest types of each pixel. The vegetation vector map of China (1 : 1 million) in 2000 year was converted into grid with 1km resolution	Continuous	km^2
Coniferous	Coniferous area of each pixel. Source is the same to forest with 1km resolution	Continuous	km^2
Deciducus	Deciduous area of each pixel. Source is the same to forest with 1km resolution	Continuous	km^2
Anthropogenic factors			
Farmland area	Farmland areas proportion of each pixel	Continuous	km^2
Road density	Total length of road in each pixel. Road includes primary high-ways, secondary highways, improved light-duty paved roads, improved light duty gravel roads, and improved light-duty dirt roads	Continuous	km
Distance to railway	Distance from the central point of each pixel to railway	Continuous	m
Distance to village	Distance from the central point of each pixel to village	Continuous	m
Distance to city	Distance from the central point of each pixel to city (including small town)	Continuous	m
Topography and river			
Elevation	Elevation grid with 1km resolution	Continuous	m
Elevation SD	Standard deviation of elevation of each pixel. Source elevation grid with 1km resolution	Continuous	m
Slope	Slope grid with 1km resolution derived from the digital elevation model above	Continuous	(°)
Aspect	Aspect grid with 1km resolution derived from the digital elevation model above	Continuous	
Viewshed	Viewshed analysis indicates not only what areas of a surface can be seen by one or more observers, but also, for any visible position, how many observers can see the position. The points are the center of each pixel	Continuous	%
River density	Total length of river in each pixel. The river includes the primary river and their branches	Continuous	km
Spatial autocorrelation (SA)	Spatial autocovariance term tested in the "spatial" models. Each sample unit observation (0,1) within 342km (in Changbaishan + Wandashan), 162km (in Changbaishan) and 94km (in Wandashan) of the reference sample unit was weighted by its inverse squared distance to the sample unit, normalized by the sum of weights for all sample units in the 342, 162 and 94km region, respectively	Continuous	%

Note: Data were processed in ArcGIS 9.0 in grid format (14km×14km pixel) based on re-sampled or interpolated measurements. Data were extracted from the Environmental and Ecological Science Data Center for West China (EESDCWC)(website: www.resdc.cn), National Natural Science Foundation of China (NSFC) and Information Center of Amur-Heilong River Basin (ICAHRB)(http://amur-heilong.net/) databases.

Anthropogenic factors included farmland encroachment indexed by farmland proportion of each pixel, the distance of the center of each pixel to the nearest road, railway, village and city. Poaching may be an important factor driving tiger distribution, but we do not have spatial data to represent this and anthropogenic factors may be correlated with poaching access, i.e., roads (Bennett and Robinson, 2000; Kerley *et al.*, 2002). All distance measurements were completed by GIS spatial analyst package with ArcGIS 9.0 software. Topography and river variables were derived from the digital elevation model with 1km resolution and river distribution GIS shape data, including elevation, standard deviation of elevation, slope, aspect and viewshed.

1.2.4 Spatial autocorrelation examination

Because of spatial dispersal and behaviors affecting distribution patterns of Amur tiger presence (Goodrich *et al.*, 2005), spatial autocorrelation problems may exist in such clustering distribution of population (José and Mariana, 2002; Haining, 2003). Amur tiger presence is locally clustered and patterns suggest the effect of movement behavior on current tiger distributions in northeastern China. There is a risk when analyzing such spatial distribution patterns that the assumption of independence will be invalidated by spatial autocorrelation (Koenig, 1999), because adjacent habitat pixels may share similar values as the dependent variable values. After the best model was selected, we calculated Moran's I statistic (Cliff and Ord, 1981) for the residuals of the models using SAM version 4.0 (Rangel *et al.*, 2010) to test for spatial autocorrelation in the dependent variables. To measure the magnitude of clustering, we created a spatial autocorrelation (SA) based on the presence or absence of Amur tigers at neighboring pixels (Beard *et al.*, 1999; Segurado and Araujo, 2004; Jiang *et al.*, 2009). The autocovariance term was derived to explicitly model spatial autocorrelation distance range of model residuals where it occurred (Augustin *et al.*, 1996) and calculated from the point pairs of the pixel centers by the inverse of the squared distance to the reference pixels, normalized by the sum of weights for all pixels in the significant autocorrelation distance identified by Moran's I value check as radius (Davis *et al.*, 2007; Jiang *et al.*, 2009).

In this study, when the best predictive models selected at different subregions were ranked and selected in terms of suitability, we diagnosed the spatial autocorrelation problems indicated with fits, then checked the spatial structure in the residuals using randomization tests (Rangel *et al.*, 2010). If there was significant spatial autocorrelation in residuals, we created an SA to model as to no significant spatial structure in the residual occurrence. Thus, Amur tiger occurrence data based on environmental variables measured at a pixel scale of 196km^2 with and without a spatial autoregressive term were modeled. For the two types of modeling, we compared habitat associations and predicted distributions of Amur tigers in both the Changbaishan and Wandashan subregions, and tested the accuracy of the models developed in each region in predicting Amur tiger habitat suitability. Finally, we compared the robustness of the spatial model (with an SA) fits with the non-spatial model (without an SA) fits, and created the prediction rank probability maps of habitat suitability in different subregions, compared and checked the observed frequency rank maps according to Amur tiger occurrence data recorded in recent years.

1.2.5 Statistical analysis

Both Wandashan and Changbaishan have different environmental, anthropogenic and climatic conditions (Chen *et al.*, 2011; Zhou and Zhang, 2011), which may lead differences in characteristics for tiger habitat selection (Goodrich *et al.*, 2010). Davis *et al.* (2007) recommended that regional variations of environmental or anthropogenic factors could be revealed by considering habitat modeling of the subregions separately. Therefore, we decided to treat Wandashan and Changbaishan independently in our modeling procedure, we recorded Amur tiger presence in 808 pixels altogether with 549 pixels in the Changbaishan and 259 pixels in Wandashan Mountains. Our sampling design for this model using a used-available design following Manly *et al.* (2002) and the limitation of this design is such that statistically only yields a relative probability of habitat rankings. The pixels without tiger information were regarded as pseudo-absence. To select the significant variables influencing habitat selection of Amur tiger into the modeling procedure, we conducted pair-wise association between habitat variables and Amur tiger presence using the Wilcoxon rank sum tests by presence and pseudo-absence. Highly skewed variables were normalized using standard transformation methods. All tests at $P<0.05$ were regarded as significant with 10 000 Monte Carlo sampled. Subsequently, all significant variables were entered into a Pearson's correlation matrix to identify co-linearity (i.e., $r>0.5$) (Ramsay *et al.*, 2003). Because the generalized additive model (GAM) can provide a close fit to the original data and produce more reliable models than an equivalent generalized linear model (GLM) with several polynomial terms, it represents the real responses of species to the environment for the data characterizes the non-linear functions in it (Suarez-Seoane *et al.*, 2002). Therefore, we modeled Amur tiger presence as a function of quantitative habitat factors using a nonlinear GAM (Hastie and Tibshirani, 1990; Guisan *et al.*, 2002). To reduce the potential impact of concurvity in the GAM, we limited the number of variables with co-linearity for stepwise selection. After this, we produced multivariate models using stepwise GAM (Hastie and Tibshirani, 1990; Guisan *et al.*, 2002). For subregional datasets, habitat variable selections were conducted as described above. All analyses were undertaken in R version 2.12.1 using the MGCY package version 1.7-2 (Wood, 2006; R Development Core Team, 2010). We used minimization of the generalized cross-validation value (an index of the model's out-of-sample predictive mean squared error, GCV) to determine the optimal roughness of the smoothing terms (Stige et al., 2006). We selected the best-fitting models from candidate models by following the rule of minimization of GCV on the condition that all variables must be statistically significant ($P < 0.05$).

We used the area under the receiver operating characteristic (ROC) curve (AUC) to evaluate the classification skill of the models (Altman and Bland, 1994; Fielding and Bell, 1997). We tested model robustness using fivefold cross validation of the fitted models. We divided both presence and absence data into five equal partitions, withholding one partition for testing, and using the remaining data to fit the model. We repeated the process five times and calculated the mean performance of the five trials (Davis et al., 2007). We compared both the model AUC and cross-validated AUC (CV AUC). Furthermore, based on AUC statistics, we examined the classification results of the best spatial or non-spatial models developed for different regions in predicting Amur tiger presence. Then, we ran the best model to quantify and plot the response curve of Amur tiger presence to environmental and anthropogenic factors, and identified the nonlinear response threshold in the two subregions. We linked the predicted relatively probabilities of Amur tiger presence of spatial or non-spatial models to the observed frequencies using linear regression to examine whether spatial models fits better than non-spatial models according to the biological meanings (Granadeiro et al., 2004).

1.3 Results

Amur tigers were present at 12.25% (99/808) pixels of the whole region, including 11.66% (64/549) pixels of Changbaishan and 13.51% (35/259) pixels of Wandashan in the northeastern China (Fig. 1.1). The longest distance of tiger presence pixel from the Sino-Russia border was 293km. The nearest distance presence point pairs between Wandashan and Changbaishan was 60km (Fig. 1.1), less than the adult male tiger movement maximum diameter of home range (Goodrich et al., 2010).

1.3.1 Changbaishan model

For Changbaishan, slope, elevation, road density, snow depth, elevation standard deviation, distance to Sino-Russia border, as well as distance to railway were significantly associated with Amur tiger presence (Wilcoxon Z, $P < 0.05$) (Table 1.2).

Table 1.2 Univariate tests of association between Amur tiger presence and environmental factors based on the Wilcoxon U-test with 10 000 Monte Carlo sampled

Variable	Changbaishan ($n = 64$)		Wandashan ($n = 35$)	
	Wilcoxon Z	P	Wilcoxon Z	P
Viewshed	−0.342	0.737	**−4.88**	**<0.001**
Aspect	−0.989	0.327	−0.575	0.571
Slope	**−1.977**	**0.047**	**−4.811**	**<0.001**
Elevation	**−2.023**	**0.043**	**−3.261**	**0.001**
Coniferous	−0.678	0.501	**−3.026**	**0.003**
Deciduous	−0.713	0.475	**−4.839**	**<0.001**
Forest	−0.47	0.633	**−5.558**	**<0.001**
Farmland area	−0.345	0.73	**−5.24**	**<0.001**
Road density	**−3.596**	**<0.001**	**−4.668**	**<0.001**
River density	−0.165	0.867	**−3.523**	**<0.001**
Snow depth	**−5.272**	**<0.001**	**−5.695**	**<0.001**
Elevation SD	**−2.143**	**0.031**	**−4.991**	**<0.001**
Distance to city	−0.132	0.895	−0.507	0.618
Distance to Sino-Russia border	**−8.991**	**<0.001**	**−6.789**	**<0.001**
Distance to village	−1.777	0.075	**−3.2**	**0.001**
Distance to railway	**−4.518**	**<0.001**	−1.461	0.142

Note: Bold face refers significant factors.

Among these significant relationships for habitat variables, Amur tiger presence was not directly associated with forest, deciduous forest or farmland (Wilcoxon Z, $P = 0.143$). However, we found that the significant relationships between slope and elevation (+), and elevation standard deviation (+), as well as elevation and elevation standard deviation (+) ($r > 0.5$) (Table 1.3).

We selected these significant associated variables, and then compared the contribution of each of these strong relationship variable pairs one by one. Thus, the elevation, snow depth, distance to Sino-Russia border and distance to railway were entered to combine for GAM and became the best model (deviation explained 29.96%, GCV score 0.0746) (Table 1.4), model residuals (observed-predicted) were significantly positively auto-correlated up 70km (5 pixels) (Fig. 1.2a), mainly because the model over predicted Amur tiger presence in eastern Changbaishan by producing negative residuals (Fig. 1.3d). When we added the spatial autocovariance, the classification skill was improved by an increase in cross-validated AUC from 0.834 to 0.997, and removed the significant spatial autocorrelation of model residuals (Table 1.5, Fig. 1.3e).

Table 1.3 Correlations among habitat variables applied to Amur tiger habitat modeling in Changbaishan subregion of the northeastern China

	Viewshed	Aspect	Slope	Elevation	Coniferous	Deciduous	Forest	Farmland area	Road density	River density	Snow depth	Elevation SD	Distance to city	Distance to Sino-Russia border	Distance to village	Distance to railway
Aspect	0.08	1														
Slope	0.16	−0.08	1													
Elevation	0.08	−0.11	**0.66**	1												
Coniferous	0.16	0.03	−0.05	0.01	1											
Deciduous	−0.13	−0.02	0.36	0.24	−0.39	1										
Forest	0.03	−0.06	0.49	**0.54**	0.17	**0.61**	1									
Farmland area	−0.08	0.06	−0.44	−0.48	−0.18	−**0.55**	−**0.92**	1								
Road density	−0.07	0.02	−0.23	−0.14	−0.06	−0.24	−0.28	0.29	1							
River density	0.17	−0.03	−0.27	−0.24	0.08	−0.29	−0.19	0.16	0.22	1						
Snow depth	−0.12	0.15	−0.26	−0.12	−0.15	−0.11	−0.22	0.25	0.17	−0.12	1					
Elevation SD	0.21	−0.02	**0.95**	**0.62**	−0.04	0.33	0.47	−0.43	−0.25	−0.26	−0.25	1				
Distance to city	0.12	−0.06	0.31	0.42	0.03	0.17	0.3	−0.29	−0.21	−0.16	0.08	0.29	1			
Distance to Sino-Russia border	0.13	0.13	−0.19	−0.27	−0.01	−0.19	−0.22	0.2	0.03	−0.03	0.49	−0.16	0.06	1		
Distance to village	0.17	−0.01	0.38	0.47	0.01	0.19	0.34	−0.33	−0.25	−0.16	−0.13	0.36	0.25	−0.1	1	
Distance to railway	0.16	−0.03	−0.12	−0.2	0.15	0.00	−0.01	−0.08	−0.44	0.07	−0.24	−0.11	−0.03	−0.19	0.01	1

Note: Bold face represents correlation coefficients are significant.

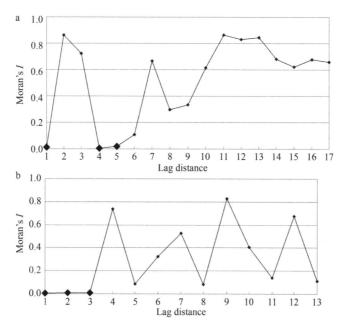

Fig. 1.2 Moran's *I* spatial correlograms of generalized additive model (GAM) residuals from the non-spatial models. The *x*-axis shows the distance between site pairs. Models include a. Changbaishan, b. Wandashan. Significant ($P < 0.05$) values of Moran's *I* are indicated with diamonds. Distance is measured as number of pixels.

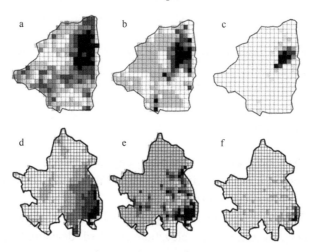

Fig. 1.3 Observed frequencies of occurrence and probabilities of occurrence predicted from generalized additive models (GAMs) for the Amur tiger. Darker squares (14km×14km) represent areas of higher observed or predicted occurrence (the values are given in Table 1.2). a. Predicted probability from non-spatial model of Wandashan; b. Predicted probability from spatial model of Wandashan; c. Observed frequency of Wandashan; d. Predicted probability from non-spatial model of Changbaishan; e. Predicted probability from spatial model of Changbaishan; f. Observed frequency of Changbaishan.

Table 1.4 The five best models resulting from the model selection procedure

Region		Model combination	Deviance explained	GCV score
Changbaishan	1	Elevation + Snow depth + Distance to Sino-Russia border + Distance to railway	0.2996	0.0746
	2	Snow depth + Distance to Sino-Russia border + Distance to railway	0.2945	0.0749
	3	Distance to Sino-Russia border + Distance to railway	0.2820	0.0755
	4	Snow depth + Distance to Sino-Russia border	0.2787	0.0761
	5	Distance to Sino-Russia border	0.2679	0.0765
Wandashan	1	Viewshed + Elevation + Farmland area + Road density + Snow depth	0.3480	0.0817
	2	Viewshed + Elevation + Farmland area + Snow depth + Distance to village	0.3509	0.0818
	3	Viewshed + Elevation + Farmland area + Snow depth	0.3327	0.0832
	4	Viewshed + Elevation + Road density + Snow depth + Distance to village	0.3312	0.0835
	5	Viewshed + Elevation + Road density + River density + Snow depth	0.3331	0.0843

Note: The five best models were used to estimate the probability of the Amur tiger's presence in southern mountains (Changbaishan) and northern mountains (Wandashan) of northeastern China with the model stepwise procedure GCV generalized cross validation.

Table 1.5 Summary of the best spatial and non-spatial generalized additive models (GAMs) to predict Amur tiger presence and absence derived using monitoring data during 2000–2012

Model and regional data	n (prevalence)	Non-spatial GAM			Spatial GAM	
		Variables	AUC	CV AUC	AUC	CV AUC
Changbaishan	64(11.66%)	Elevation, snow depth, distance to Sino-Russia border, distance to railway	0.835	0.834	0.998	0.997
Wandashan	35(13.51%)	Viewshed, elevation, farmland area, road density, snow depth	0.895	0.895	0.998	0.998

Note: AUC. area under the curve; CV AUC. the mean AUC of cross validation of five folds.

1.3.2 Wandashan model

For Wandashan, only aspect, distance to city and distance to railway were not significantly associated with Amur tiger presence (Wilcoxon Z, $P > 0.05$), but 13 other habitat variables were associated (Table 1.2).

We found that significant relationships existed between slope and elevation(+), deciduous forest(+), forest(+) and elevation standard deviation(+); elevation and deciduous forest(+), forest (+) and elevation standard deviation (+); deciduous forest and forest (+), farmland area (−) and elevation standard deviation (+); forest and farmland area (−), and elevation standard deviation (+); distance to Sino-Russia border and road density (+), snow depth (−)($r > 0.5$) (Table 1.6).

We selected these significantly associated variable pairs, and then compared the magnitude of contribution of each variable one by one. Thus, elevation, coniferous

Table 1.6 Correlations among habitat variables applied to Amur tiger habitat modeling in Wandashan subregion of the northeastern China

	Viewshed	Aspect	Slope	Elevation	Coniferous	Deciduous	Forest	Farmland area	Road density	River density	Snow depth	Elevation SD	Distance to city	Distance to Sino-Russia border	Distance to village	Distance to railway
Viewshed	1															
Aspect	0.08	1														
Slope	0.49	0.14	1													
Elevation	0.36	0.16	**0.90**	1												
Coniferous	0.09	0.08	0.34	0.40	1											
Deciduous	0.34	0.22	**0.74**	**0.72**	0.13	1										
Forest	0.38	0.24	**0.81**	**0.78**	0.30	**0.95**	1									
Farmland area	-0.24	-0.03	-0.49	-0.47	-0.21	**-0.61**	**-0.64**	1								
Road density	-0.08	0.04	-0.08	0.02	0.07	-0.21	-0.20	0.46	1							
River density	-0.13	-0.09	-0.31	-0.25	-0.22	-0.23	-0.28	0.23	0.14	1						
Snow depth	0.19	-0.02	0.09	-0.09	-0.03	0.03	0.06	-0.16	-0.20	-0.04	1					
Elevation SD	0.48	0.13	**0.97**	**0.85**	0.32	**0.74**	**0.81**	-0.45	-0.08	-0.29	0.10	1				
Distance to city	0.01	0.03	0.26	0.37	0.19	0.27	0.28	-0.35	-0.22	-0.24	0.03	0.22	1			
Distance to Sino-Russia border	-0.06	0.19	0.18	0.41	0.22	0.15	0.17	0.16	**0.50**	-0.10	**-0.65**	0.16	0.12	1		
Distance to village	0.17	-0.12	0.38	0.31	0.05	0.29	0.32	-0.42	-0.31	-0.23	0.24	0.30	0.28	-0.28	1	
Distance to railway	-0.17	-0.19	-0.10	-0.18	-0.16	0.00	-0.04	-0.30	**-0.68**	-0.05	0.01	-0.10	0.04	-0.43	0.21	1

Note: Bold face represents correlation coefficients are significant.

forest, farmland area, snow depth and distance to village were entered to combine for GAM models and the best model included viewshed, elevation, farmland area, road density and snow depth (deviation explained 34.80%, GCV score 0.0817) (Table 1.4). The model residuals (observed-predicted) were significantly positively auto-correlated up to 42km (3 pixels) (Fig. 1.2b), mainly because the model over predicted Amur tiger presence in western and southern Wandashan producing negative residuals (Fig. 1.3a). When we added the spatial autocovariance, classification skill was improved by an increase in cross-validated AUC from 0.895 to 0.998, and removed significant autocorrelation of model residuals (Table 1.5, Fig. 1.3b).

1.3.3 Subregional comparison and quantitative responses

For univariate tests, we found that the same habitat variable may have different associations in different subregions (Table 1.2). For example, viewshed was significantly associated with Amur tiger presence in Wandashan, but not in Changbaishan. Interestingly, the relative probability response of Amur tiger presence to snow depth is different in the two subregions: its response to snow depth is positive in Changbaishan, but convex in Wandashan (Fig. 1.4a, b).

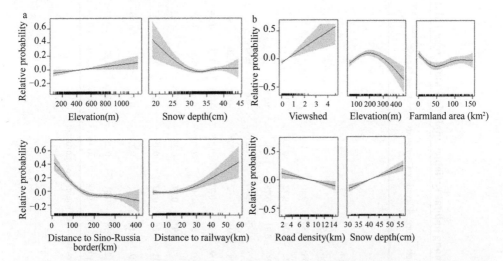

Fig. 1.4 a. Partial probability response curves of generalized additive model (GAM) for the Changbaishan region of northeastern China based on topography (elevation), climate (snow depth), the relationship to the Russia tiger population (distance to Sino-Russia border) and distance to railway. b. Partial probability response curves of generalized additive model (GAM) for the Wandashan region of northeastern China based on topography (viewshed, elevation), climate (snow depth), farmland area, and road density. The x-axis is the value of the model independent variable, and the y-axis is the additive contribution of the variable to the nonparametric GAM smoothing function. Shaded areas are two standard errors about the estimated function.

According to the models, Amur tigers preferred areas with high viewshed in Wandashan (Fig. 1.4b). The Sino-Russia border Amur tiger population showed effective diffusion into Changbaishan within 150km (Fig. 1.4a). The areas within 15km from railways seriously limited Amur tiger presence (Fig. 1.4a). Farmland encroachment and road density negatively influenced the Amur tiger presence in Wandashan (Fig. 1.4b). The lowest relative probability of tiger presence occurred at the farmland area proportion of 50km^2 per pixel in Wandashan (Fig. 1.4b).

Different autocorrelation distances were detected for different subregions (Fig. 1.2a, b). When the spatial autocorrelation effects of these best models were removed, the two best subregion models were improved for robustness (Table 1.5, Fig. 1.3). In

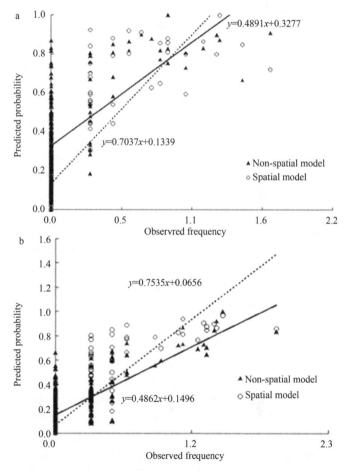

Fig. 1.5 Linear correlation between predicted probability from spatial model or non-spatial model and observed frequency in Wandashan (a) and Changbaishan (b). Observed frequency (n) for x-axis has been transformed into ln(n + 1). Higher values of R^2 of spatial and non-spatial models have been labeled by their own equations near their own modeling straight lines, respectively. Dash lines represent the spatial models.

addition, we found that relative predicted probability of the spatial models was significantly linearly correlated with the relative frequency of Amur occurrence observed (Wandashan, $R^2 = 0.7042$ vs 0.3693; Changbaishan $R^2 = 0.7212$ vs 0.4945) (Fig. 1.5a, b). Therefore, the probability of spatial models predicted is better than that of non-spatial models relating to real observed frequency for the two subregions.

1.4 Discussion

1.4.1 Effects of environmental factors

As far as the role of Russia as a big source population was concerned, we indeed found that a significantly responsive relationship between distance to Sino-Russia border and Amur tiger presence pixels existed within 150km (Fig. 1.4a, Table 1.4). In addition, frequently information, including tiger snow tracking in China or information from the Russian Far East, has shown that some wild Amur tiger individuals share a home range across the Sino-Russia border areas. Russia tiger population immigration into Chinese habitat is key task of conservation work for maintaining stable populations by opening or creating international migration corridors.

Our results suggested that viewshed and elevation were the two key environmental factors for Amur tiger presence. Amur tiger used areas with higher viewshed and elevation, which might be associated with tiger strategies to locate ungulate and other prey (Karanth and Sunquist, 2000; Valeix et al., 2009) or this might also be a strategy to avoid areas of human disturbance (e.g., see moose habitat selection strategy in Jiang et al., 2009). Amur tigers preferred areas with lower elevations ranging from 150m to 250m in Wandashan. The median range of this study is lower than empirical results of median elevations from 400m to 800m in Changbaishan (Hebblewhite et al., 2012).

Heptner and Sludskii (1992) reported that Amur tigers are unable to hunt in deep snow, and have been known to starve to death in their current range in winters with heavy and persistent snow cover. Both snow depth and the length of time that snow remains may affect tiger mortality (Kitchener and Dugmore, 2000). In fact, different snow depths directly influence their prey(i.e., ungulates), as well as their own survival and distribution due to restricted mobility and increased energy costs from travel in deep snow (Pauley et al., 1993; Jiang et al., 2008, 2010). For example, basal metabolic rates of white-tailed deer *Odocoileus virginianus* were depressed when snow depths exceeded 40cm (Pauley et al., 1993). Jiang et al. (2008) suggested that the smaller-bodied roe deer may select sites with less snow than red deer that, due to their larger body size, can move more freely and with less energetic expenditures in deeper snow. In this study area, sika deer were the abundant species in the prey community in eastern Changbaishan (Chen et al., 2011), which have a relatively smaller body size and prefer shallower snow. Whereas, in Wandashan, the red deer or wild boar are the abundant species and adapt to relatively deeper snow (Zhou and Zhang, 2011). Furthermore, there are areas of higher latitude in Wandashan, meaning that snow cover remains longer and lies more deeply there. However, our results suggested a depressive effect of an annual maximum snow depth of more than 30cm on Amur tiger presence occurred in Changbaishan (Fig. 1.4a), but a positive effect on them in Wandashan (Fig. 1.4b). Miquelle *et al.* (1999) reported that there was no evidence of tigers avoiding deep snow in the Russian Far East, but both predation and snow cover were often noted as mortality factors in the ungulate community of northern world (Jędrzejewski *et al.*, 1992). Hence, we conclude that the Amur tiger shows different responses to snow depth in different subregions because of different prey community distributions dependent on snow depth. However, the interaction between prey community and Amur tiger as related to snow cover needs further research to be better understood in China.

1.4.2   Effects of anthropogenic factors

Human disturbance may directly result in habitat loss and increased avoidance behaviors by affected wildlife, which in turn also changes spatial distribution (Stevens and Boness, 2003; Jiang *et al.*, 2007, 2009; Proulx and MacKenzie, 2012; Mushtaq *et al.*, 2013). Amur tigers are known to keep away from human disturbances (Tian *et al.*, 2009; Kang *et al.*, 2010), and this causes additional habitat loss resulting from areas that are seemingly suitable that are in fact avoided (Mace and Waller, 1996; Stevens and Boness, 2003; Johnson *et al.*, 2005). Some experts suggest that there is a threshold distance for normal wildlife activity of 500m to 5km from human disturbance (Mahoney and Schaefer, 2002; Frid, 2003; Jiang *et al.*, 2009, 2010). However, at present, most researches identify the avoidance threshold distance of wildlife from human disturbance using generalized linear models (Mahoney and Schaefer, 2002; Frid, 2003; Jiang *et al.*, 2009, 2010) by predefining the unreliable modeled functions with several polynomial terms (Suarez-Seoane *et al.*, 2002). Our results showed that the effects of human disturbance factors were nonlinearly impacting tiger occurrence. When farmland area accounted for more than 50km^2 per pixel, the probability of Amur tiger presence decreased rapidly, showing strong avoidance (Fig. 1.4b). When areas were within 15km to the railway in Changbaishan, tiger presence was depressed (Fig. 1.4a). These response curves of GAMs reflected the quantitative threshold distances of avoidance extensity for anthropogenic factors. Furthermore, increased road density in Wandashan posed dramatically linear negative effects on Amur tiger presence (Fig. 1.4b). Kerley *et al.* (2002) reported that the construction of a road in Russia reduced a source population to a sink population that no longer provided supplementary tigers to neighboring areas. In other species, roads detrimentally affect lynx by increasing unnatural mortality rates through road kill (Kramer-Schadt et al., 2004) and affect wolf by restricting movement across their ranges (Whittington et al., 2004). Moreover, roads provide farmers with better quality (Linkie et al., 2004) and poachers with greater access to remote habitats containing greater amounts of bushmeat (Bennett and Robinson, 2000).

1.4.3 Effects of spatial autocorrelation

Spatial autocorrelation (SA) is a very challenging issue of which ecologists should be aware (José and Mariana, 2002; Haining, 2003). Dormann (2007) and it is recommended that both spatial and non-spatial models should be presented to aid future comparisons on the effect of incorporating spatial autocorrelation on model estimates. Although the pervasive effects of SA on distribution models are still widely debated (Dormann et al., 2007), it is indeed a potential issue in our study because the Amur tiger presence data were spatially auto-correlated. Does distance with significantly spatial autocorrelation reflect differences in different regions or is there difference between habitat suitability predicted by spatial and non-spatial models?

Our results indicated the probability that the spatial models predicted were more closely related to observed frequency than that of non-spatial models, demonstrating the spatial models owned what biological meaning can be explained (Fig. 1.5). Furthermore, different scale subregional models indeed reflect different occurrence point clustering pattern in focal subregional. This could be due to both the fact that we used presence-/pseudo-absence data to calibrate the model and the fact that the tiger information has not been collected in many apparently suitable areas. Indeed, absolute residuals for absence points are naturally higher where the habitat suitability is high, conversely, residual value for presence data decreases as prediction value increases (Lyet et al., 2013). Our results suggested that we might feel quite confident about the results of the habitat predicted on the eastern parts of Changbaishan and northern Wandashan subregions. However, more caution is required in the interpretation of the model on the western part of Changbaishan mountains, where presence pixel with low or once-only occurrence was obviously left by a wandering tiger family or an individual. Thus, as potential tiger habitat, the western habitat suitability of Changbaishan may be underestimated for absence of neighboring tiger population and dispersive ability restriction.

Until now, few studies have attempted to evaluate whether the spatial models are actually less biased than non-spatial models (Dormann, 2007). This is a question deserving further research, since model predictions depend strongly on bias introduced by the model structure (Reineking and Schröder, 2006). Both Augustin et al. (1996)

and Betts *et al.* (2009) argued that autocovariate models were best suited for prediction in ecology rather than necessarily being useful for inference. In this study, we compared the predication of habitat suitability in different subregional scales and indicated that habitat suitability predicated using spatial models is more meaningful biologically than those of non-spatial models.

1.4.4 Conservation implications

Our research firstly quantitatively assessed the effects of environmental and anthropogenic factors on Amur tiger presence in northeastern China. These quantitative findings are essential for taking practical measures for Amur tiger habitat recovery. Thus, we should consider how to recover key area ranges by alleviating human disturbance, for example, by closing roads, returning farmland to forest, building smart green infrastructure by constructing tunnel or bridges for railways (Quintero *et al.*, 2010; Wikramanayake *et al.*, 2011). We found Amur tiger tracks in the snow, passing through a road tunnel and heading west of Changbaishan on 28 December 2012. This indicated that smart infrastructure design might possibly play key role in coordinating human development and Amur tiger coexistence in the future (Quintero *et al.*, 2010). In field conservation practice, we should focus on establishing a corridor at the Sino-Russia border, and among fragmented suitable habitat patches, by building a nature reserve network, in order to mitigate negative effects of human disturbance (Tian *et al.*, 2011).

Furthermore, modeling tool used is crucial to identify threat factors and new potential habitats (Li and Wang, 2013). Although multiple scale modeling is prevailing (Riitters *et al.*, 1997), both spatial model and multiple subregional scale methods should also be considered for large-scale conservation. Spatial models and multiple subregional scale methods may provide new insights on endangered species habitat prediction and should be used according to ecological characteristics of target animals and actual future conservation requirements.

Despite the useful conservation information of this study, data collection of this study was not designed by a true used-unused sampling design based on systematic surveys. Therefore, in the future, inherently flaws and weaknesses within the survey design must be overcome. We should consider different animal population distribution patterns, and then adopt different methods to collect necessary information. For example, in high presence frequency and high suitability habitat areas, systematic survey route survey should be chosen, because resident tigers may exist; large-scale areas with wandering individual information may be used as an information network to monitor distribution dynamics opportunistically (e.g. the wolf network monitoring system used by Duchamp *et al.*, 2012). Although opportunistically collected information has inherent shortcomings, it still provides vital information for some rare, elusive and wandering animal populations, in order to monitor and assess habitats.

1.5 Summary

We examined environmental and anthropogenic factors drive range loss in large mammals, using distribution data of Amur tigers opportunistically collected between 2000 and 2012, and anthropogenic and environmental variables to model the distribution of the Amur tiger in northeastern China. Our results suggested that population distribution models of different subregions showed different habitat factors determining tiger population distribution patterns. Where farmland cover was over 50km^2 per pixel (196km^2), distance was within 15km to the railway in Changbaishan and road density (length per pixel) increased in Wandashan, the relative probability of Amur tiger occurrence exhibited monotonic avoidance responses; however, where distance was within 150km of the Sino-Russia border, the occurrence probability of Amur tiger was relatively high. We analyzed the avoidance or preference responses of Amur tiger distribution to elevation, snow depth and viewshed. Furthermore, different subregional models detected a variety of spatial autocorrelation distances due to different population clustering patterns. We found that spatial models significantly improved model fits for non-spatial models and made more robust habitat suitability predications than that of non-spatial models. Consequently, these findings provide useful guidance for habitat conservation and management.

References

Altman D G, Bland M. 1994. Diagnostic tests 2: predictive values. British Medical Journal, 309(6947): 102.

Augustin N H, Mugglestone M A, Buckland S T. 1996. An autologistic model for the spatial distribution of wildlife. Journal of Applied Ecology, 33: 339-347.

Beard K H, Hengartner N, Skelly D K. 1999. Effectiveness of predicting breeding bird distributions using probabilistic models. Conservation Biology, 13: 1108-1116.

Bennett E L, Robinson J G. 2000. Hunting of wildlife in tropical forests: implications for biodiversity and forest peoples // Environment Department working papers, no.76. Biodiversity series. Washington DC: The World Bank.

Betts M G, Ganio L M, Huso M M P, et al. 2009. Comment on "Methods to account for spatial autocorrelation in the analysis of species distributional data: a review". Ecography, 30(5): 609-628.

Carroll C, Miquelle D G. 2006. Spatial viability analysis of Amur tiger *Panthera tigris altaica* in the Russian Far East: the role of protected areas and landscape matrix in population persistence. Journal of Applied Ecology, 43: 1056-1068.

Channell R, Lomolino M V. 2000. Dynamic biogeography and conservation of endangered species. Nature, 403: 84-86.

Chen J, Na S, Shun Q, *et al*. 2011. Amur tiger and prey in Jilin Hunchun National Nature Reserve, China. Chinese Journal of Zoology, 46: 46-52. (in Chinese)
Cliff A, Ord J K. 1981. Spatial Processes-models and Applications. London: Pion.
Davis F W, Seo, Zielinski W J. 2007. Regional variation in home range scale habitat models for fisher (*Martes pennanti*) in California. Ecological Application, 17: 2195-2213.
Dormann C F, McPherson J M, Araujo M B, *et al*. 2007. Methods to account for spatial autocorrelation in the analysis of species distributional data: a review. Ecography, 30: 609-628.
Dormann C F. 2007. Effects of incorporating spatial autocorrelation into the analysis of species distribution data. Global Ecology and Biogeography, 16: 129-138.
Duchamp C, Boyer J, Briaudet P E, *et al*. 2012. A dual frame survey to assess time-and space-related changes of the colonizing wolf population in France. Hystrix the Italian Journal of Mammalogy, 23: 14-28.
Fielding A H, Bell J F. 1997. A review of methods for the assessment of prediction errors in conservation presence/absence models. Environmental Conservation, 24: 38-49.
Frid A. 2003. Dall's sheep responses to overflights by helicopter and fixed-wing aircraft. Biological Conservation, 110: 387-399.
Goodrich J M, Kerley L L, Miquelle D G, *et al*. 2005. Social structure of Amur tigers on Sikhote-Alin Biosphere Zapovednik // Miquelle D G, Smirnov E N, Goodrich J M. Tigers in Sikhote-Alin Zapovednik. Vladivostok, Russia (in Russian): PSP, 50-60.
Goodrich J M, Miquelle D G, Smirnov E N, *et al*. 2010. Spatial structure of Amur (Siberian) tigers (*Panthera tigris altaica*) on Sikhote-Alin Biosphere Zapovednik, Russia. Journal of Mammalogy, 91: 737-748.
Granadeiro J P, Andrade J, Palmeirim J M. 2004. Modelling the distribution of shorebirds in estuarine areas using generalised additive models. Journal of Sea Research, 52: 227-240.
Guisan A, Edwards T C, Hastie T. 2002. Generalized linear and generalized additive models in studies of species distributions: setting the scene. Ecological Modelling, 157: 89-100.
Haining R. 2003. Spatial Data Analysis—Theory and Practice. Cambridge: Cambridge University Press.
Hastie T J, Tibshirani R J. 1990. Generalized Additive Models. London: Chapman, Hall/CRC.
Hebblewhite M, Zimmermann F, Li Z, *et al*. 2012. Is there a future for Amur tigers in a restored tiger conservation landscape in Northeast China. Animal Conservation, 15(6): 1-14.
Hemmer H. 1976. Fossil history of the living Felidae // The Carnivore Research Institute. The World's Cats. Seattle: Burke Museum, 111(2): 1-14.
Heptner V G, Sludskii A A. 1992. Mammals of the Soviet Union. Volume II, Part 2. Carnivora (hyenas and cats). Leiden, Brill.
Hurlbert S H. 1984. Pseudoreplication and the design of ecological field experiments. Wildlife Monographs, 54: 187-211.
Jędrzejewski W, Jędrzejewska B, Okarma H, *et al*. 1992. Wolf predation and snow cover as mortality factors in the ungulate community of the Białowieża National Park, Poland. Oecologia, 90: 27-36.
Jiang G, Ma J, Zhang M, *et al*. 2009. Multiple spatial scale resource selection function models in relation to human disturbance for moose in northeastern China. Ecological Research, 24: 423-440.
Jiang G, Ma J, Zhang M, *et al*. 2010. Multi-scale foraging habitat use and interactions by sympatric cervids in northeastern China. Journal of Wildlife Management, 74: 678-689.
Jiang G, Ma J, Zhang M. 2006. Spatial distribution of distribution of ungulate responses to habitat factors in Wandashan, northeastern China. Journal of Wildlife Management, 70: 1470-1476.
Jiang G, Zhang M, Ma J. 2007. Effects of human disturbance on movement, foraging and bed site

selection of red deer *Cervus elaphus xanthopygus* in the Wandashan Mountains, northeastern China. Acta Theriologica, 52: 435-446.

Jiang G, Zhang M, Ma J. 2008. Habitat use and separation between red deer *Cervus elaphus xanthopygus* and roe deer *Capreolus pygargus bedfordi* in relation to human disturbance in the Wandashan Mountains, northeastern China. Wildlife Biology, 14(1): 92-100.

Johnson C J, Boyce M S, Case R L, *et al*. 2005. Cumulative effects of human developments on Arctic wildlife. Wildlife Monographs, 160: 1-36.

José A F D-F, Mariana O D C T. 2002. Spatial autocorrelation analysis and the identification of operational units for conservation in continuous populations. Conservation Biology, 16: 924-935.

Kang A, Xie Y, Tang J, *et al*. 2010. Historic distribution and recent loss of tigers in China. Integrative Zoology, 5: 335-341.

Karanth K U, Nichols J D, Kumar N S, *et al*. 2004. Tigers and their prey: predicting carnivore densities from prey abundance. Proceedings of the National Academy of Sciences of the United States of America, 101: 4854-4858.

Karanth K U, Sunquist M E. 2000. Behavioural correlates of predation by tiger (*Panthera tigris*), leopard (*Panthera pardus*) and dhole (*Cuon alpinus*) in Nagarahole, India. Journal of Zoology, 250: 255-265.

Kerley L L, Goodrich J M, Miquelle D G, *et al*. 2002. Effects of roads and human disturbance on Amur tigers. Animal Conservation, 16: 97-108.

Kitchener A C, Dugmore A J. 2000. Biogeographical change in the tiger, *Panthera tigris*. Animal Conservation, 3: 113-124.

Kitchener A C, Yamaguchi N. 2010. What is a tiger? Biogeography, morphology, and taxonomy// Tilson R and Nyhus P. Tigers of the World: The Science, Politics and Conservation of *Panthera tigris*. 2nd ed. Oxford: Elsevier: 53-85.

Koenig W D. 1999. Spatial autocorrelation of ecological phenomena. Trends in Ecology and Evolution, 14: 22-26.

Kramer-Schadt S, Revilla E, Wiegand T, *et al*. 2004. Fragmented landscapes, road mortality and patch connectivity: modelling influences on the dispersal of Eurasian lynx. Journal of Applied Ecology, 41: 711-723.

Latimer A M, Wu S S, Gelfand A E, *et al*. 2006. Building statistical models to analyze species distributions. Ecological Application, 16: 33-50.

Lawton J H. 1993. Range, population abundance and conservation. Trends in Ecology and Evolution, 8: 409-413.

Li B, Zhang E, Liu Z. 2009. Livestock depredation by Amur tigers in Hunchun Nature Reserve, Jilin, China. Acta Theriologica Sinica, 29: 231-238.

Li B, Zhang E, Zhang Z, *et al*. 2008. Preliminary monitoring of Amur tiger population in Jilin Hunchun National Nature Reserve. Acta Theriologica Sinica, 28: 333-334.

Li T, Jiang J, Wu Z, *et al*. 2001. Survey on Amur tigers in Jilin Province. Acta Theriologica Sinica, 21: 1-6.

Li X, Wang Y. 2013. Applying various algorithms for species distribution modeling. Integrative Zoology, 8: 124-135.

Li Z, Zimmerman F, Hebblewhite M, *et al*. 2010. Study on the Potential Tiger Habitat in the Changbaishan Area. Beijing: China Forestry Publishing House.

Linkie M, Smith R J, Leader-Williams N. 2004. Mapping and predicting deforestation patterns in the lowlands of Sumatra. Biodiversity and Conservation, 13: 1809-1818.

Luan X, Qu Y, Li D, *et al*. 2011. Habitat evaluation of wild Amur tiger (*Panthera tigris altaica*) and

conservation priority setting in north-eastern China. Journal of Environmental Management, 92: 31-42.
Luo S J, Kim J H, Johnson W E, *et al*. 2004. Phylogeography and genetic ancestry of tigers (*Panthera tigris*). PLoS Biology, 2: 2275-2293.
Lyet A, Thuiller W, Cheylan M, *et al*. 2013. Fine-scale regional distribution modelling of rare and threatened species: bridging GIS Tools and conservation in practice. Diversity and Distributions, 19(7): 651-663.
Ma Y. 2005. Changes in numbers and distribution of the Amur tiger in Northeast China in the past century: a summary report // Zhang E D, Miquelle D G, Wang T H. Recovery of the Wild Amur Tiger Population in China: Process and Prospect. Beijing: China Forestry Publishing House.
Mace R D, Waller J S. 1996. Grizzly bear distribution and human conflicts in Jewel Basin Hiking Area, Swan Mountains, Montana. Wildlife Society, 24(3): 461-467.
Mahoney S P, Schaefer J A. 2002. Hydroelectric development and the disruption of migration in caribou. Biological Conservation, 107: 147-153.
Manly B F J, McDonald L L, Thomas D L, *et al*. 2002. Resource Selection by Animals: Statistical Analysis and Design for Field Studies. Boston: Kluwer: 2.
Miquelle D G, Smirnov E N, Merrill T W, *et al*. 1999. Hierarchical spatial analysis of Amur tiger relationships to habitat and prey//Seidensticker J, Christie S, Jackson P. Riding the Tiger: Tiger Conservation in Human-Dominated Landscapes. Cambridge: Cambridge University Press: 71-99.
Miquelle D G, Smirnov E N, Quigley H G, *et al*. 1996. Food habits of Amur tigers in Sikhote-Ali Zapovednik and Russian Far East, and implication for conservation. Journal of Wildlife Research, 1: 138-147.
Mushtaq M, Hussain I, Mian A, *et al*. 2013. Field evaluation of some bait additives against Indian crested porcupine (*Hystrix indica*) (Rodentia: Hystricidae). Integrative Zoology, 8: 285-292.
Pauley G R, Peek J M, Zager P. 1993. Predicting white-tailed deer habitat use in northern Idaho. Journal of Wildlife Management, 57: 904-913.
Proulx G, MacKenzie N. 2012. Relative abundance of American badger (*Taxidea taxus*) and red fox (*Vulpes vulpes*) in landscapes with high and low rodenticide poisoning levels. Integrative Zoology, 7: 41-47.
Quintero J, Roca R, Morgan A, *et al*. 2010. Smart Green Infrastructure in Tiger Range Countries. Washington DC: The World Bank.
R Development Core Team. 2010. R: A language and environment for statistical computing. Vienna, Austria: R Foundation for Statistical Computing.
Ramsay T, Burnett R, Krewski D. 2003. Exploring bias in a generalized additive model for spatial air pollution data. Environmental Health Perspectives, 111: 1283-1288.
Rangel T F, Diniz-Filho J A, Bini L M. 2010. SAM: a comprehensive application for spatial analysis in macroecology. Ecography, 33: 46-50.
Reineking B, Schröder B. 2006. Constrain to perform: regularization of habitat models. Ecological Modelling, 193: 675-690.
Riitters K H, O'neill R, Jones K B. 1997. Assessing habitat suitability at multiple scales: a landscape-level approach. Biological Conservation, 81: 191-202.
Segurado P, Araujo M B. 2004. An evaluation of methods for modelling species distributions. Journal of Biogeography, 31: 1555-1568.
Seidensticker J. 1996. Tigers. Vancouver: Voyageur Press.
Shenko A N, Bien W F, Spotila J R, *et al*. 2012. Effects of disturbance on small mammal community structure in the New Jersey Pinelands, USA. Integrative Zoology, 7: 16-29.

Stevens M A, Boness D J. 2003. Influences of habitat features and human disturbance on use of breeding sites by a declining population of southern fur seals (*Arctocephalus australis*). Journal of Zoology, 260: 145-152.

Stige L C, Ottersen G, Brander K, et al. 2006. Cod and climate: effect of the North Atlantic Oscillation on recruitment in the North Atlantic. Marine Ecology Progress, 325: 227-241.

Suarez-Seoane S, Osborne P E, Aloneso J C. 2002. Large-scale habitat selection by agricultural steppe birds in Spain: identifying species-habitat responses using generalized additive models. Journal of Applied Ecology, 39: 755-771.

Sunquist M. 2010. What is a tiger ecology and behavior // Tilson R, Nyhus P. Tigers of the World: The Science, Politics and Conservation of *Panthera tigris*. 2nd ed. Oxford: Elsevier: 19-34.

Tian Y, Wu J, Kou X, et al. 2009. Spatiotemporal pattern and major causes of the Amur tiger population dynamics. Biodiversity Science, 17(3): 211-225. (in Chinese)

Tian Y, Wu J, Smith A T, et al. 2011. Population viability of the Siberian Tiger in a changing landscape: going, going and gone? Ecological Modelling, 222: 3166-3180.

Valeix M, Loveridge A J, Jammes C, et al. 2009. Behavioral adjustments of African herbivores to predation risk by lions: spatiotemporal variations influence habitat use. Ecology, 90: 23-30.

Whittington J, St Clair C C, Mercer G. 2004. Path tortuosity and the permeability of roads and trails to wolf movement. Ecology and Society, 99(1): 1759-1763.

Wikramanayake E, Dinerstein E, Seidensticker J, et al. 2011. A landscape-based conservation strategy to double the wild tiger population. Conservation Letters, 4: 219-227.

Wood S N. 2006. Generalized Additive Models: An Introduction with R. Boca Raton: CRC Press.

Yackulic C B, Sanderson E W, Uriartea M. 2011. Anthropogenic and environmental drivers of modern range loss in large mammals. Proceedings of the National Academy of Sciences of the United States of America, 108(10): 4024-4029.

Zhang C, Zhang M, Jiang G. 2012. Assessment of monitoring methods for population abundance of Amur tiger in Northeast China. Acta Ecologica Sinica, 32: 5943-5952.

Zhou S, Sun H, Zhang M, et al. 2008. Regional distribution and population size fluctuation of wild Amur tiger (*Panthera tigris altaica*) in Heilongjiang Province. Acta Theriologica Sinica, 28: 165-173.

Zhou S, Zhang M. 2011. An integrated analysis into the causes of ungulate mortality in the Wanda Mountains (Heilongjiang Province, China) and an evaluation of habitat quality. Biological Conservation, 144(10): 2517-2523.

Chapter 2   The synergy between ecology and society in Amur tiger conservation

2.1   Introduction

Top predators have been used as surrogate, umbrella and flagship species in biodiversity conservation due to their indispensable ecological and socioeconomic roles in ecosystems (Ripple *et al.*, 2014). However, many carnivores have suffered substantial population declines, geographic range contractions, habitat loss and fragmentation (Ceballos and Ehrlich, 2002; Morrison *et al.*, 2007). For wide-ranging large carnivores like tigers and leopards, conservation of viable populations is complex and challenging because of the need to maintain extensive permeable landscapes to facilitate movement and involvement of political and socioeconomic issues arising from high habitat restoration costs and human-carnivore conflicts (Wikramanayake *et al.*, 2004; Athreya *et al.*, 2013). Some large carnivore populations have shown recovery due to forest restoration, increased ungulate prey, and reduced anthropogenic disturbance (e.g., poaching, livestock grazing, habitat destruction). Recovering large carnivores either coexist with humans in modern, crowded landscapes in Europe (Chapron *et al.*, 2014) or live in separation from humans in preserves or wilderness areas in Africa (Packer *et al.*, 2013) and North America (Gompper *et al.*, 2015). These different successes represent an ongoing debate about land-sharing and land-sparing models for conservation. Both models or strategies are key for understanding how landscape management and conservation protect and help the recovery of populations of key species. Conservation practice will benefit from understanding the mechanisms, by which each model contributes to population recovery, or whether a combination of strategies is required.

During the past six decades, the Chinese government has committed to eradicating poverty and improving the lives of more than one billion people with rapid economic growth, particularly after 1978. China has become the world's second largest economy since 2010. Prior to the 1990s in northeastern China a high density of forest workers congregated in several regions. Large-scale forest logging lasted for almost 50 years (From the founding of the People's Republic of China until 1998) across China including the Northeast. Consequently, forest resources were over-exploited with serious impacts on regional biodiversity. Ungulates, such as sika deer(*Cervus nippon*) and red deer (*C. elaphus*) became endangered or locally extinct in northeast China (Jiang *et al.*, 2015). During this period, the majority of Amur tiger habitat (*Panthera tigris altaica*) was lost

(Yu et al., 2009), and severe drought and massive flooding shocked the national economy and compromised human safety (Zong and Chen, 2000; Xu et al., 2006). The Amur tiger population declined from 200 tigers of the 1950s to 14 tigers of 1999 (Yu et al., 2009). In China, <10 free-ranging Amur leopards (*Panthera pardus orientalis*) were found in 1998 (Yang et al., 1998). The ensuing socioeconomic and ecological losses made the Chinese government to initiate Natural Forest Protection Program (NFPP) in 1998, with aims of protecting and restoring forests. This program represents one of the largest government investments and spatially extensive environmental rehabilitation efforts in the world (Hyde et al., 2003). It covers all natural forest regions including the historical range of the endangered big cats in north-eastern China.

Recent reports have demonstrated improvement in the populations and habitats of the Amur tiger and the Amur leopard in China over the past two decades (Jiang et al., 2015; Wang et al., 2016). Both forest management and natural preserves were involved in the restoration. Recent studies have found the correlates of recent big cat population and habitat increases, but were restricted to only big cat habitat and population distributions. Specifically, Wang et al. (2016) used camera trap data to assess tiger and leopard populations and distributions, and Jiang et al. (2014, 2015) investigated habitat suitability and potential habitats of Amur leopards and Amur tigers. Yet, no studies have considered the long-term effects of human and socioeconomic factors on the conservation of these two big cats. It would be beneficial to compare the long-term effects of temporal changes in forest management practice, human population changes, the costs of management actions, and subsequent effects on the big cat and prey populations. Doing so would help elucidate mechanisms that inform the roles of land sparing vs. land sharing, and the extent to which these strategies are mutually exclusive vs. synergistic.

Quantitative analyses of the relationships between human livelihood and biodiversity conservation are necessary to test for synergies between social and ecological aims of conservation investments (Persha et al., 2011). Trade-offs or synergies between biodiversity conservation and eradication of poverty may also reveal conservation needs and opportunities (Naughton-Treves et al., 2005). However, few studies of top predators have focused on these interactive relationships across both social and ecological dimensions (Persha et al., 2011). Here we investigated the relationships among landscape conservation investments, forest logging volume, number of relocated foresters (i.e., shifts in livelihoods), and habitat areas occupied by Amur tigers and Amur leopards in the current core range of the Amur tiger and Amur leopard in northeastern China.

The government actions of relocating humans together with basic predator-prey theory led us to hypothesize several steps towards recovery of Amur tigers and Amur leopards. First, relocation of forestry workers (land sparing) led to increases in forest standing biomass (Hypothesis 1). Second, increases in forest biomass and decrease in human disturbance (i.e., poaching, cow grazing, non-timber collection etc.) led to

Chapter 2   The synergy between ecology and society in Amur tiger conservation | 197

subsequent increases in ungulate prey abundance (supporting land sharing or sparing; Hypothesis 2). Third, reductions in human density led to decreased poaching of big cats and ungulate prey (supporting land sparing; Hypothesis 3). Last, we predicted a subsequent increase in the population size and distributional range of the big cat species following both prey increases and reduced human disturbance (Hypothesis 4). We also discussed the extent, to which the timing of events allows us to separate out the effects of removal of human effects vs. recovery of forest biomass, and hence determine whether land sharing, land sparing models, or some mix of the two are involved.

2.2   Methods

2.2.1   Collection and analysis of forest management data

To test the relationship between conservation costs and human population removal, and between human population decline and increases in forest biomass, we selected 31 forest bureaus (69 605km^2) and 10 nature reserves (4050km^2) for data collection and surveys within the current range of the Amur tiger and Amur leopard in northeast China. The total survey area of 73 655km^2 comprised the current range of the tiger and leopard (Table 2.1, Fig. 2.1). Most of the forest bureaus were established in the 1950s and have had commercial logging for > 60 years (Table 2.1). Apart from one nature reserve, most natural reserves were established after 2000 (Table 2.1). We compiled the annual records of relevant variables from 1950 to 2015. Those variables included conservation investments, forest logging volumes, the amount by which forest stock

Table 2.1 Names, geographic locations, areas, and years of establishment of 31 forest bureaus and 10 nature reserves in the current range of Amur tigers and Amur leopards in northeastern China

No.	Name of forest bureau	Province	East longitude(°)	North latitude(°)	Forest area (km^2)	Year of establishment
1	Chaihe	Heilongjiang	129.460 2	45.238 83	3 452.72	1947
2	Dongjingcheng	Heilongjiang	129.134 7	43.936 18	4 184.68	1948
3	Fangzheng	Heilongjiang	129.425 9	45.761 73	2 035.82	1958
4	Hailin	Heilongjiang	129.110 5	44.783 35	1 566.21	1958
5	Shanhetun	Heilongjiang	127.870 7	44.336 3	2 064.09	1948
6	Suiyang	Heilongjiang	130.799 6	44.029 39	5 160	1948
7	Likou	Heilongjiang	130.061 8	45.532 36	2 730.25	1963
8	Boli County	Heilongjiang	130.569 7	45.752 32	785.07	1958
9	Jidong County	Heilongjiang	131.122 5	45.243 25	1 640.99	1965
10	Linkou	Heilongjiang	130.270 6	45.274 11	1 939.04	1950
11	Muling City	Heilongjiang	130.514 4	44.915 73	335.316	1958

Table 2.1 (continued)

No.	Name of forest bureau	Province	East longitude(°)	North latitude(°)	Forest area (km²)	Year of establishment
12	Ning'an County	Heilongjiang	129.477 7	44.344 62	1 487.369	1958
13	Antu County	Jilin	128.912 7	43.110 6	1 237.65	1962
14	Dunhua City	Jilin	128.226 2	43.369 63	1 743.51	1951
15	Antu Forest Management	Jilin	128.912 7	43.110 6	1 092.41	1962
16	Baihe	Jilin	128.220 2	42.405 72	1 904.7	1971
17	Dashitou	Jilin	128.579 8	43.140 93	2 634.38	1952
18	Daxinggou	Jilin	129.577 8	43.476 87	1 272	1960
19	Dunhua	Jilin	127.976 5	43.122 54	2 373.55	1958
20	Helong	Jilin	128.664 7	42.213 04	1 704.89	1958
21	Helong City	Jilin	129.003 4	42.535 29	2 708.13	1950
22	Huangnihe	Jilin	128.106 4	43.833 15	1 965.96	1953
23	Hunchun City	Jilin	130.363 2	42.874 27	880.04	1950
24	Tianqiaoling	Jilin	130.077 8	43.768 04	1 935.06	1959
25	Tunmen City	Jilin	129.844 1	42.966 02	886.36	1974
26	Wangqing	Jilin	130.472 1	43.392 19	3 040	1947
27	Wangqing County	Jilin	129.756 4	43.309 12	3 290.11	1956
28	Hunchun	Jilin	130.501 7	43.034 8	4 000.4	1994
29	Dahailin	Heilongjiang	128.474 4	44.378 39	2 663.1	1947
30	Muling	Heilongjiang	130.212 2	44.266 63	2 675.3	1947
31	Dongfanghong	Heilongjiang	133.627 8	46.523 41	4 216.76	1963
Total					69 605.87	

No.	Name of nature reserve	Province	East longitude(°)	North latitude(°)	Forest area (km²)	Year of establishment
1	Daxiagu	Heilongjiang	127.917 3	44.183 4	249.98	2004
2	Hudieling	Heilongjiang	129.447 7	45.607 9	108.02	2004
3	Laoyeling	Heilongjiang	131.045 4	43.575 4	781.278	2011
4	Mulingzishan	Heilongjiang	130.153 7	43.969 3	356.48	2007
5	Fenghuangshan	Heilongjiang	131.163 2	44.977 4	265.7	2006
6	Mudanfeng	Heilongjiang	129.745 1	44.447 8	194.68	1981
7	Niaoqingshan	Heilongjiang	131.011 1	44.540 4	180.02	2007
8	Huangnihe	Jilin	128.127	43.975	234.76	2000
9	Wangqing	Jilin	130.894	43.397	674.34	2002
10	Hunchun	Jilin	130.775 6	42.932 5	1 008	2002
Total					4 053.26	

Note: The longitudes and latitudes represent the geographic centroids of forest bureaus and nature reserves.

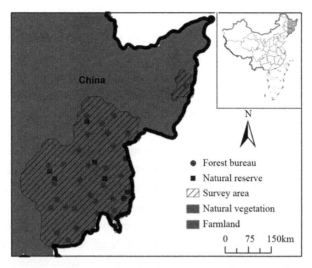

Fig. 2.1 Distribution of forest bureaus (red points) and nature reserves (blue squares) surveyed for forest management, forest worker populations, natural forest protection investments, and occurrence information collected for Amur tigers and Amur leopards from 1950 to 2015. Please scan the QR code on the back cover to see the color image.

growth exceeded logging volume (hereafter net forest stock growth), plantation areas, number and density of forest workers, compensation costs of wildlife-human conflicts, and, human welfare benefits provided by resettlement-assistance projects for forester relocations from each forest bureau and nature reserve.

We used linear or nonlinear regression to test the relationships between annual forest management investment amounts, forest logging volumes, net forest stock growth, density of forest workers, and habitat areas occupied by Amur tigers and Amur leopards after the initiation of the natural forest protection project in 1998. We regressed logarithmically transformed poaching numbers on time to quantify the linear increase rates of the Amur tigers poaching from 1998 to 2006 and from 2007 to 2015. All statistical tests were two-sided tests at the significance level of 0.05 and carried out using by Prism (GraphPad Prism, Prism 5.0; www.graphpad.com).

2.2.2 Population size and habitat distribution of Amur tiger and Amur leopard

We obtained data on the abundance and habitat distribution of the Amur tigers and Amur leopards from the literature published in China since the 1950s (Yu *et al.*, 2009), which we summarized below. Historical population size was assessed using snow-track line transect methods based on big cat footprint identification from 1998 to 2015.

In the core habitat of Laoyeling, Zhangguangcailing and Wandashan, we deployed >1500 camera traps covering almost 2900km^2 covering this core big cat distribution area since 2013. Camera traps were set up at a density of 1 pair per 10km^2 and were checked every 3 months from 2013 to 2015.

Based on population size from both camera trap data and historical reports (Yu et al., 2009), we estimated the realized population growth rate of the Amur tiger population as $R = [\ln(N_{2015}) - \ln(N_{1999})]/t$, where N_{1999} is tiger population in 1999, N_{2015} is population size in 2015, and $t = 16$ (years) (Berryman and Turchin, 2001). We converted annual population growth rate R to finite rate of increase (λ) using the equation $\lambda = eR$. With relatively low survival rates (e.g. 85%), at least 100 individuals of a tiger population should be required for ensuring long-term persistence (Chapron et al., 2008). We predicted how many years 100 individuals of Amur tigers will be realized according to this increase rate (Berryman and Turchin, 2001). We also conducted similar calculations for Amur leopards.

From 1998 to 2015, we recorded 779 occurrences of Amur tiger in this region, including 355 killings of prey or livestock, 51 fecal samples (determined by DNA analysis), and 345 sets of footprints and 71 photographs from camera traps. Also, 643 occurrences of Amur leopard were found, including 24 killings of prey or livestock, 36 fecal samples (determined by DNA analysis), 133 sets of footprints, and 459 photos from camera traps. We established the national tiger and leopard information database in the Feline Research Center of National Forestry and Grassland Administration. Habitat loss is normally a strong indicator of population declines (Dinerstein et al., 2007). To estimate habitat areas occupied by the big cats, we first created two gridded polygons, covering the entire study region, at the spatial resolutions of 20km × 20km and 10km × 10km, respectively, using ArcGIS software (Environmental Systems Resource Institute, ArcGIS 10.0; www.esri.com). The two spatial resolutions were chosen according to average home range sizes of female Amur tigers or Amur leopards (Goodrich et al., 2010; Hebblewhite et al., 2011). Then we determined the occurrence frequencies of Amur tigers within a 400km^2 grid cell and those of Amur leopards within a 100km^2 grid cell, respectively, using Hawth's Analysis Tools for ArcGIS (http://www.spatialecology.com/htools/download.php). A grid cell with a non-zero (> 0) occurrence frequency was considered occupied. The annual sum of all occupied grid cells was used to estimate the annual total habitat area occupied by each big cat species. The total areas occupied from 1999 to 2014 are shown in Fig. 2.2 (a) for Amur tigers and Fig. 2.2 (b) for Amur leopards. We tested for relationships between annual total habitat areas occupied by big cats and socioeconomic and forest variables using linear or non-linear regression models.

We collected ungulate prey density data from 2010 to 2014 over an area of 878km^2 in Jilin Wangqing Nature Reserve using snow track sample plots (Qi et al., 2015). We did a total of 33 plots in 2010–2011, 14 plots in 2012–2013, and 10 plots in 2013–2014. Each plot with the area of approximately 10km^2 (i.e., 5km × 2km) consisted of five parallel 5km transect lines separated by 500m. We only used animal tracks within 24h to calculate the number of individuals of each prey species in each

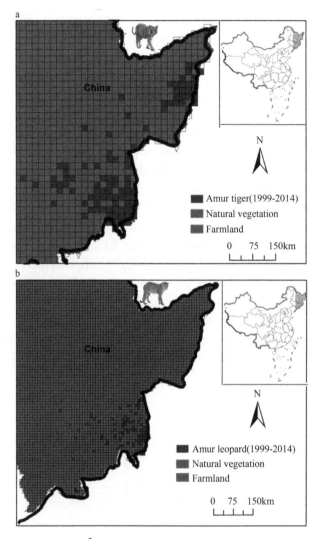

Fig. 2.2 Habitat areas (41 200km^2) occupied by the Amur tigers (a) and habitat areas (10 200km^2) occupied by Amur leopards from 1999 to 2014 (b). Red squares represent the habitat areas occupied by big cats. Please scan the QR code on the back cover to see the color image.

plot (Qi *et al.*, 2015) and hence, we calculated ungulates prey density. Prey abundance data were not available across the whole study area for the entire study period. Furthermore, to estimate the efforts to reduce poaching on tiger and prey and numbers of snares encountered, we collected the records of patrols for removals of steel snares as well as data on steel snare density in the Jilin Wangqing Nature Reserve in both 2009 and 2015.

2.3 Results

Most Amur tigers in China live near the Sino-Russia border, but we obtained the camera-trap image of Amur tiger individuals 270km west of the border, indicating the successful dispersal of Amur tigers into the interior of northeastern China. We also recorded breeding Amur tigers and breeding Amur leopards (Jiang, 2014; Shi et al., 2015). Furthermore, we found the current range of Amur leopards to be 48 000km^2 and to contain 37 suitable habitat patches for Amur leopards, which may harbor about 195 Amur leopards in northeast China (Jiang et al., 2015).

The Chinese government invested $4.476 billion of U.S. dollars (USD) within the current range of big cats, including $2.723 billion in NFPP (Fig. 2.3a), $0.013 billion in human-wildlife conflict compensation (Fig. 2.3b), $1.712 billion in human settlement improvement for relocations (Fig. 2.3c), and $0.027 billion in nature reserve protecttion. The total amount of annual NFPP investment steadily increased from 1998 to 2015 ($R^2 = 0.78$, $n = 17$, $P < 0.001$; Fig. 2.3a). From 2007 to 2014, annual rates of livestock killing by big cats increased by 31%. To mitigate the conflict between the livelihoods of local people and big cats or forest recovery (Fig. 2.3b), the Chinese government initiated the forester resettlement project in 2008. The government

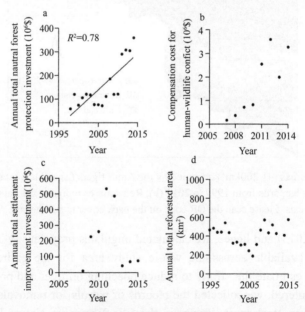

Fig. 2.3 Annual total costs of natural forest investment for 31forest bureaus from 1998 to 2015 (a), annual total cost of compensation for human-wildlife conflicts from 2007 to 2015 (b), annual total cost of human settlement improvements for employees migrating out of forest farms from 2008 to 2015 (c), annual total planting areas in the forest bureaus from 1998 to 2015 (d).

constructed welfare housing to relocate foresters from forest farms or villages to the nearby towns or cities and helped most relocated foresters to shift to other careers and livelihoods independent of forest resources (Fig. 2.3c).

Annual NFPP investment per square km averaged $2927 and linearly increased from $1098 to $3930 ($R^2 = 0.76$, $n = 17$, $P < 0.001$; 2.4a). Consistent with Hypothesis 1 and land sparing, the NFPP policy initiated in 1998 caused human population to fall from 17.77×10^4 in 1999 to 7.18×10^4 in 2015(Fig. 2.4b, Fig. 2.5a), and almost 100, 000 people changed careers during this period (Fig. 2.5a). From 1999 to 2015, annual mean population density of foresters decreased from 2.73 km^{-2} to 1.07km^{-2} ($R^2 = 0.95$, $n = 17$, $P < 0.001$; Fig. 2.4b). Moreover, there were no human residents in the forest areas of 2861km^2 within 9 of 31 forest bureaus inside the current range of the big cats from 2008 to 2015. Concurrent with human population declines, annual total forest logging volume linearly declined from 5.51×10^6 to 2.06×10^6m^3 from 1999 to 2015 ($R^2 = 0.88$, $n = 17$, $P < 0.001$; Fig. 2.5b). Annual mean logging volume decreased from 117m^3/km^2 to 30.3m^3/km^2 from 1998 to 2015 ($R^2 = 0.90$, $n = 17$, $P < 0.001$; Fig. 2.4c),

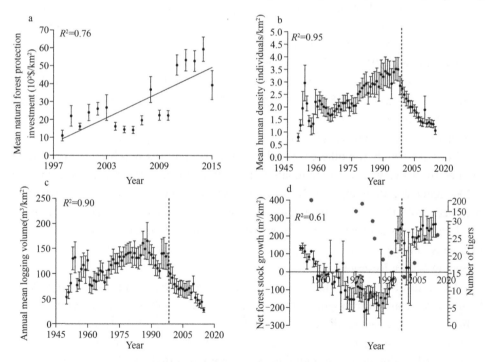

Fig. 2.4 Annual mean natural forest protection investments per square kilometer from 1998 to 2015 (a); annual mean human population density (b); annual mean logging volume (c); dynamics of the Amur tiger population and annual mean net forest stock growth (d) from 1950 to 2015. Dashed line represents the year 1998 when natural forest protection project was implemented. Points represent the number of Amur tigers.

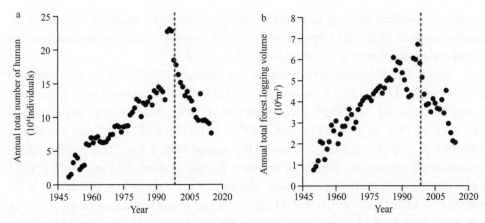

Fig. 2.5 Dynamics of human density (a) and annual total forest logging volume (m³) (b) in Amur tiger habitat from 1950 to 2015. Dashed line represents the initiation of natural forest protection project in 1998.

and forest logging entirely ceased in northeastern China in April 2015. Consistent with Hypothesis 1, annual mean net forest stock growth increased from 66.62m³/km² in 1999 to 232.45m³/km² in 2014 ($R^2 = 0.61$, $n = 16$, $P = 0.004$; Fig. 2.4d). Forest area increased at a rate of 368km²/year from 1995 to 2015, and the total area reforested between 1995 and 2015 was 7736km² (Fig. 2.3d).

Annual mean costs of NFPP per km² were inversely related to the number of foresters per km² ($R^2 = 0.68$, $n = 17$, $P = 0.002$; Fig. 2.6a) and mean forest logging volume per km² ($R^2 = 0.74$, $n = 17$, $P < 0.001$; Fig. 2.6b), but were positively related to mean net forest stock growth ($R^2 = 0.45$, $n = 16$, $P = 0.069$; Fig. 2.6c).

Consistent with Hypotheses 2 and 3, the total recorded number of Amur tigers poached by snares was five individuals from 1998 to 2015: four individuals from 1998 to 2006 and one individual from 2007 to 2015, suggesting that management approaches improved protection of the tiger and may facilitate population growth. In addition, patrolling records showed a snare density of 0.4 snares/km in 2009 compared to only 0.1 snares/km for prey or tiger by catch during line transect survey in 2015 in Jilin Wangqing Nature Reserve.

The rate of tiger population decline paralleled increases in human population density and mean forest logging volume and decreases in net forest stock growth (Fig. 2.4b-d). However, as predicted by Hypothesis 4, the Amur tiger population size recovered from 14 tigers in 1999 to 27 animals based on our camera trap data from 2013 to 2015, similar to 26 individuals or more reported by Wang et al. (2016). Furthermore, the number of Amur leopards increased from approximately 10 leopards in 1998 (Yang et al., 1998) to 42 individuals in 2015 (Wang et al., 2016). Thus, the annual finite rate of increase (λ) of the Amur tiger population from 1999 to 2015 was 1.04, so that China's wild Amur tiger population could potentially grow to 100

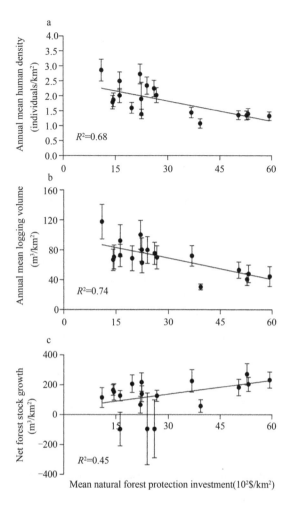

Fig. 2.6 Relationships between mean natural forest protection investment (the abscissa), annual mean human density (a), annual mean forest logging volume (b), and net forest stock growth (c) in 31 forest bureaus of Northeastern China from 1998 to 2015.

individuals by 2050. In addition, the rate λ for increase of the Amur leopard population from 1999 to 2015 was 1.08, and the wild Amur leopard population could potentially grow to 100 individuals by 2025.

For Hypothesis 4, we did see the predicted effects of changes in human density, forest growth, logging volume and prey on habitat areas occupied and population sizes of big cats. A total area of 41 200km^2 was occupied by Amur tigers since 1999 (Fig. 2.2). A total area of 10 200km^2 was occupied by Amur leopards from 1999 to 2014 (Fig. 2.2). Furthermore, annual habitat areas occupied by Amur tigers and Amur leopards increased rapidly and linearly from 1999 to 2015 (R^2= 0.52, n = 14, P = 0.049 for tigers; R^2 = 0.86, n = 19, P< 0.001 for leopards; Fig. 2.7).

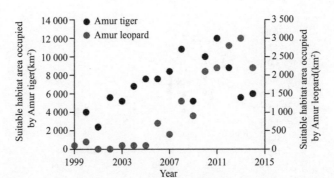

Fig. 2.7 Estimates of habitat areas occupied by Amur tigers and Amur leopards in northeastern China after the implementation of natural forest protection project in 1998. Please scan the QR code on the back cover to see the color image.

Habitat area occupied by the Amur tiger was inversely related to human density (R^2 = 0.55, n = 14, P = 0.032; Fig. 2.8a), positively related to mean net forest stock growth (R^2 = 0.63, n = 14, P = 0.011; Fig. 2.8b), but not related to mean forest logging volume (R^2 = 0.24, n = 14, P = 0.387; Fig. 2.8c). Likewise, habitat area occupied by the Amur leopard also was inversely related to human density (R^2 = 0.80, n = 15, P =

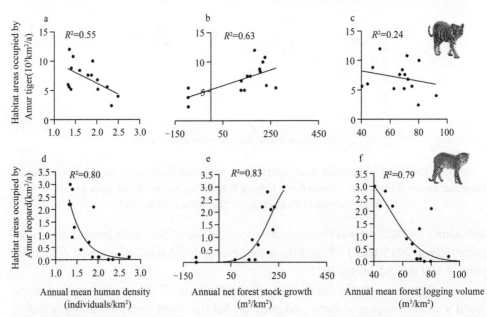

Fig. 2.8 Linear relationships between habitat areas occupied by big cats, annual mean human density (a), mean net forest stock growth (b), and mean forest logging volume (c), and nonlinear relationships between habitat areas occupied by big cats (the ordinate), annual mean human density (d), mean net forest stock growth (e), and mean forest logging volume (f) since the beginning of natural forest protection in 1998.

0.001; Fig. 2.8d), positively related to mean net forest stock growth ($R^2 = 0.83$, $n = 15$, $P = 0.002$; Fig. 2.8e), and inversely related to mean forest logging volumes per km^2 ($R^2 = 0.79$, $n = 15$, $P < 0.001$; Fig. 2.8f). Ungulate prey surveys showed that population density of key prey species increased from 2010 to 2014 over an area of 878km^2 in Jilin Wangqing Nature Reserve, supporting Hypothesis 3. Roe deer (*Capreolus pygargus*) density increased from (0.67 ± 0.13) individuals/km^2 in 2010 to (1.88 ± 0.22) individuals/km^2 in 2013, and then to (2.33 ± 0.52) individuals/km^2 in 2014. Wild boar (*Sus scrofa*) density increased from (0.32 ± 0.13) individuals/km^2 in 2010 to (0.40 ± 0.02) individuals/km^2 in 2013, and to (0.93 ± 0.66) individuals/km^2 in 2014. Sika deer density arose from (0.03 ± 0.022) individuals/km^2 in 2010 to (0.06 ± 0.03) individuals/km^2 in 2013, and to (0.34 ± 0.34) individuals/km^2 in 2014. Ungulate prey population data also provided evidence for the hypothesis concerning the relationship between tiger or leopard habitat areas or population recovery and prey population density growths.

2.4 Discussion

We found a transition from unsustainable forest management to land sharing and to land sparing for forest conservation in northeast China. During the same period, substantial decreases in regional human population density, forest volume logged, and known number of Amur tigers poached coincided with increases in three key prey species. Both increasing trends in big cat population sizes and increases in habitat areas occupied were also observed during the transition to land sharing and land sparing for forest conservation of northeast China.

We provided evidence of the lack of land sharing before 1999. We found that Amur tiger population size linearly decreased from 1950 to 1998 and has then increased since 1999 (Fig. 2.4d), while annual forest logging volume increased till 1998 and then decreased with an increasing trend of net forest stocks from 1999 to 2015 (Fig. 2.4d). The timings and temporal sequence of the changes in forest management, population dynamics of big cats during the past 65 years provided evidence for ineffective land sharing before 1999. Specifically, the more overexploitation of forest resources led to lower forest biomass and ungulate prey and higher poaching of prey and cats. Subsequently, since the beginning of NFPP in 1998, land sparing has promoted forest biomass recovery followed by prey recovery and increases in both habitat area and abundance of big cats.

Forest management should focus on both forest resource sustainability and humane social development (Carpenter *et al*., 2009). Our study provides partial support for successful big cat conservation in China, although populations were still small. We cannot simply attribute the success in the recovery of the two big cats to either the land sparing or land sharing model, because it might be caused by the role of ecological synergy. The lack of land sharing between the big cats and humans led to the dramatic declines in the abundance of the two large carnivores before 1998. By contrast, implementation of land sparing, through setting the majority of forestlands aside from logging, and land sharing, through maintaining the minimal forest logging and other forest use, has increased the abundance and habitat distribution of the two big cats. This represents a step towards the goals suggested by Chapron et al. (2014), that the conservation of large carnivores should focus on preserving the ecological processes driven by large carnivores in human-dominated landscapes and the functionality of forest ecosystems with different levels of completeness (Chapron et al., 2014). Hence, we should simultaneously emphasize the social processes of conservation investment with measurable social indicators (Mace, 2014); otherwise, we may not be able to ensure ecological and conservation outcomes of conservation investments.

On the other hand, styles of land ownership also influence the effectiveness of conservation investments due to the will of land owners (Wilcove et al., 1998). One big challenge western countries face in biodiversity conservation is persuading landowners to agree to protect wildlife on the privately owned lands (McShane et al., 2011). However, land in China is state-owned. State-owned land management provides benefits for "top-down" initiatives and was shown to be beneficial despite its complexity. Considering the potential risks of nationally uniform policies and their large-scale impacts, the Chinese government first conducted a pilot project and then elevated the annual amount of investments gradually, according to the program's progress (Fig. 2.4a, Fig. 2.9). In addition, welfare housing, increasing forest stock volume, and wildlife compensation policies played an important role in resolving human-tiger conflicts and protecting human wellbeing (Jiang et al., 2014). As far as the level of conservation investment is concerned, it is remarkably similar to the funding levels (> $3000/km^2) for successful lion conservation in unfenced African reserves (Packer et al., 2013).

Large carnivores have exhibited species-specific sensitivities to changes in human population density, allowing for successful adaptation to human-dominated landscapes (Chapron et al., 2014). Our results indicate that the decline of human density enhanced the distributional expansion of both Amur tigers and leopards (Fig. 2.4b, Fig. 2.6a, Fig. 2.8a, d). The control of human density in the core range of big cats not only directly reduced forest harvesting volume, but also mitigated human disturbances such as forest harvesting, farmland reclamation, infrastructure construction, poaching, and human activity (Giller et al., 2008; Qi et al., 2015). Disentangling the relationship between big cat poaching and changes in human densities is more complex because poaching effects may have a time lag. For example, once steel snares have been set up, they may impose risks to wild animals for many years. With increases in the frequency of big cat

Chapter 2 The synergy between ecology and society in Amur tiger conservation | 209

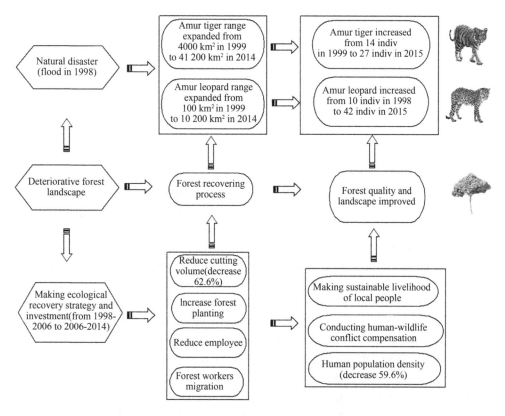

Fig. 2.9 Conservation solutions and actions (causes and synchrony roadmap for China from 1998 to 2015).

occurrence, local governments have conducted patrols in the key tiger and leopard habitats to locate and remove old snares to ensure big cat safety. Consequently, our results show that snare density decreased from 2009 to 2015 in the Jilin Wangqing Nature Reserve. Coincident with this, from 2007 to 2015, the recorded number of Amur tigers poached was significantly lower than that from 1998 to 2006. Although fewer and fewer people will live in the forests, local governments will need to continue to remove old snares through patrolling.

Evidence suggests that long-term forest restoration and simultaneous controls of illegal ungulate hunting and other human disturbances have increased ungulate abundance and distribution range (Jiang et al., 2015). For example, our ungulate surveys suggested that ungulate prey density (i.e., roe deer, wild boar and sika deer) increased from 2010 to 2014 over the 878km^2 Jilin Wangqing Nature Reserve. Based on the evidence of the prey driven distribution of big cats in previous studies (Jiang et al., 2015; Qi et al., 2015) and this study, we conclude that forest restoration contributed to the recovery of ungulate populations and subsequently increased the distribution range of big cats in northeastern China. Our findings provide a good example, in which big cat populations expanded in both size and habitat area used when humans retreated from forests through a combination of land sparing and sharing. Like other conservation studies, there is need to consider social problems caused by relocating people, such as the needs for infrastructure and new job opportunities. Conservationists have often ignored the social context and implications of conservation actions (Jiang et al., 2014). Our study reveals that the controlling of human density from a conservation investment standpoint may bring out synergy for social or ecological developments (Persha et al., 2011).

Several European studies have shown that the land sharing model for large carnivores to coexist with humans at high densities can be successful on a continental scale, and demonstrated that four large carnivore species live largely outside of protected areas (Chapron et al., 2014). However, on smaller regional scales in southwest Ghana and northern India, land sparing was a more promising strategy for minimizing negative effects of food production on tree and bird species biodiversity (Phalan et al., 2011). Similar to the European findings, we also found that the conservation of large carnivores may not merely rely on local nature reserve systems in China, which is the case for giant panda conservation (Liu et al., 2001; Loucks et al., 2001). The Changbaishan Nature Reserve was established for the conservation of Amur tigers in the 1960s. However, the reserve was insufficient to sustain the tiger population because the tiger has not been found in this reserve over the last 30 years. In this study, our results have demonstrated the effectiveness of landscape conservation on Amur tigers and leopards in China on a large scale. After the NFPP was initiated in 1998, forestland management in northeastern China incorporated not only the large-scale land sharing model to reduce human densities, but also regional-scale land sparing model to set aside 2861km^2 of unpopulated forests within the 9 forest bureaus. More generally, there is a need to integrate the land sparing and land sharing models for the conservation of large carnivores at multiple spatiotemporal scales (Hurlbert, 1978; Carter et al., 2012; Chapron et al., 2014).

The recovery of big cats, like other large carnivores, is limited by their low annual population growth rates and by diseases. The annual finite rate of increase of China's Amur tiger population is similar to that (1.046) of the Russian Amur tiger population during its increasing phase over 41 years (Miquelle et al., 2015). Long-term monitoring is needed for the assessments of the success or failure of the recovery (Miquelle et al., 2015). Short-term monitoring may not be adequate owing to the lack of data for sensitive ecological indicators and slowly changing populations of long-lived species. At the early stages of big cat conservation, we can use indirect social, economic, primary production (i.e., habitat vegetation improvement, etc.) or prey indicators to measure the success of conservation activities. Only after a relatively long period of conservation may we link the habitat area or population abundance of big cats to conservation investment indicators. Up to now, there is not a published time limit on the life-span of the NFPP from the Chinese government. China's central government still continues to fund the forest protection under the on-going forest harvest ban in northeastern China. Moreover, a series of centrally controlled large national parks (14 600km^2) for tigers and leopards has been scheduled to be established step by step across the big cat habitat landscape (Li *et al.*, 2016). The new National Park Initiative will ensure the long-term protection and conservation of big cats in northeastern China (Kathleen, 2016). Except the Amur tiger and Amur leopard national park pilot, the Chinese government is planning to establish other national parks, including Sanjiang-yuan national park pilot (123 100km^2), giant panda (Ailuropoda melanoleuca) national park pilot (27 000km^2), Hubei province Shennongjia national park pilot (1170km^2), Zhejiang province Qianjiangyuan national park pilot (252km^2), Hunan province Nanshan national park pilot (635.94km^2), Fujian province Wuyishan national park pilot (982.59km^2), Qilianshan national park pilot(50 200km^2), Beijing Great Wall national park pilot (59.91km^2), and Yunnan Shangri-La Pudacuo (602.1km^2). It is urgent to assess how much areas of both national parks and international or national ecological corridors are needed to maintain extensive permeable land-scapes for Amur tiger and leopard survival (Chapron *et al.*, 2014; Ekroos *et al.*, 2016). Based on this study, historical declines of the big cat population were accompanied with human encroachment, forest harvest intensification, and then the occurrence of natural disasters and conservation decision-making (Fig. 2.9). With the progress of conservation, some new social or ecological problems may occur. Increasing tiger-human conflicts (mainly killing livestock) also occurred simultaneously with the tiger recovery (Fig. 2.3b). Consequently, attention should be paid to large-scale synergistic effects between big cats or other large carnivores and human livelihoods to improve conservation and future recovery potential with sustainable forestland management. At the same time, we also should pay attention to differences in habitat and prey requirements between the two big cats. Critically endangered Amur leopards are more vulnerable while competing with sympatric Amur tigers (Jiang *et al.*, 2015). Hence, the balanced conservation of sympatric big predator guild may be a future challenge for conservationists. The findings of this study provide great lessons of land management strategies not only for big cats, but also for other large carnivore guilds and habitat conservation under complex social and ecological synergy worldwide.

2.5 Summary

Global biodiversity conservation has recently focused on the roles of land management strategies of land sharing vs. land sparing. However, few studies have evaluated the roles of social and ecological interactions in modifying the effectiveness of land management for top predator conservation. Using a 65-year dataset about the ecology and society economic conditions in northeastern China, we evaluated the roles of government social policies in resolving human-wildlife conflicts and improving human livelihood. From 1998 to 2015, both big cat populations and their habitats have increased significantly. Concurrently, regional human population density decreased by 59.6%, forest volume logged was reduced by 62.6%. Consequently, increases of key prey species were observed during the same periods. Although populations remained small, the annual finite rate of increase was 1.04 for the Amur tiger population and 1.08 for Amur leopards from 1999 to 2015. Habitat areas occupied by big cats increased significantly. Overexploitation of forest resources and big cat declines under previous unsustainable land use are progressively being reversed under land sparing. Large economic investment and intense human-relocation projects coupled with efforts to reduce poaching and illegal hunting and trapping demonstrate a complex social and ecological synergy in big cat conservation in China.

References

Athreya V, Odden M, Linnell J D, *et al*. 2013. Big cats in our backyards: persistence of large carnivores in a human dominated landscape in India. PLoS ONE, 8: e57872.

Berryman A, Turchin P. 2001. Identifying the density-dependent structure underlyingecological time series. Oikos, 92: 265-270.

Carpenter S R, Mooney H A, Agard J, *et al*. 2009. Science for managing ecosystemservices: beyond the millennium ecosystem assessment. Proceedings of the National Academy of Sciences of the United States of America, 106: 1305-1312.

Carter N H, Shrestha B K, Karki, J B, *et al*. 2012. Coexistence between wildlife and humans at fine spatial scales. Proceedings of the National Academy of Sciences of the United States of America, 109: 15360-15365.

Ceballos G, Ehrlich P R. 2002. Mammal population losses and the extinction crisis. Science, 296: 904-907.

Chapron G, Kaczensky P, Linnell J D C, *et al*. 2014. Recovery of large carnivores in Europe's modern human-dominated landscapes. Science, 346: 1517-1519.

Chapron G, Miquelle D G, Lambert A, *et al*. 2008. The impact on tigers of poaching versus prey depletion. Journal of Applied Ecology, 45: 1667-1674.

Dinerstein E, Loucks C, Wikramanayake E, *et al*. 2007. The fate of wild tigers. Bioscience, 57: 508-514.

Ekroos J, Ödman A M, Andersson G K S, *et al*. 2016. Sparing land for biodiversity at multiple spatial scales. Frontiers in Ecology and Evolution, 3: 145.

Giller K E, Leeuwis C, Andersson J A, *et al*. 2008. Competing Claims on Natural Resources: What Role for Science? Ecology and Society, 13(2): 34.

Gompper M E, Belant J L, Kays R. 2015. Carnivore coexistence: America's recovery. Science, 347: 382-383.

Goodrich J M, Miquelle D G, Smirnov E N, *et al*. 2010. Spatial structure of Amur (Siberian) tigers

(*Panthera tigris altaica*) on Sikhote-Alin Biosphere Zapovednik, Russia. Journal of Mammalogy, 91: 737-748.

Hebblewhite M, Miquelle D G, Murzin A A, *et al.* 2011. Predicting potential habitat and population size for reintroduction of the Far Eastern leopards in the Russian Far East. Biological Conservation, 144: 2403-2413.

Hurlbert S H. 1978. The measurement of niche overlap and some relatives. Ecology, 59: 67-77.

Hyde W F, Belcher B M, Xu J. 2003. China's forests: global lessons from market reforms. Washington DC: Resource for the Future Press.

Jiang G, Qi J, Wang G, *et al.* 2015. New hope for the survival of the Amur leopard in China. Scientific Reports, 5.

Jiang G, Sun H, Lang J, *et al.* 2014. Effects of environmental and anthropogenic drivers on Amur tiger distribution in northeastern China. Ecological Research, 29: 801-813.

Jiang G. 2014. New evidence of wild Amur tigers and leopards breeding in China. Oryx, 48: 326.

Kathleen M. 2016. Can a new park save China's big cats? Science. http://dx.doi.org/10.1126/science.aah 7201[2017-3-6].

Li J, Wang W, Axmacher J C, *et al.* 2016. Streamlining China's protected areas. Science, 351: 1160.

Liu J, Linderman M, Ouyang Z, *et al.* 2001. Ecological degradation in protected areas: the case of Wolong Nature Reserve for giant pandas. Science, 292: 98-101.

Loucks C J, Lü Z, Dinerstein E, *et al.* 2001. Giant pandas in a changing landscape. Science, 294: 1465.

Ma Z, Melville D S, Liu J, *et al.* 2014. Rethinking China's new great wall. Science, 346(6212): 912-914.

Mace G M. 2014. Whose conservation? Science, 345: 1558-1560.

McShane T O, Hirsch P D, Trung T C, *et al.* 2011. Hard choices: making trade-offs between biodiversity conservation and human well-being. Biological Conservation, 144: 966-972.

Miquelle D G, Smirnov E N, Zaumyslova O Y, *et al.* 2015. Population dynamics of Amur tigers (*Panthera tigris altaica*) in Sikhote-Alin Biosphere Zapovednik: 1966-2012. Integrative Zoology, 10: 315-328.

Morrison J C, Sechrest W, Dinerstein E, *et al.* 2007. Persistence of large mammal faunas as indicators of global human impacts. Journal of Mammalogy, 88: 1363-1380.

Naughton-Treves L, Holland M B, Brandon K. 2005. The role of protected areas in conserving biodiversity and sustaining local livelihoods. Annual Review of Environment and Resources, 30: 219-252.

Packer C, Loveridge A, Canney S, *et al.* 2013. Conserving large carnivores: dollars and fence. Ecology Letters, 16: 635-641.

Persha L, Agrawal A, Chhatre A. 2011. Social and ecological synergy: local rule making, forest livelihoods, and biodiversity conservation. Science, 331: 1606-1608.

Phalan B, Onial M, Balmford A, *et al.* 2011. Reconciling food production and biodiversity conservation: land sharing and land sparing compared. Science, 333: 1289-1291.

Qi J, Shi Q, Wang G, *et al.* 2015. Spatial distribution drivers of Amur leopard density in northeast China. Biological Conservation, 191: 258-265.

Ripple W J, Estes J A, Beschta R L, *et al.* 2014. Status and ecological effects of the world's largest carnivores. Science, 343: 1241484.

Shi Q, Li Q, Zhang M. 2015. First camera-trap video evidence of the Amur tiger breeding in China. Oryx, 49: 205-206.

Sikes RS, Gannon WL. 2011. Guidelines of the American Society of Mammalogists for the use of wild mammals in research. Journal of Mammalogy, 92: 235-253.

Tian Y, Wu J, Kou X, *et al*. 2009. Spatial and temporal dynamics of the amur tiger population and its cause analysis. Biodiversity, 17(3): 211-225. (in Chinese)

Wang T, Feng L, Mou P, *et al*. 2016. Amur tigers and leopards returning to China: direct evidence and a landscape conservation plan. Land scape Ecology, 31: 491-503.

Wikramanayake E, McKnight M, Dinerstein E, *et al*. 2004. Designing a conservation landscape for tigers in human-dominated environments. Conservation Biology, 18: 839-844.

Wilcove D S, Rothstein D, Dubow J. 1998. Quantifying threats to imperiled species in the United States. Bioscience, 48: 607-615.

Xu J, Yin R, Li Z, *et al*. 2006. China's ecological rehabilitation: Unprecedented efforts, dramatic impacts, and requisite policies. Ecological Economics, 57: 595-607.

Yang S, Jiang J, Wu Z, *et al*. 1998. Report on the Sino-Russian joint survey of Far Eastern leopards and Siberian tigers and their habitat in the Sino-Russian boundary area, eastern Jilin Province, China, winter 1998 // A Final Report to the UNDP and the Wildlife Conservation Society.

Chapter 3 An integrated analysis of the causes of ungulate mortality in the Wandashan Mountains (Heilongjiang Province, China) and an evaluation of habitat quality

3.1 Introduction

In many parts of the world, large wild ungulates are a significant economic resource for local and regional communities (Loibooki *et al.*, 2003; Corlett, 2007); however, illegal hunting for wild bushmeat is a real threat to their survival (IUCN, 2010). Numerous reports have shown that poaching has reduced some ungulate populations to crisis levels(Milner-Gulland and Bennett, 2003; Baillie *et al.*, 2004; Sodhi *et al.*, 2004; Corlett, 2007). This crisis is particularly noticeable in the Wandashan Mountains of Heilongjiang Province, China. Increasing development pressure on natural resources in the Wandashan Mountains has presented local researchers and management authorities with a unique dilemma: on the one hand, road construction and logging has boosted economic development, on the other, it has also provided easier access for poachers (Blake *et al.*, 2007). In an effort to find the most effective measures for preserving large ungulates (Zhang and Liu, 2008), conservationist studies have focused on habitat selection and spatial distribution (Jiang and Zhang, 2008; Jiang *et al.*, 2011). Habitat selection models by ungulate species have been developed, providing an effective conservation plan. However, the analyses, as with many habitat suitability models, are limited due to lack of data on human-caused ungulate mortality, leading to erroneous conclusions if species occurrence does not correspond with positive reproductive and survival rates (Garshelis and David, 2000).

Some conservationists believe that high quality habitats, identified within habitat models and generated by species occurrence alone, might actually be located in ecological sink areas where reproductive and/or survival rates are too low and the mortality risk is too high to sustain a viable population. The attractive sink as a high quality habitat is a particular case where animals perceive an area as a good habitat, even though human-caused habitat modification will ultimately reduce demographic performance(Delibes *et al.*, 2001; Naves *et al.*, 2010). The identification of attractive sinks as high quality habitat will lead to erroneous conclusion in management and conservation plans. (Naves et al., 2010) originally proposed the identification of attractive sink-like habitats using a two-dimensional habitat model, while(Nielsen et al., 2008) considered both variables in occurrence and human-caused mortality models. Following this approach, (Falcucci et al., 2009), using ecological niche factor analysis, evaluated habitat quality for conservation via an occurrence-mortality model. They built a two-dimensional model, i.e., an occurrence and human-caused mortality model, and then integrated the reclassified models. (Falcucci et al., 2009) defined seven habitat states based on the interactions between suitability and mortality risk categories: unsuitable habitats (unsuitable class in the occurrence model and all mortality risk values), attractive sink-like habitats (medium and high mortality), and source-like or secure habitats (no and low mortality risk). Although the integrated occurrence-mortality model performed better for habitat analysis, considering demographic features, it failed to resolve how species respond to a single habitat factor, analyzing human-caused mortality only in the process of habitat selection for occurrence and survival.

Knowledge of mortality rates and causes of mortality are important in understanding ungulate population dynamics. Datasets on mortality and survival have yielded a real opportunity to analyze the causes of mortality and insights into ways of conserving current populations by reducing mortality (Groom et al., 2006). The habitat selection index (HSI), for example, is an effective way of analyzing the effects of specific factors on animal habitat selection(Johnson, 1980; Underwood, 2014). The HSI is usually produced using species presence to infer the preference of a consumer for a particular habitat component(Li et al., 2006). Habitat suitability index models, based on HSI, are usually generated to infer habitat quality (Duberstein et al., 2008; Guisan and Thuiller, 2010). The integrated occurrence-mortality model is based on both the occurrence and mortality models, and can be used to develop suitable strategies for conservation and management(Falcucci et al., 2009).

In analyzing habitat selection and quality evaluation using the two datasets on occurrence and human-caused mortality, our goals were: ① to analyze the causes of mortality in wild boar(*Sus scrofa*), red deer(*Cervus elaphus*) and roe deer(*Capreolus pygargus*), ② to analyze habitat factors affecting ungulate occurrence and mortality, ③ to generate an integrated occurrence-mortality model which can be used to distinguish between attractive sink-like and source-like habitats, and provide effective management strategies.

3.2 Study area and background

Our study area covered 3692.06km^2, located between 46°07′55″N–47°01′41″N and 132°48′52″E–133°56′55″E, where the population of wild boar, roe deer and red deer have been recorded over the last 37 years(Ma *et al.*, 1986). The study area bordered the Russia Far East and was dominated by mountains with mean elevations of –200m. Vegetation types included coniferous deciduous forest, coniferous/broad leaved mixed forest, broad-leaved forest, shrub land, forest wetland, and farmland. Mean annual rainfall ranges 500-800mm. Temperatures, however, vary considerably throughout the year, with a monthly maximum temperature of 34.6 ℃ and monthly minimum temperature of –34.8 ℃. Snow begins to fall in late November, to a maximum depth of 110cm, and does not begin to thaw until the end of April. Human density (higher at lower elevations) is 32 inhabitants/km^2 at township level.

The diversity of habitat reflects the biodiversity, supporting animals such as wild boar, red deer, roe deer, goral (*Naemorhedus goral*), snow hare (*Lepus timidus*), and amur tiger (*Panthera tigris altaica*)(Li *et al.*, 2008). However, forest clearance and human subsistence poaching has reduced the population number of wild boar, red deer and roe deer(Zhou *et al.*, 2010). The Chinese government has passed several laws and regulations in the fight against poaching, and civilian-held guns were confiscated by the Government in the mid-1990s. Population recovery of these three ungulate species is an urgent issue for wildlife conservation authorities in the study area.

3.3　Materials and methods

3.3.1　Mortality data

We obtained two datasets on ungulate mortality in the Wandashan Mountains. The first dataset included indirect or direct signs of ungulate occurence from line-transect surveys of ungulate population abundance in later winter 2008 to early spring 2009. The second included indirect or direct signs obtained from a long-term monitoring program of the amur tiger population between 2002 and 2009(Zhou *et al.*, 2008). A total of 160 cases (58 wild boar, 40 red deer and 62 roe deer) was collected. We edited the datasets for spatial accuracies and spatial–temporal independence, with selection locations of at least 500m and 24h apart. From these, we excluded nine cases with inaccuracies or missing coordinates to account for anthropogenic ungulate mortality. We therefore used 151 human-caused mortality cases (56 wild boar, 37 red deer and 58 roe deer) for analysis. Two datasets were pooled and the inputs used for the mortality model of the three ungulate species.

3.3.2　Cause of mortality in the three ungulate species

We combined the mortality datasets from the line-transect surveys of ungulate population abundance and long-term monitoring of the amur tiger population. The causes of ungulate mortality were examined and the principle causes grouped into four categories: ① natural mortality, ② human-caused mortality by cable snares, ③ human-caused mortality by poison, ④ human-caused mortality by hunting with hounds. The year was split into two periods from November to April (snowfall period), and May to October (non-snowfall period).

3.3.3 Survey design and data collection

Ungulates were difficult to observe directly due to low population density and high sensitivity to human disturbance. The information on ungulate occurrence and locations is based on snow–tracks surveys. Datasets were collected over two time-periods. The first dataset included GPS locations of ungulate activity collected along 105 line transects (each line transect was 5km in length) in later winter 2003 and 2006, as well as in earlier spring 2005 (135 activity locations for wild boar, 165 for red deer and 258 for roe deer). The second dataset included 234 activity locations collected along 240 line transects in later winter 2008, and early spring and later winter 2009 (63 activity locations for wild boar, 54 for red deer and 117 for roe deer). A total of 240 line transects, 5km in length, were place in forest stand maps with the beginning and end locations demarcated. Line transects were randomly assigned to a sub-sample of these predetermined locations, which were accessible to the surveyors. Surveyors walked slowly (<0.8km/h) along the line transect noting animal signs (tracks, dung, fecal material, evidence of foraging), recording habitat factors and collecting mortality events. Participants either worked for ungulate conservation organizations for many years or had experience in the forestry areas, and were specially trained in field survey techniques.

3.3.4 Analysis of habitat selection

3.3.4.1 Habitat selection index based on occurrence locations: ungulate occurrence model

In this study "habitat type" represents a category within a "habitat class", which in turn signifies a geographical or biological category. For statistical analysis of habitat selection using occurrence location data, six habitat factors were selected based on previous studies(Zhou et al., 2006; Zhou et al., 2010; Jiang et al., 2011) (Table 3.1). The habitat selection index (E_i^*) method was used to analyze habitat selection by three ungulate species (Han et al., 2004):

$$E_i^* = \frac{W_i - 1/n}{W_i + 1/n} \qquad (3.1)$$

$$W_i = \frac{r_i / p_i}{\sum(r_i / p_i)} \quad (3.2)$$

where n is the rank of each habitat factor, W_i is the selection coefficient, p_i is the percentage of available habitat area with an i rank character versus the whole study area, r_i is the percentage of activity locations (usage habitat) within the available habitat of i hierarchy character versus the total number of activity locations within whole study area.

Table 3.1 Rating category of single habitat factors used in analyzing habitat selection by ungulates

Number	Aspect	Slope (°)	Elevation(m)	Vegetation	Settlement(m)	Road (m)
1	Plain	<5	0–150	Coniferous forest	0–1500	0–1000
2	N	5–15	150–300	Coniferous broad leaved mixed forest	1500–3000	1000–2000
3	NE	15–25	300–450	Broad leaved forest	3000–4500	2000–3000
4	E	>25	450–830	Forest wetland	4500–6000	3000–4000
5	SE			Farmland	>6000	>4000
6	S					
7	SW					
8	W					
9	NW					

E_i^* has values from -1 to $+1$, which show the preference of animal for special components of the habitat. Habitats that were not selected by consumers were denoted by $E_i^* = -1$ and those that were preferentially selected were denoted by $E_i^* = 1$. Habitats that were randomly selected by ungulate were denoted by $E_i^* = 0$. To gain a better insight into the preference of habitat selection, a χ^2 test was performed to determine if the i ranked habitat for each variable used by the consumer was statistically significant by comparing the number of actual activity locations in the i ranked habitats with the number of random sampling locations (Table 3.2). All preferred categories were classed as $E_i^* > 0$ with significant difference (high suitability habitat), those avoided as $E_i^* < 0$ with significant difference (unsuitable habitat), the third habitat was selected randomly as $E_i^* > 0$ without significant difference (medium suitability habitat), and the final habitat was selected randomly as $E_i^* < 0$ without significant difference (low suitability habitat) (Jiang, 2006). Values (0-3) were assigned to each rank of significant factors (0 for unsuitable habitats, 1 for low suitability habitats, 2 for medium suitability habitats and 3 for high suitability habitats). Finally, the habitat suitability index (*HIS*) model was used, incorporating the four habitat factors (aspect, elevation, vegetation type and settlements) to construct an ungulate occurrence model. If the suitability index for any component variable was zero, then there is no suitable habitat for ungulates, i.e., $HSI=0$. It was assumed that the four components played an equal role in determining the probability of occurrence of the three ungulate species(Thomasma, 1981; Jiang et al., 2011). A geometric mean of the four suitability indices ($HSI_{geometric}$) was used, such that:

$$HSI_{geometric} = (V_1 V_2 V_3 V_4)^{1/4} \quad (3.3)$$

The values of V_1, V_2, V_3, V_4 were awarded to the appropriate ranks of habitat factors. *HSI* values were rated 0-3 (0 unsuitable habitat, 0-1 low suitability habitat, 1-2 medium suitability habitat and 2-3 high suitability habitat). Finally, the three occurrence model maps for wild boar, red deer and roe deer were combined to generate an integrated predictive occurrence model map. The degree of habitat overlap among species was calculated and the integrated predictive occurrence model map reclassified.

Table 3.2 Selective index (E_{ij}) for wild boar, red deer and roe deer

Number	Aspect	Elevation(m)	Vegetation	settlements
1	$-0.22^a, 0.18^b, -0.10^c$	$-0.08^a, -0.05^b, -0.16^c$	$-0.16^a, -0.15^b, -0.13^c$	$-0.16^a, -0.23^b, -0.11^c$
2	$-0.12^a, -0.25^b, -0.29^c$	$0.32^{a*}, 0.18^b, ^c0.41^{c*}$	$0.45^{a*}, 0.25,^b, 0.21^c$	$0.15^a, 0.05,^b, 0.01^c$
3	$0.08^a, 0.10^b, 0.31^c$	$0.11^a, 0.36^{b*}, 0.06^c$	$0.36^a, 0.38^{b*}, 0.57^{c*}$	$0.43^{a*}, 0.22^b, 0.36^{c*}$
4	$0.40^{a*}, 0.32^{b*}, 0.45^{c*}$	$-0.15^a, -1^{b*}, -1^{c*}$	$-0.29^a, -0.12^b, -0.28^c$	$0.19^a, 0.47^{b*}, 0.18^c$
5	$0.19^a, 0.13^b, 0.12^c$		$-1^{a*}, -1^{b*}, -0.31^c$	$0.02^a, 0.06^b, -0.11^c$
6	$-0.23^a, 0.19^b, -0.22^c$			
7	$-0.14^a, -0.07^b, -0.11^c$			
8	$-0.65^{a*}, -0.37^b, -0.58^{c*}$			
9	$-0.15^a, -0.55^{b*}, -0.26^c$			

Note: a. wild boar; b. red deer; c. roe deer. (*$P<0.05$).

Slope and aspect were derived from 10-m contour lines. Aspect was defined by how many degrees the slope (rotated into the horizontal in the normal plane) would vary from true north. The distances to human disturbance of settlements and roads were calculated using the Arcview 3.3 spatial analyst program.

Although water resource are a principle factor affecting habitat selection (Simcharoen et al., 2008), they were not considered as limiting factors in the Wandashan Mountains. Even though rivers were frozen during the study period, snow was freely available; thus the variable "water" was excluded.

3.3.4.2 Habitat selection index: human-caused mortality risk and an integrated occurrence-mortality model

Using mortality datasets on the three ungulate species, the procedure described above was used to analyze the relationship between single habitat factors and mortality locations and construct a human-caused mortality risk model for each species. The three models were then combined to create an integrated model of human-caused mortality and the integrated model of human-caused mortality reclassified for the three ungulate species.

To develop the integrated occurrence-mortality habitat model, the reclassified integrated occurrence and mortality models for the three ungulate species were combined. Following(Falcucci *et al.*, 2009), seven habitat states were defined, based on the interactions between suitability and mortality risk categories and obtained habitat size at various levels.

3.4 Results

3.4.1 Causes of mortality

Endemic poaching in forest areas contributed 87.42% to ungulate mortality. Total natural mortality for the three ungulate was typically low and accounted for only 12.58%. Poaching via poisoning accounted for 40.07% and steel cable snaring for 27.79% of wild boar mortalities. For red deer, 51.35% of the mortalities were due to poisoning and 40.54% to steel cable snaring. For roe deer, the figures were 29.31% from poisoning and 56.90% by steel cable snaring (Fig. 3.1). Analysis of seasonal trends in ungulate mortality show that mortality in the three ungulate species was concentrated during late winter and early spring (together accounting for 86.00% of mortality).

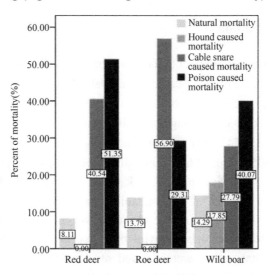

Fig. 3.1   Causes of mortality in 56 wild boar, 37 red deer and 58 roe deer in the Wandashan Mountains between 2007 and 2009.

3.4.2 Habitat selection based on the occurrence and mortality locations of ungulate

We excluded "road" and "slope" factors, which had no statistical significance; the remaining four habitat factors were used to construct an occurrence model. To generate mortality models, slope and settlements without statistical significance were excluded from the six habitat factors. Thus, four habitat factors with statistical significance were used to generate a mortality model (Table 3.3).

Table 3.3 Mortality index (E_{ij}) for wild boar, red deer and roe deer

No.	Aspect	Elevation	Vegetation	Roads
1	-0.11^a, 0.05^b, -0.10^c	-0.04^a, -0.06^b, -0.05^c	-0.06^a, -0.08^b, -0.15	0.36^{a*}, 0.31^{b*}, 0.42^{c*}
2	-0.13^a, -0.18^b, -0.16^c	0.02^a, 0.05^b, 0.09^c	0.38^{a*}, 0.16^b, 0.11	0.21^a, 0.22^b, 0.21^c
3	0.23^a, 0.17^b, 0.13^c	0.21^{a*}, 0.39^{b*}, 0.28^{c*}	0.15^a, 0.43^{b*}, 0.45^*	0.06^a, 0.09^b, 0.06^c
4	0.30^{a*}, 0.25^{b*}, 0.33^{c*}	-0.15^a, -1^{b*}, -1^{c*}	-1^{a*}, -1^{b*}, -1^{c*}	-0.08^a, -0.05^b, -0.10^c
5	0.09^a, 0.02^b, 0.02^c		-1^{a*}, -1^{b*}, -1^{c*}	-0.21^a, -0.18^b, -0.05
6	-0.14^a, 0.06^b, -0.18^c			
7	-0.34^a, -0.07^b, -0.11^c			
8	-0.18^a, -0.37^b, -0.15^c			
9	-0.11^a, -0.10^b, -0.15^c			

Note: a. wild boar; b. red deer; c. roe deer (*$P<0.05$).

3.4.3 Habitat evaluation

A *HSI* model map, with respect to the occurrence of each species, was created using the GIS tool (Fig. 3.2a). Species-specific habitat values were derived as: ① 573.38km² of unsuitable habitat, 852.25km² of low suitability habitat, 1101.5km² of medium suitability habitat, 1164.93km² of high suitability habitat for wild boar; ② 954.06km² of unsuitable habitat, 954.87km² of low suitability habitat, 1177.95km² of medium suitability habitat, 605.18km² of high suitability habitat for red deer; ③ 303.49km² of unsuitable habitat, 374.06km² of low suitability habitat, 1596.07km² of medium suitability habitat, 1418.44km² of high suitability habitat for roe deer. In the mortality analysis, four habitat factors, i.e., aspect, elevation, vegetation and roads, were imputed to generate the mortality models for wild boar, red deer and roe deer (Fig. 3.2b). Species-specific habitat values were derived as: ① 511.35km² of no risk habitat, 957.35km² of low risk habitat, 846.59km² of medium risk area, 1376.77km² of high risk habitat for wild boar; ② 557.87km² of no risk habitat, 1247.92km² of low risk habitat, 1149.33km² of medium risk habitat, 736.94km² of high risk habitat for red deer; ③ 1667.19km² of no risk habitat, 663.46km² of low risk habitat, 285.39km² of medium risk habitat, 1076.02km² of high risk habitat for roe deer.

Fig. 3.2 Occurrence model, mortality model and integrated occurrence-mortality model for wild boar, red deer and roe deer.

By creating the integrated multi-species occurrence model for the three ungulate species, 8.06% (297.49km²) of the study area was designated as unsuitable, 14.30% (528.06km²) as low suitability habitat, 63.30% (2337.07km²) as medium suitability habitat and 14.34% (529.44km^2) as high suitability habitat (Fig. 3.2c). In the integrated mortality model map (Fig. 3.2c), the risk class areas were 12.13% (448.06km^2) as no risk, 41.47% (1531.00km^2) as low risk, 27.01% (997.30km^2) as medium risk and 19.39% (715.70km^2) as high risk habitat.

In the integrated occurrence-mortality model, 8.06% (297.49km^2) of the study area was designated as unsuitable for ungulate species, with 5.43% (200.44km^2), 32.76% (1209.85km^2) and 8.20% (302.77km^2) as the first, second and third attractive sink-like habitats, respectively, and 8.91% (329.00km^2), 30.53% (1127.22km^2) and 6.11% (225.29km^2) as the first, second and third source-like habitat, respectively.

3.5 Discussion

3.5.1 Causes and percentage of mortality

Mortality data collected from line-transect surveys was low because poaching occurs in secret. The main data sources on mortality for the three ungulate species were collected from our long-term amur tiger-monitoring program(Zhou *et al.*, 2010). The combined natural mortality for all three ungulate species was 12.58% versus 87.42% for human-caused mortality. The natural death of the three wild boars was due primarily to food shortage during a severe winter; with red and roe deer the probable cause was old age. Extensive poaching did contribute significantly to mortalities; specifically, the pattern of steel cable snares and poisoning were found to be serious threats to survival of the ungulate populations in the Wandashan Mountains. The most alarming fact was that poaching accounted for 87.42% of combined ungulate mortality (85.71% for wild boar, 91.89% for red deer and 87.21% for roe deer). This study has substantiated previous surveys on ungulate populations as to harmful effect and extent of poaching in the Wandashan Mountains(Zhang and Zhang, 2011). It is very important to understand the reasons for ungulate exploitation(Baillie *et al.*, 2004; Kühl *et al.*, 2009). Exploitation of bushmeat by forest-dwellers has increased in recent years, which is strongly linked to poverty and unemployment (Kühl *et al.*, 2009). Our study would indicate that forest-dwellers are feeding their families by hunting large ungulates (wild boar, red deer and roe deer). Furthermore, some studies have shown that man's attitude towards wildlife conservation in human-dominated landscape is a key factor influencing the survival of ungulates (Linnell *et al.*, 2001). The long-term viability of ungulate populations requires an effective education program that positively changes the attitude of the public and poachers to ungulates, an aggressive road-management program (forest road deactivation) to reduce human-caused ungulate mortality, and increased employment opportunities to boost the income of residents of the Wandashan Mountains. A positive change in the public's attitude has resulted in increased survival and a recovery of the wolf population (Linnell et al., 2001). Moreover, since poachers operate in secret, government-funded anti-poaching patrols need to be established to prevent the killing of ungulates, specifically during late winter and early spring.

3.5.2 Habitat selection and quality evaluation

Identifying the resource requirements of a species is a very important part of conserving wildlife populations (Freitas et al., 2010). However, some problems are associated with two of the main underlying assumptions: that recorded data can be used to infer habitat selection, and that evidence of habitat selection is related to fitness and population growth (Alldredge and Ratti, 1986; Porter and Church, 1987). Nevertheless, investigating how individuals select different habitats is a necessary step in identifying resources that influence the fitness of individuals and population viability(Ratcliffe and Crowe, 2001; Simcharoen et al., 2008).

Our study found that many factors were relevant in habitat selection by ungulates, although some factors were more influential than others in winter (Podchong et al., 2009). The results indicated that the three ungulate species showed a preference for east-facing slopes, presumably because there are warmer and have more available food (Singer et al., 1981; Gao and Zhang, 1995). Vegetation type was also an important factor in habitat selection. Previous studies on wild boar found a preference for coniferous forests, presumably because they provide abundant food resources throughout winter (i.e., nutritious conifer seeds) (Grodziński and Sawicka-Kapusta, 1970). Our findings also showed that human disturbance is an important factor in habitat selection. Wild boar avoided coniferous forests and settlements, possible due to the collecting of conifer seeds by the local population. Although the "road" factor was not statistically significant, as noted previously, roads provided easy access for poachers. A human presence can lead to chronic stress in deer(Sauerwein et al., 2004), which lead to avoidance behavior through reduced fitness(Bèchet et al., 2004). Agricultural land and forest wetlands are two extreme habitats, because of their exposure and lack of food in winter. The three ungulate species avoided agricultural land and forest wetlands, preferring coniferous/broad-leaf and broad-leaf forests in winter (Zhou et al., 2010).

Ungulate species usually occurred at elevations ranging between 150 and 450m, areas we consider essential for the survival and long-term preservation of these important ungulate species. In addition, habitats at these elevations are dominated by coniferous/broad-leaf and broad-leaf forests, which not only provided adequate food and shelter, but also reduce the impact of human activity on wildlife (Zhou et al., 2010).

Despite the similarities in slope selection, the three ungulate species showed a clear difference in tolerance of human-associated disturbance.

Mortality records for the three ungulate species were difficult to collect along the line transects, with a number of locations in later winter 2008 and early spring 2009. The long-term amur tiger-monitoring program, which began in 2002 and also monitors population fluctuations in wild boar, red deer and roe deer, provided better data on ungulate mortality. Furthermore, some of the amur tiger-monitoring team work at the local Forestry Bureau, which keeps records of mortality events. Locations of mortality events related to east-facing aspects, coniferous/broad-leaf and broad-leaf forests may indicate a potential relationship between ungulate occurrence and mortality in these areas. Our study found that disturbance from human activities increased the survival threat to ungulates, and the settlements and roads increased the mortality risk (Stankowich, 2008; Laurance et al., 2010). In the study area, the loss of forest cover via increased forest clearance led to increased road density to access and remove timber.

In a human-dominated landscape, habitat evaluation is important and should be linked with human demographic data(Naves et al., 2010; Thomas and Kunin, 2010). Habitat evaluation can be used to understand the biological needs of a species and to create conservation strategies for the effective management of ungulates(Underwood, 2014). However, evaluation of high quality habitat using animal occurrence alone is questionable and often provides misleading information(Falcucci et al., 2009). We used both environment and anthropogenic factors to create a habitat evaluation model, which avoided some potential errors in conservation planning. Although habitat loss represents the main threats to ungulate survival(Li et al., 2008) in the study area, habitat availability does not appear to be a limiting factor (Zhang and Liu, 2008). This conclusion is well supported by the integrated occurrence-mortality model map for the three ungulate species. Our integrated model indicated that the traditional habitat suitability model can provide misleading indications for conservation and management planning, with 46.39% of suitable habitat in the occurrence model being associated with mortality risk. There are areas of suitable source-like (secure) habitat (far from roads, settlements and farmland) for the three ungulate species around the Shendingfeng Mountains, and we believe that the effective control of human-caused mortality in these source-like habitat areas, in addition to controlling habitat loss, should be a priority for the conservation of ungulates.

The respective occurrence and mortality models relating to demographic performance lead to the development of an integrated occurrence-mortality model based on the assumption that human-caused mortality can be used to model the effect of anthropogenic factors on ungulate' survival and that ungulate occurrence can be used to generate an appropriate model.

A potential weakness in the process occurs when entire datasets are utilized from ungulate data collected only in winter. Snow tracking in winter provides more accurate and precise information than data from other seasons (Zhang and Liu, 2008; Zhou et al., 2010; Alexander et al., 2011). Moreover, the weakness is perpetuated when mortality data from the entire year is used in the mortality model. Nevertheless, 86.00% of ungulate mortality occurred in winter, thus reducing the disparity between the two datasets.

In terms of overall habitat quality, our integrated occurrence-mortality model indicated that many areas characterized as high suitability habitats were commonly interspersed with attractive sink-like habitats (Fig. 3.2c). Analysis of habitat factors, which were inputted into the mortality model and then reclassified with the occurrence-mortality model, confirmed that human-induced mortality is a major threat to ungulate survival. In addition, the "road" factor resulted in reduced high suitability habitats and increased attractive sink-like habitats.

3.6 Summary

Through this study, we consider that steel cable snares and poison, used to kill ungulates, were the primary threats to survival of wild boar, red deer and roe deer. Furthermore, these factors, including aspect, slope, elevation and forest type, are important factors in determining ungulate habitat preference. The integrated occurrence-mortality model indicated that 46.39% of suitable habitat was associated with mortality risk. The attractive sink-like habitats account for a large proportion of study area. So we suggest that source-like habitats should be preserved to prevent habitat loss and degradation, while attractive sink-like habitats should be managed effectively to mitigate mortality risks. In particular, the various authorities need to be more proactive (increase patrolling, thereby providing employment, educational opportunities and increasing income) to reduce human-caused ungulate mortality.

References

Alexander S M, Paquet P C, Logan T B, et al. 2011. Snow-tracking versus radiotelemetry for predicting wolf-environment relationships in the Rocky Mountains of Canada. Wildl.soc.bull, 33(4): 1216-1224.

Alldredge J R, Ratti J T. 1986. Comparison of some statistical techniques for analysis of resource selection. The Journal of Wildlife Management, 50: 157-165.

Baillie J E M, Hiltontaylor C, Stuart S N. 2004. 2004 IUCN red list of threatened species.

Bèchet A, Giroux J-F, Gauthier G. 2004. The effects of disturbance on behaviour, habitat use and energy of spring staging snow geese. Journal of Applied Ecology, 41(4): 689-700.

Blake S, Strindberg S, Boudjan P, et al. 2007. Forest elephant crisis in the Congo Basin. Plos Biology, 5(4): e111.

Corlett R T. 2007. The impact of hunting on the mammalian fauna of tropical Asian forests. Biotropica, 39(3): 292-303.

Delibes M, Gaona P, Ferreras P. 2001. Effects of an Attractive Sink Leading into Maladaptive Habitat Selection. American Naturalist, 158(3): 277-285.

Duberstein C A, Simmons M A, Sackschewsky M R, et al. 2008. Development of a Habitat Suitability Index Model for the Sage Sparrow on the Hanford Site. Revista Colombiana De Entomología, 37(2): 228-233.

Falcucci A, Ciucci P, Maiorano L, et al. 2009. Assessing Habitat Quality for Conservation Using an Integrated Occurrence-Mortality Model. Journal of Applied Ecology, 46(3): 600-609.

Freitas C, Kovacs K, Lydersen C, et al. 2010. A novel method for quantifying habitat selection and predicting habitat use. Journal of Applied Ecology, 45(4): 1213-1220.

Gao Z, Zhang M, Hu R. 1995. Winter bedding site selection of ussurian wild pig in the lesser khing-an mountains. Acta Theriologica Sinica, 15(1): 25-30. (in Chinese)

Garshelis David. 2000. Delusions in Habitat Evaluation: Measuring Use, Selection, and Importance. New York: Colomia University Press: 111-164.

Grodziński W, Sawicka-Kapusta K. 1970. Energy values of tree-seeds eaten by small mammals. Oikos, 21(1): 52-58.

Groom M J, Meffe G K, Carroll C R, et al. 2006. Principles of conservation biology. Sinauer associates, Sunderland, USA: 63-109.

Guisan A, Thuiller W. 2010. Predicting species distribution: offering more than simple habitat models. Ecology Letters, 8(9): 993-1009.

Han Z X, Wei F W, Zhang Z J. 2004. Habitat Selection by Red Pandas in Fengtongzhai Natural Reserve. Acta Theriologica Sinica, 24(3): 185-192. (in Chinese)

IUCN Red list of threatened species. 2010. http: //www.iucnredlist.org[2017-3-6].

Jiang G S, Ma J Z, Zhang M H. 2011. Spatial Distribution of Ungulate Responses to Habitat Factors in Wandashan Forest Region, Northeastern China. Journal of Wildlife Management, 70(5): 1470-1476.

Jiang G S, Zhang M H, Ma J Z. 2008. Habitat use and separation between red deer *Cervus elaphus xanthopygus* and roe deer *Capreolus pygargus bedfordi* in relation to human disturbance in the Wandashan Mountains, northeastern China. Wildlife Biology, 14(1): 92-100.

Jiang Z G. 2006. Biodiversity in the Laoxiancheng Nature Reserve, Shanxi, China. Beijing: Tsinghua University Press: 78-79. (in Chinese)

Johnson D H. 1980. The Comparison of Usage and Availability Measurements for Evaluating Resource Preference. Ecology, 61(1): 65-71.

Kühl A, Balinova N, Bykova E, et al. 2009. The role of saiga poaching in rural communities: Linkages between attitudes, socio-economic circumstances and behaviour. Biological Conservation, 142(7): 1442-1449.

Laurance W F, Croes B M, Guissouegou N, et al. 2010. Impacts of roads, hunting, and habitat alteration on nocturnal mammals in African rainforests. Conservation Biology the Journal of the Society for Conservation Biology, 22(3): 721-732.

Li Y K, Zhang M H, Jiang Z G. 2008. Habitat selection by wapiti *Cervus elaphus xanthopygus* in Wandashan Mountains based habitat availability. Acta Ecologica Sinica, 28(10): 4619-4628. (in Chinese)

Li Z S, Jiang Z G, Li C W, et al. 2006. Biodiversity in the Laoxiancheng Nature Reserve, Shanxi,

China. Beijing: Tsinghua University Press: 78-79.

Linnell J D C, Swenson J E, Anderson R. 2001. Predators and people: conservation of large carnivores is possible at high human densities if management policy is favourable. Animal Conservation, 4: 345-350.

Loibooki M, Hofer H, Campbell K L I, *et al.* 2003. Bushmeat hunting by communities adjacent to the Serengeti National Park, Tanzania: the importance of livestock ownership and alternative sources of protein and income. Environmental Conservation, 29(3): 391-398.

Ma Y Q, *et al.* 1986. Mammals of Heilongjiang Province. Harbin: Heilongjiang Science and Technology Press. (in Chinese)

Milner-Gulland E J, Bennett E L. 2003. Wild meat: the bigger picture. Trends in Ecology & Evolution, 18(7): 351-357.

Naves J, Wiegand T, Revilla E, *et al.* 2010. Endangered Species Constrained by Natural and Human Factors: The Case of Brown Bears in Northern Spain. Conservation Biology, 17(5): 1276-1289.

Nielsen S E, Stenhouse G B, Beyer H L, *et al.* 2008. Can natural disturbance-based forestry rescue a declining population of grizzly bears? Biological Conservation, 141(9): 2193-2207.

Podchong S, Schmidt-Vogt D, Honda K. 2009. An improved approach for identifying suitable habitat of Sambar Deer (Cervus unicolor Kerr) using ecological niche analysis and environmental categorization: Case study at Phu-Khieo Wildlife Sanctuary, Thailand. Ecological Modelling, 220(17): 2103-2114.

Porter W F, Church K E. 1987. Effects of Environmental Pattern on Habitat Preference Analysis. Journal of Wildlife Management, 51(3): 681-685.

Ratcliffe C S, Crowe T M. 2001. Habitat utilisation and home range size of helmeted guineafowl (Numida meleagris) in the Midlands of KwaZulu-Natal province, South Africa. Biological Conservation, 98(3): 333-345.

Sauerwein H, Müller U, Brüssel H, *et al.* 2004. Establishing baseline values of parameters potentially indicative of chronic stress in red deer (*Cervus elaphus*) from different habitats in western Germany. European Journal of Wildlife Research, 50(4): 168-172.

Simcharoen S, Barlow A C D, Simcharoen A, *et al.* 2008. Home range size and daytime habitat selection of leopards in Huai Kha Khaeng Wildlife Sanctuary, Thailand. Biological Conservation, 141(9): 2242-2250.

Singer F J, Otto D K, Tipton A R, *et al.* 1981. Home Ranges, Movements, and Habitat Use of European Wild Boar in Tennessee. Journal of Wildlife Management, 45(2): 343-353.

Sodhi N S, Pin K L, Brook B W, *et al.* 2004. Southeast Asian biodiversity: an impending disaster. Trends in Ecology & Evolution, 19(12): 654-660.

Stankowich T. 2008. Ungulate flight responses to human disturbance: A review and meta-analysis. Biological Conservation, 141(9): 2159-2173.

Thomas C D, Kunin W E. 2010. The Spatial Structure of Populations. Journal of Animal Ecology, 68(4): 647-657.

Thomasma L E. 1981. Standards for the Development of Habitat Suitability Index Models. Wildlife Society Bulletin, 19: 1-171.

Underwood F. 2014. Resource selection by animals: Statistical design and analysis for field studies. Journal of Animal Ecology, 63(3): 351.

Zhang C Z, Zhang M H. 2011. Population status and dynamic trends of Amur tiger's prey in Eastern Wandashan Mountain, Heilongjiang Province. Acta Ecologica Sinica, 31(21): 6481-6487. (in Chinese)

Zhang M H, Liu Q X. 2008. Estimation of winter carrying capacity of wapiti in the Eastern

Wandashan Mountains, Heilongjiang Province, China. Acta Theriologica Sinica, 28(1): 56-64.

Zhou S C, Sun H Y, Zhang M H, et al. 2008. Regional distribution and population size fluctuation of wild Amur tiger (*Panthera tigris altaica*) in Heilongjiang Province. Acta Therioligica Sinica, 28(1): 165-173. (in Chinese)

Zhou S C, Zhang M H, Sun H Y, et al. 2010. Population size and habitat of wild boar (*Sus scrofa*) in the eastern Wanda Mountains. Acta Theriologica Sinica, 30(1): 28-34. (in Chinese)

Zhou S C, Zhang M H, Wang S L. 2006. Habitat Selection of Red Deer (*Cervus elaphus*) and Roe Deer (*Capreolus capreolus*) in Winter in Logged and Unlogged Forest of the Wandashan Mountains, Heilongjiang. Zoological Research, 27(6): 575-580.

Chapter 4 Effects of human disturbance on vegetation, prey and Amur tigers in Hunchun Nature Reserve, China

4.1 Introduction

The effects of human disturbance on vegetation and wildlife in human-dominated landscapes is a critical issue in wildlife management. In recent decades, conflict between people and wildlife has become more frequent because of the exponential increase in the human population and spread of human activity (Woodroffe, 2000; Conover, 2001; Pettigrew *et al.*, 2012). Disturbance results in habitat loss, either directly or indirectly as a consequence of avoidance behavior by affected wildlife (Mace and Waller, 1996). Human disturbances may cause wildlife to shift habitat use and reduce forage quality (Hernandez and Laundre, 2005). Large carnivores (>40kg) are more susceptible to human disturbance and changes in the configuration and connectivity of habitats such as habitat fragmentation and loss (Ngoprasert *et al.*, 2007). Some identified human disturbances effecting tigers (*Panthera tigris Linnaeus*, 1758) include roads, settlements, farmlands, logging, poaching, grazing and quarrying (Kerley *et al.*, 2002; Barbermeyer *et al.*, 2013; Bhattarai and Kindlmann, 2013). Some studies have suggested that prey abundance and human disturbance are the most important parameters for tiger occupancy, and serious disturbances can cause prey depletion and tiger extinction (Bhattarai and Kindlmann, 2013). However, most research into human disturbances is descriptive, explores only a few causes of disturbance, and rarely yields mechanisms. The majority of wild tigers remain affected by human disturbance, even in conservation zones (Linkie *et al.*, 2003). Further, little research has compared the contribution of different disturbances affecting wildlife and vegetation. In nature, different factors often have different contributions to the survival of taxa (Tisseuil *et al.*, 2013) and many disturbance factors have interactions that affect taxa (Soh *et al.*, 2014). Therefore, just focusing on particular disturbances may cause one-sided problems, and comparing different types of human disturbances allows an understanding of the deep and complex mechanisms threatening wildlife.

As one of the few areas where wild tigers (*Panthera tiger*) and leopards (*Panthera pardus*) coexist in China, Hunchun Nature Reserve (HNR) is a key area for Amur tiger (*Panthera tigris altaica*) and leopard conservation. The main human disturbances around

HNR include roads, grazing, ungulate poaching (snares) and crop cultivation (ginseng and other farmland). A few studies have looked at disturbances in HNR and tiger abundance, leopards and ungulates (Li et al., 2006; Li, 2009; Chen et al., 2011). Due to historical problems, many pastures remain in HNR and grazing is a typical human disturbance across the Hunchun region. Research on grazing has mainly focused on habitat selection of livestock depredation(Liu et al., 2005; Li et al., 2009) and not assessed the real effect of disturbance on Amur tigers and habitat. Soh et al. (2014) showed that the probability of livestock depredation by Amur tigers increases in areas close to ungulate snares and there was an overall lower density of ungulate prey closer to snare sites. Ginseng cultivation is another typical disturbance in Hunchun and requires the cleaning of all trees and shrubs and establishment of blue greenhouses that results in large habitat loss. Ginseng planting also requires chemical fertilizer and pesticides which may affect vegetation. Roads are a serious factor for wildlife and tigers in HNR (Smirnov and Miquelle, 1999). In Far East Russia, roads decrease survivorship and reproductive success of Amur tigers (Kerley et al., 2002). Although logging ceased years ago in HNR, many logging roads are still used and remolded for national defense, the collection of non-timber forest products and tourism. There are three kinds of roads within or around HNR: tertiary, secondary and primary roads (Fig. 4.1). Most tertiary roads are abandoned logging roads used by local people, possessing rugged condition and soil paved only the farm vehicle available.; secondary roads are used for connecting villages around HNR with cement paved all year-round; and primary roads with asphalt paved are not within HNR but divide it into two sections, connecting Jilin and Heilongjiang Province and allowing traffic to move at high speed (Fig. 4.1).

The purpose of the statistical model is to provide a mathematical basis for interpretation, examining such parameters as fit (do the measured predictors adequately explain the response?), strength association (is the relationship between the response and the predictors significant?), and to ascertain the contributions and roles of different variables (Guisan et al., 2002). Different kinds of models provide different insights into the role of statistical modeling in ecology (Guisan et al., 2002) and aims to provide insights into the ecological processes that produce patterns (Austin et al., 1990). As the mathematical extensions of linear modeling (LM), GLMs do not force data into unnatural scales, and thereby allow for nonlinearity and non-constant variance structures in data (Hasties and Tibshirani, 1990). Data assumed to be from several families of probability distributions, including the normal, binomial, Poisson, negative binomial, or gamma distribution, many of which better fit the non-normal error structures of most ecological data (Guisan et al., 2002). GLMs are more flexible and better suited for analyzing ecological relationships, which can be poorly represented by classical Gaussian distributions (Austin, 1987). GAMs (Hastie and Tibshirani, 1986; Hasties and Tibshirani, 1990) are semi-parametric extensions of GLMs; the only underlying assumption made is

Chapter 4 Effects of human disturbance on vegetation, prey and Amur tigers in... | 233

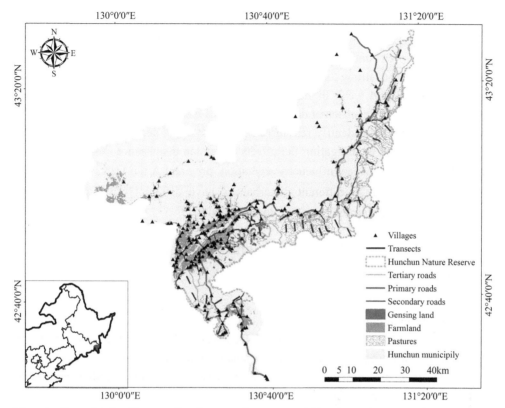

Fig. 4.1 Seven kinds of human disturbance and transects within HNR. Please scan the QR code on the back cover to see the color image.

that the functions are additive and that the components are smooth. The strength of GAMs is their ability to deal with highly nonlinear and non-monotonic relationships between the response and the set of explanatory variables for which the GLMs do not have strong capacity to handle (Guisan et al., 2002). GAMs represent the real responses of species to the environment for data characteristics and nonlinear functions (Suarezseoane et al., 2002). One advantage of SEM is to estimate and test relationships among constructs while GAMs and GLMs only reveal correlations between the response and explanatory variables alternately. Compared with other general linear models, SEM, which estimate causal effects through the study of path relations (Grace et al., 2010), allows for the use of multiple measures to represent constructs and addresses the issue of measure-specific error (Weston and Gore, 2006).

Top-down and bottom-up effects are classic mechanisms known to affect prey and predator systems (Sabatier, 1986; Hunter and Price, 1992; Suarez and Case, 2002; Aryal et al., 2014). Bottom-up effects combined with climate change, virology and fear ecology can better reveal interactions between prey and predators (Wilmers et al., 2006; Laundre et al., 2014) and formed the focus of the present study. Specifically, understanding the relative importance of bottom-up effects is critical for predicting impacts on top predators (Frederiksen et al., 2006). Predicted patterns and correlations between predators and prey under various control mechanisms differ markedly: positive correlations suggest strong bottom-up control, negative correlations indicate strong top-down control and weak or no correlations between predators and prey suggest weak trophic links or insufficient data (Worm and Myers, 2003).

Here, we aimed to identify real threats to Amur tigers and provide recommendations for eliminating or mitigating the effects of human disturbance. We hypothesized that ① human disturbance influences vegetation, prey and Amur tigers via bottom-up effects in HNR, and ② different disturbances have different effect intensities on specific biological hierarchies along bottom-up chains.

4.2 Methods and techniques

4.2.1 Study area

Hunchun municipality is a 5145km^2 area in Jilin province, northeast China. The actual reserve, HNR, covers 1087km^2. HNR borders Russia to the east and North Korea to the south west and includes four management zones (core area, experimental area, buffer zone and community co-management area) (Li, 2009). HNR is connected to three Amur tiger and Amur leopard protected areas in Russia, and is a key corridor for movement of Amur tigers and Amur leopards between China, Russia and North Korea. As part of the Changbai Mountains, HNR is in a temperate zone and has an average rainfall of 661mm concentrated between July and September (50% of yearly precipitation). The main vegetation forms in HNR are mixed broad-leaved forests and secondary Mongolian oak (*Quercus mongolica* Fisch. ex Ledeb) forest and the main animals include Amur tigers, Amur leopard (*Panthera pardus*), red deer (*Cervus elaphus*), sika deer (*Cervus nippon*), wild boar (*Sus scrofa*), roe deer (*Capreolus pygargus*) and musk deer (*Moschus moschiferus*). There were 29 villages and 14 953 people within HNR before it was established in 2001 (Li, 2009); currently there are 98 villages and four towns (Soh et al., 2014).

4.2.2 Data collection

In order to confirm pasture distribution and grazing intensities, field surveys and questionnaires were used (Hasselfinnegan et al., 2013; Zimmermann et al., 2013). We interviewed every rancher within HNR to survey the distribution of pastures and cows. For huge and stable pastures which were difficult to walk around to get their ranges and locations precisely, we calculated their positions and areas using ArcGIS; for small and private ones we used field measurements with tracking recorders (HOLUX M241).

Stratified sampling methods were used for designing transect lines with a 10% sampling ratio (10km/100km^2) and 32 transects (total length = 96km) were established. The following factors were considered when designing transects: ① each transect was 3 km in length and the interval between adjacent transects was larger than 3km. ② Forest types: we selected four kinds of main forest forms (coniferous forests, broad-leaved forests, mixed coniferous-broad leaf forests and shrub lands) with an area ratio of 22 ∶ 127 ∶ 15 ∶ 1. The total transects length for the four kinds of forest were 11.4km, 70.3km, 10.7km and 3.6km, respectively. ③ Areas of management zones: the area ratio for buffer zone, core zone and experimental zone in HNR is 6 ∶ 5 ∶ 1.7 and the transect length in the three zones was 47km, 37km and 12km, respectively.

In order to assess vegetation structure and quality, 512 tree survey quadrats and 2560 shrub survey plots were made along transects with an interval of 200m. For tree quadrats, the size was 10m × 10m and data on hiding cover, tree species, quantity and diameter at breast height (DBH) were recorded. For shrub quadrats, we sampled five 2m × 2m plots at the four corners and the center of each tree quadrat. We measured shrub density, type, quantity and density of newborn edible shrub twigs. Besides, geographic parameters such as the slope, aspect and elevation of each quadrats were also recorded.

We used the line transect method to measure wild ungulate and cattle relative abundance. We defined grazing intensities as cow track numbers per kilometer, and for wild ungulates we calculate presence/absence (1/0) in each tree quadrat and track numbers per kilometer. Considering the surveys were in the snow free period of northeast China, we identified the calculate presence/absence of prey by multiple parameters such as footprints, food trace, feces and so on to calculate the occupancies of prey.

We used a tiger presence recordings database for the assessment of tiger distribution. The database included tiger presences such as footprints, kill sites, feces and so on from 2000 to 2012 within HNR. We hope to detect the prey level effects on the distributions of Amur tigers based on the location discriminations of recordings database in the period in discussion. Considering that female Amur tigers average a total distance of 7km/day (Yudakov and Nikolaev, 1987), we calculated tiger occurrence frequency using a 196km^2 grid (14km × 14km) (Jiang et al., 2014) and then calculated Amur tiger occurrence probabilities for each 10m × 10m quadrat. All calculations used the spatial analysis tool Zonal statistics in ArcGIS.

For human disturbance factors we measured the nearest distance from the central point in each tree quadrat to each kind of disturbance. Seven kinds of human disturbance including grazing, ginseng farming, cropland, primary roads, secondary roads, tertiary roads and settlements were considered. We obtained distribution shape data for ginseng farms, cropland, primary roads, secondary roads, tertiary roads and settlements using ArcGIS 9.3.

4.2.3 Data analysis

Pearson correlation coefficients were used to calculate correlation coefficients among independent variables, such as tree species, tree density, DBH of tree, shrub density, density of newborn edible shrub twigs, aromatic wiener index, grazing intensities, ungulate presence/absence (1/0) in each tree quadrat, the nearest distance to ginseng land, farmland, villages, tertiary roads, secondary roads and primary road.

We considered a strong correlation between independent variables if Pearson correlation coefficients were over 0.5 and in these cases one of the independent variables was deleted to make sure all variables entering the models had low correlations (Ramsay et al., 2003).

In order to explore specific bottom-up mechanisms related to people, we divided mechanisms into three hierarchies: the human effects on vegetation, prey and Amur tigers. In the vegetation hierarchies, only human disturbance was considered; human disturbances and vegetation parameters were used to test the effect on prey in second hierarchies. Then, disturbance, vegetation and prey parameters were used together to determine the effects on Amur tigers. In each hierarchy we used generalized additive models (GAM), generalized liner models (GLM) and structural equation models (SEM) to test specific effect mechanisms. We then established an interaction network among human disturbances and three hierarchies to test our hypotheses whether human disturbance influences vegetation, prey and Amur tigers along a bottom-up chain. Of particular note is that we used ungulate presence/absence (1/0) in each tree quadrat as their relative abundance in the GLM because ungulate tracks were very few. For the GAM, we took ungulate track numbers per kilometer as their relative abundance because this model does not permit response variables that contain many zero values. We applied a hierarchical partitioning model (HPM) approach (Chevan and Sutherland, 1991) to quantify the explanatory power of each biological parameter from the three hierarchies in explaining their effect on vegetation, prey and tigers, respectively. SEM analyses were carried out in Amos v21 and others analyses were carried out in R v2.12.1 using the mgcv and hier. part packages (Chevan and Sutherland, 1991; R-DevelopmentCoreTeam, 2010; Li and Wang, 2013). All parameters in models have been standardized by SPSS 17.0.

4.3 Results

Based on our questionnaire survey, 63 pastures are located in or around HNR and 3066 cows were recorded. Of these, there were 42 pastures with an area of 355km^2 within HNR making up 32.7% of the total reserve area; 19 pastures were completely or partially located in the core zone, accounting for 53.3% of the core zone area. The average cow density was 11 cows per km^2. Pearson correlation coefficients show that cropland and settlements have higher correlations with secondary and primary roads and that grazing, ginseng farming, primary roads and secondary roads were considered last from the 7 disturbances at first. The effect network predicted by GAM, GLM and SEM showed that human disturbances have spread into every hierarchy of the Amur tiger food chain and are having a bottom-up effect on Amur tigers (Fig. 4.2). The HPM showed that different disturbances have different effect intensities on specific biological hierarchies (Fig. 4.3). The HPM results were based on our four focal disturbances.

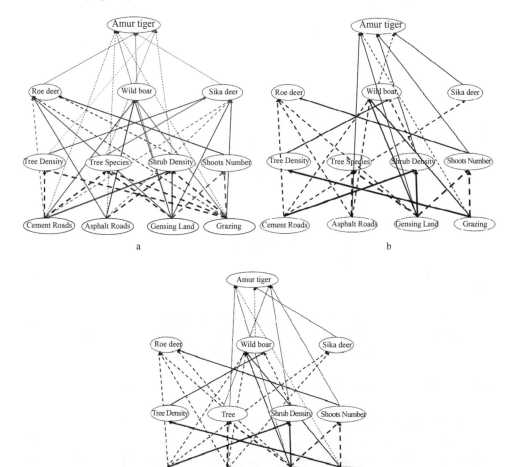

Fig. 4.2 Effect of secondary roads, primary roads, ginseng land and grazing on vegetation, prey and Amur tigers according to GAM (a), GLM (b) and SEM (c). "Primary" and "Second" mean primary roads and second roads respectively. Red solid lines are positive correlations; red dotted lines are negative correlations; black solid lines are n-shape correlations; black dotted lines are u-shape correlations; and blue lines are wave shape correlations in 2(a). Black solid lines are positive correlations; and black dotted lines are negative correlations in 2(b) and 2(c).

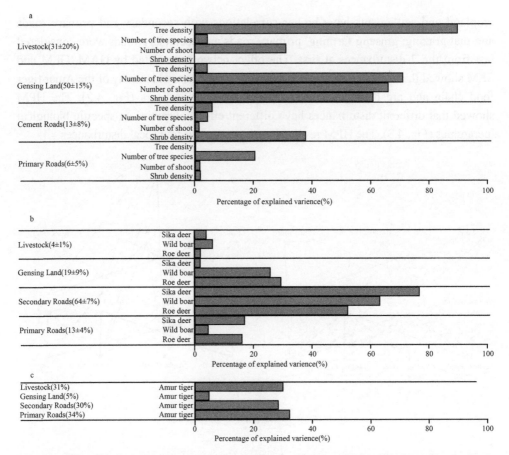

Fig. 4.3 Hierarchical partitioning models quantified the total contribution (given as the percentage of the total explained deviance based on Pseudo R^2) of livestock, ginseng land, secondary roads and primary roads on vegetation, prey and Amur tigers.

4.3.1 Human disturbance on vegetation

The GAM revealed 11, the GLM seven and the SEM six significant effects of disturbance on vegetation hierarchies, of a total of 16 effects (Fig. 4.2, Table 4.1). All models showed that both ginseng planting and grazing activities had negative effects on the number of shoots (Fig. 4.2, Table 4.1) with a contribution of 65.4% and 30.9% (Fig. 4.3), respectively. The GAM showed more non-monotonic significant effects with n, u and w (wave) shapes, but the GLM and SEM test liner correlations only. The HPM showed that grazing activities (31%±20%) and ginseng planting (50%±15%) have a greater contribution than secondary roads (13%±8%) and primary roads (6%± 5%). Grazing activities mainly affected tree density (u-shape, GAM; positive, GLM) and shoot numbers while ginseng planting has a similar main contribution on tree species (u-shape, GAM; negative, GLM and SEM), shoot number (negative, GAM, GLM and SEM) and shrub density (wave shape, GAM; positive, GLM and SEM) (Fig. 4.2, Table 4.1).

Table 4.1 Results of GAM, GLM and SEM six effects were similar across the three models

Code	Correlations	GAM	GLM	SEM	The same effects or not
1	Secondary - Tree density	u			×
2	Secondary - Tree species	u			×
3	Secondary - Shrub density	w	+	+	×
4	Secondary - Shoot numbers				
5	Secondary - Roe deer	−	−	−	√
6	Secondary - Wild boar	w	−	−	×
7	Secondary - Sike deer	−	−	−	√
8	Secondary - Amur tiger	u			×
9	Primary - Tree density				
10	Primary - Tree species		−	−	×
11	Primary - Shrub density	u			×
12	Primary - Shoot numbers				
13	Primary - Roe deer	w	−	−	×
14	Primary - Wild boar	n	−	−	×
15	Primary - Sike deer	−		−	
16	Primary - Amur tiger	u	+	+	×
17	Ginseng land - Tree density				
18	Ginseng land - Tree species	u	−	−	×
19	Ginseng land - Shrub density	w	+	+	×
20	Ginseng land - Shoot numbers	−	−	−	√
21	Ginseng land - Roe deer	w	−	−	×
22	Ginseng land - Wild boar	w	+	+	×
23	Ginseng land - Sike deer	w			×
24	Ginseng land - Amur tiger	w	+	+	×
25	Grazing - Tree density	u	+	+	×
26	Grazing - Tree species	u			×
27	Grazing - Shrub density	u			×
28	Grazing - Shoot numbers	−	−	−	√
29	Grazing - Roe deer	u			
30	Grazing - Wild boar	+	+	+	√

Table 4.1 (continued)

Code	Correlations	GAM	GLM	SEM	The same effects or not
31	Grazing - Sike deer	w			×
32	Grazing - Amur tiger	u	–	–	×
33	Tree density - Roe deer				
34	Tree density - Wild boar		+	+	×
35	Tree density - Sike deer	w			×
36	Tree density - Amur tiger	–			×
37	Tree species - Roe deer			–	×
38	Tree species - Wild boar				
39	Tree species - Sike deer	+			×
40	Tree species - Amur tiger	+		+	×
41	Shrub density - Roe deer	u			×
42	Shrub density - Wild boar		–	–	×
43	Shrub density - Sike deer				
44	Shrub density - Amur tiger	u	+	+	×
45	Shoot numbers - Roe deer	n	+	+	×
46	Shoot numbers - Wild boar				
47	Shoot numbers - Sike deer				
48	Shoot numbers - Amur tiger		+	+	×
49	Roe deer - Amur tiger	w			×
50	Wild boar - Amur tiger	w	–	–	×
51	Sike deer - Amur tiger	+	+	+	√

Note: Plus signs, minus signs, n signs, u signs and w signs represent positive, negative, n-shape, u-shape and wave shape correlations, respectively. Check marks indicates that all three models show the same effects; otherwise, cross marks. GAM, GLM and SEM exhibited 37, 25 and 28 significant effects from a total 51 correlations between three hierarchies, respectively.

4.3.2 Human disturbance on prey

The GAM revealed 12, the GLM eight and SEM nine significant direct effects of disturbance on prey hierarchies from a total of 12 effects (Fig. 4.2, Table 4.1). The HPM showed that secondary roads explained 64%±7% of the variance and had preponderant effects on prey over other disturbances (Fig. 4.3); GAM indicated that the four kinds of disturbance have significant effects on sika deer, wild boar and roe deer (Fig. 4.2, Table 4.1). Secondary roads and primary roads showed negative effects on sika deer in GAM and all three models indicated that secondary roads have negative effects on roe deer and sika deer, with a contribution of 51.9% and 76.3% respectively. Human disturbance also affects prey in indirect ways. For example, the GAM showed that secondary roads and ginseng planting have u-shape effects on tree species, while tree species have a positive effect on sika deer presence (Fig. 4.2).

4.3.3 Human disturbance on Amur tigers

The GAM revealed four direct effects, GLM three and SEM three significant direct effects of disturbance on prey hierarchies, from a total of four effects (Fig. 4.2, Table 4.1). The GLM and SEM showed that primary roads and ginseng planting had a positive effect on Amur tigers but grazing had a negative effect, while the GAM showed that these effects may not be linear (Fig. 4.2). The HPM showed that grazing (31%), secondary roads (30%) and primary roads (34%) have almost the same main contribution to Amur tiger presence (Fig. 4.3).

We found that human disturbance may affect Amur tiger presence by affecting vegetation and prey (Fig. 4.2). The GAM and SEM indicated that secondary and primary roads had a negative effect on sika deer presence (sika deer are preferred by Amur tigers as prey). All three models showed that secondary roads had negative effects on the probability of roe deer presence. The probability of Amur tiger occurrence was affected significantly by roe deer according to GAM. All three models showed that both grazing activities and ginseng planting may reduce the number of shoots (twigs), which is preferred and browsed upon by ungulates; shoot number had a positive effect on the probability of roe deer occurrence according to GLM and SEM. Roe deer presence had a significant n-shaped effect on Amur tiger presence.

4.4 Discussion

4.4.1 Effects of grazing

The effect of pastures on Amur tiger presence is multi-leveled. The indirect effect of pasture shows a bottom-up mechanism whereby grazing activities affect Amur tigers through vegetation and prey hierarchies. One of the most important findings was that grazing activities can reduce the number of shoots preferred by ungulates. Research shows that prey avoid grazing areas (Bhattarai and Kindlmann, 2013) similar to roe deer responses to grazing in our results (Fig. 4.2b,c). This phenomena may be explained by the behavioral ecology of roe deer because they are small to intermediate sized ungulates that favor dense shrub habitat(Jiang et al., 2008; Jiang et al., 2010). Roe deer absence may affect the probability of Amur tiger presence by predation loss. Soh et al. (2014) points out that snares are a significant cause in the decline of tiger prey abundance and that the negative effect of hunting snares on prey exacerbates tiger depredation on cattle.

However, Soh et al. (2014) failed to explore the deeper relationships between pasture and the distribution of ungulates and habitat. We deem that prey loss caused by grazing through resource competition is another serious threat to Amur tigers besides seniors. The GAM showed that grazing had some direct nonlinear effects on prey, but the HPM showed that the contribution of grazing to vegetation was 31%±20% and grazing contribution to prey was just 4%±1%. Therefore, we argue that compared to effects on prey, grazing may mainly affect vegetation through effects on prey.

The direct effect of grazing on Amur tiger presence was via changes in prey distribution. More than 30% of HNR is covered with pastures and the average grazing density is 11 cows per km^2. In HNR, the predicted densities of red deer, sika deer, wild boar and roe deer were 0-0.16, 0-0.19, 0-2.9 and 0-5.5 per km^2, respectively (Soh et al., 2014). Because the selection of wild and domestic ungulate prey is influenced mainly by local abundance and accessibility (Meriggi and Lovari, 1996), the huge density discrimination between livestock and wild ungulates means that tigers have more chance of encountering livestock and cause serious depredation. Serious livestock depredation results in conflict between locals and wildlife and agricultural losses, and possible bi-directional disease infection at the wildlife/livestock interface (Bengis et al., 2002). For example, in HNR foot and mouth disease is the main illness among livestock (Bengis et al., 2002) and Amur tigers may suffer from diseases that threaten livestock. Grazing is one important economic income source for local people, but the contribution of grazing to Amur tiger presence was 31% of the total main effective disturbance (Fig. 4.3). Dealing with grazing problems in this area correctly is critical for wildlife conservation and local agricultural development (Fig. 4.4).

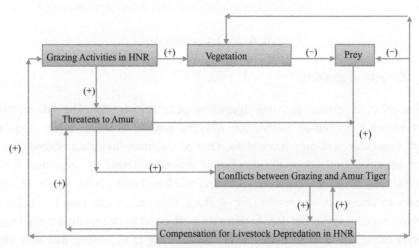

Fig. 4.4 Compensation policy for livestock within HNR implemented by the government caused one vicious spiral in which vegetation, prey and Amur Tiger may get continuous threatens. "(+)" means promoting effects and "(−)" means restrain effects.

4.4.2 Effects of roads

In eastern Russia, radio collared data shows that roads reduce tiger survival rates. Cub survival was lower in areas with roads than in areas without roads and undisturbed tigers also spent more time at each kill site than disturbed tigers (Kerley et al., 2002). In contrast to previous research, our GLM and SEM results show that primary roads had a positive effect on Amur tigers, while secondary roads did not have a significant effect on tigers. The GAM showed both secondary and primary roads had nonlinearity effects on tigers and that roads had a nonlinearity effect on prey. In a review of the empirical literature on the effects of roads on animal population abundance and distribution, most effects (59%) are negative (Fahrig and Rytwinski, 2009) and there are many neutral and positive responses (Fahrig and Rytwinski, 2009; Rytwinski and Fahrig, 2012). Our comparison of the GAM, GLM and SEM found that the linear effect tests by the GLM and SEM were nonlinear in GAM. In fact, the strength of GAMs are their ability to deal with highly nonlinear and non-monotonic relationships between the response and set of explanatory variables and hence increase our understanding of ecological systems (Guisan et al., 2002). Rytwinski and Fahrig (2013) showed that large-bodied prey under predation pressure may exhibit nonlinearity relationships with increasing road density because there was a balance of prey between the predator-loss in roads and traffic avoidance disturbance. We propose there may also be a balance between the activity of convenience and traffic avoiding that results in nonlinearity correlation.

The contribution of secondary and primary roads to Amur tigers was 30% and 34%, respectively, accounting for two thirds of the four kinds of disturbance; road contribution to prey reached 77%±11%. However, regarding vegetation and prey, these contributions was only 19% (Fig. 4.3). All of the three models showed that secondary roads had negative effects on sika and roe deer, while sika deer had positive effects on the probability of Amur tiger presence and therefore forms a bottom-up effective mode (Fig. 4.2).

4.4.3 Effects of ginseng planting

The HPM results showed that the contribution of ginseng cultivation on Amur tigers and prey was only 5% and 19% respectively, much lower than for other disturbances (Fig. 4.3). However, the contribution of ginseng planting was as high as 50%. We suggest that ginseng planting may still be an indirect potential threat to tigers because of habitat loss and pesticide use. All three models showed that ginseng planting had negative effects on the shoot number and had some significant nonlinearity effects on shrub density and tree species (Fig. 4.2, Table 4.1). Pesticides may cause protein denaturation, enzyme inactivation, membrane injury and restrain vegetation planting (Ciccotelli *et al.*, 1998). Ginseng planting means that pesticides enter the Amur tiger food chain and result in bottom-up effective impacts via bio magnification.

4.4.4 Results discrimination of GAMs, GLMs and SEMs

In our results, 22 significant effect correlations were revealed by the GAM and GLM (Table 1). Only six correlations were the same and the other 16 linear effects revealed by the GLM were proved to be non-monotonic by GAM. We set the mechanism nets of GAM and GLM (Fig. 4.2,b) according to each regression by GAM and GLM (Appendix) but obtained the mechanism net of SEM (Fig. 4.2c) directly because one advantage of SEM is to estimate and test relationships among constructs while GAMs and GLMs only reveal correlations between the response and explanatory variables alternately. Our results show that almost 90% of the results from the SEM are the same as for the GLM; and that the GAM had a stronger prediction ability than the GLM and SEM (Table 4.1).

4.5 Conclusions

Human disturbance is spreading into every hierarchy of the Amur tiger food chain and has formed a bottom-up effect on Amur tigers in HNR. Grazing, secondary roads and primary roads were the main disturbances, with direct and indirect forms; potential threats from ginseng cultivation to Amur tigers cannot be ignored. Our results reveal the effective mechanisms of four kinds of human disturbance on three bio-hierarchies. However, our study only covered major sources of human disturbance within the reserve and specific mechanisms underlying other disturbances require further investigation.

4.6 Summary

Multiple human disturbances influence vegetation, ungulates and Amur tigers (Panthera tigris altaica) in Hunchun Nature Reserve (HNR) in northeastern China. In order to understand the influence and relative contribution of human disturbance on Amur tigers, prey and vegetation, we conducted transect lines and plot surveys of human disturbance inside HNR from August to October 2013. We used generalized additive models, generalized liner models and structural equation models to explore the effects of human disturbance on vegetation, prey and Amur tigers. We then used hierarchical partitioning models to quantify the contribution of four main kinds of human disturbance. Our results suggest that all three models indicate that human disturbance can directly and indirectly affect prey and Amur tigers via 'bottom-up' chains. Among human disturbances, grazing activity and ginseng land encroachment impacted vegetation more than roads did; for prey, secondary roads had the greatest impact. Grazing activity, secondary roads and primary roads were the main factors disturbing Amur tigers. The generalized additive model had a stronger detection ability for disturbance prediction than generalized liner and structural equation models. The generalized additive model detected more complex nonlinear interaction relationships between predator and prey or predator, prey and habitat factors. Reducing or eliminating specific types of disturbance will be essential to recover tiger populations and habitat.

References

Aryal A, Brunton D H, Weihong J I, et al. 2014. Blue sheep in the Annapurna Conservation Area, Nepal: habitat use, population biomass and their contribution to the carrying capacity of snow leopards. Integrative Zoology, 9(1): 34-45.

Austin M P, Nicholls A O, Margules C R. 1990. Measurement of the realized qualitative niche: environmental niches of five Eucalyptus species. Ecological Monographs, 60(2): 161-177.

Austin M P. 1987. Models for the analysis of species' response to environmental gradients. Plant Ecology, 69: 35-45.

Barbermeyer S M, Jnawali S R, Karki J B, et al. 2013. Influence of prey depletion and human disturbance on tiger occupancy in Nepal. Journal of Zoology, 289(1): 10-18.

Bengis R G, Kock R A, Fischer J. 2002. Infectious animal diseases: the wildlife/livestock interface. Revue Scientifique Et Technique (Office International Des Epizooties), 21(1): 53-65.

Bhattarai B P, Kindlmann P. 2013. Effect of human disturbance on the prey of tiger in the Chitwan National Park-Implications for park management. Journal of Environmental Management, 131: 343-350.

Chen J Y, Nasen D, Sun Q H, et al. 2011. Amur tiger and prey in Jilin Hunchun National Nature Reserve. China J Zoology, 46: 46-52.

Chevan A, Sutherland M. 1991. Hierarchical partitioning. The American Statistician, 45(2): 90-96.

Ciccotelli M, Crippa S, Colombo A. 1998. Bioindicators for toxicity assesment of effluents from a wastewater treatment plant. Chemosphere, 37: 2823-2832.

Conover M R. 2002. Resolving human-wildlife conflicts: the science of wildlife damage management. New York: CRC Press.

Dusit N, Antony J L, George A G. 2007. Human disturbance affects habitat use and behaviour of Asiatic leopard *Panthera pardus* in Kaeng Krachan National Park, Thailand. Oryx, 41(3): 343-351.

Fahrig L, Rytwinski T. 2009. Effects of Roads on Animal Abundance: an Empirical Review and Synthesis. Ecology and Society, 14(1): 21.

Frederiksen M, Edwards M, Richardson A J, et al. 2006. From plankton to top predators: bottom-up control of a marine food web across four trophic levels. Journal of Animal Ecology, 75(6): 1259-1268.

Grace J B, Anderson T M, Olff H, et al. 2010. On the specification of structural equation models for ecological systems. Ecological Monographs, 80(1): 67-87.

Guisan A, Edwards T C, Hastie T. 2002. Generalized linear and generalized additive models in studies of species distributions: setting the scene. Ecological Modelling, 157(2): 89-100.

Hasselfinnegan H M, Borries C, Zhao Q, et al. 2013. Southeast Asian primate communities: the effects of ecology and Pleistocene refuges on species richness. Integrative Zoology, 8(4): 417-426.

Hastie T, Tibshirani R. 1986. Generalized Additive Models. Statistical Science, 1(3): 297-310.

Hastie T, Tibshirani R. 1990. Generalized additive models. London: Champion & Hall: 137-173.

Hernandez L, Laundre J W. 2005. Foraging in the 'landscape of fear' and its implications for habitat use and diet quality of elk Cervus elaphus and bison *Bison bison*. Wildlife Biology, 11(3): 215-220.

Hunter M D, Price P W. 1992. Playing Chutes and Ladders: Heterogeneity and the Relative Roles of Bottom-Up and Top-Down Forces in Natural Communities. Ecology, 73(3): 724-732.

Jiang G, Ma J, Zhang M, et al. 2010. Multi-scale foraging habitat use and interactions by sympatric cervids in Northeastern China. Journal of Wildlife Management, 74(4): 678-689.

Jiang G, Zhang M, Ma J. 2008. Habitat use and separation between red deer Cervus elaphus xanthopygus and roe deer Capreolus pygargus bedfordi in relation to human disturbance in the Wandashan Mountains, northeastern China. Wildlife Biology, 14(1): 92-100.

Kerley L L, Goodrich J M, Miquelle D G, et al. 2002. Effects of Roads and Human Disturbance on Amur Tigers. Conservation Biology, 16(1): 97-108.

Laundre J W, Hernandez L, Medina P L, et al. 2014. The landscape of fear: the missing link to understand top-down and bottom-up controls of prey abundance? Ecology, 95(5): 1141-1152.

Li B. 2009. Status of Amur Tiger and Prey Population in Hunchun Nature Reserve, China and Conservation Research. Shanghai, China.: East China Normal University.

Li B B, Wu Y, Zhang E D. 2006. Estimating Carrying Capacity of Red Deer (*Cervus elaphus*) at Qinglongtai Forestry in Hunchun Nature Reserve, Jilin. Sich J Zool, 25: 519-523.

Li B, Zhang E D, Liu Z. 2009. Livestock depredation by Amur tigers in Hunchun Nature Reserve, Jilin, China. Acta Theriol Sin, 29: 231-238.

Li X, Wang Y. 2013. Applying various algorithms for species distribution modeling. Integr Zool, (8): 124-135.

Linkie M, Martyr D J, Holden J, et al. 2003. Habitat destruction and poaching threaten the Sumatran tiger in Kerinci Seblat National Park, Sumatra. Oryx, 37(1): 41-48.

Liu Y, Zhang E, Li Z, et al. 2005. Amur tiger (*Panthera tigris altaica*) predation on livestock in Hunchun Nature Reserve, Jilin, China. Acta Theriol Sin, (26): 213-220.

Mace R D, Waller J S. 1996. Grizzly bear distribution and human conflicts in Jewel Basin Hiking Area, Swan Mountains, Montana. Wildlife Society Bulletin, 24(3): 461-467.

Meriggi A, Lovari S. 1996. A review of wolf predation in southern Europe : does the wolf prefer wild prey to livestock? Journal of Applied Ecology, 33(6): 1561-1571.

Ngoprasert D, Lynam A J, Gale G A. 2007. Human disturbance affects habitat use and behaviour of Asiatic leopard Panthera pardus in Kaeng Krachan National Park, Thailand. Oryx, 41(3): 343-351.

Pettigrew M, Xie Y, Kang A, et al. 2012. Human-carnivore conflict in China: a review of current approaches with recommendations for improved management. Integrative Zoology, 7(2): 210-226.

Ramsay T, Burnett R T, Krewski D. 2003. Exploring Bias in a Generalized Additive Model for Spatial Air Pollution Data. Environmental Health Perspectives, 111(10): 1283-1288.

R Development Core Team. 2010. R: A language and environment for statistical computing. Vienna, Austria: R Foundation for Statistical Computing.

Rytwinski T, Fahrig L. 2012. Do species life history traits explain population responses to roads? A meta-analysis. Biological Conservation, 147(1): 87-98.

Rytwinski T, Fahrig L. 2013. Why are some animal populations unaffected or positively affected by roads? Oecologia, 173(3): 1143-1156.

Sabatier P A. 1986. Top-down and bottom-up approaches to implementation research: a critical analysis and suggested synthesis. Journal of Public Policy, 6(1): 21-48.

Smirnov E N, Miquelle D G. 1999. Population dynamics of the Amur tiger in Sikhote-Alin State Biosphere Reserve // Christie J S, Jackson P. Riding the tiger: meeting the needs of people and wildlife in Asia. Cambridge: Cambridge University Press: 61-70.

Soh Y H, Carrasco L R, Miquelle D G, *et al.* 2014. Spatial correlates of livestock depredation by Amur tigers in Hunchun, China: Relevance of prey density and implications for protected area management. Biological Conservation, 169: 117-127.

Suarez A V, Case T J. 2002. Bottom-up effects on persistence of a specialist predator: ant invasions and horned lizards. Ecological Applications, 12(1): 291-298.

Suarezseoane S, Osborne P E, Alonso J C. 2002. Large-scale habitat selection by agricultural steppe birds in Spain: identifying species–habitat responses using generalized additive models. Journal of Applied Ecology, 39(5): 755-771.

Tisseuil C, Cornu J, Beauchard O, *et al.* 2013. Global diversity patterns and cross‐taxa convergence in freshwater systems. Journal of Animal Ecology, 82(2): 365-376.

Weston R, Gore P A. 2006. A brief guide to structural equation modeling. The Counseling Psychologist, 34(5): 719-751.

Wilmers C C, Post E, Peterson R O, *et al.* 2006. Predator disease out-break modulates top-down, bottom-up and climatic effects on herbivore population dynamics. Ecology Letters, 9(4): 383-389.

Woodroffe R. 2000. Predators and people: using human densities to interpret declines of large carnivores. Animal Conservation, 3(2): 165-173.

Worm B, Myers R A. 2003. Meta-analysis of cod–shrimp interactions reveals top-down control in oceanic food webs. Ecology, 84(1): 162-173.

Yudakov A G, Nikolaev I G. 1987. Ecology of the Amur tiger: winter observations during 1970–1973 in the western section of Central Sikhote-Alin. Moscow: Nauka Press.

Zimmermann F, Breitenmoserwursten C, Molinarijobin A, *et al.* 2013. Optimizing the size of the area surveyed for monitoring a Eurasian lynx (*Lynx lynx*) population in the Swiss Alps by means of photographic capture-recapture. Integrative Zoology, 8(3): 232-243.

Chapter 5 A comparison of food habits and prey preferences of Amur tiger at the southwest Primorskii Krai in Russia and Hunchun in China

5.1 Introduction

As advanced technologies like camera trapping and molecular genetic analysis have been used to monitor Amur tigers in China over the past 10 years, the whole picture of this endangered species has become clearer. The Amur tiger is now confined in 2 separate areas: the Sikhote-Alin Mountains of Russia connected with the Wandashan Mountains in China and contains nearly 90% of the tiger population (Miquelle *et al.*, 2006; Tian *et al.*, 2009; Jiang *et al.*, 2014), and a small isolated population occurring in the joint region of southwest Primorskii Krai, Russia and Hunchun, China (Miquelle *et al.*, 2006; Henry *et al.*, 2009; Sugimoto *et al.*, 2016b). These 2 populations have long been separated by urban development and wetlands (Hebblewhite *et al.*, 2011, 2014). The small and isolated population of tigers in the southern joint region is at risk of extinction from genetic, demographic and environmental stochasticity (Uphyrkina *et al.*, 2002; Henry *et al.*, 2009; Sugimoto *et al.*, 2014). Russia established a single management system known as the "Land of the Leopard National Park" in 2012 in this area for more efficient conservation. Under the North-East Asian Subregional Programme for Environmental Cooperation (NEASPEC) project *"Study of transboundary movements of the Amur tiger and Amur leopard using camera traps and molecular genetic analysis,"* the Land of the Leopard National Park and the Feline Research Center of the State Forestry Administration of China shared the camera trapping data for 2013–2015. At least 45 Amur tigers (adults only) were recorded, and 19 tigers (adults only) were recorded in both countries (southwest Primorskii Krai, Russia and Hunchun, China) during this research period (http://www.wwf.ru/resources/news/article/eng/14752). Tiger populations of such small size, especially after being isolated for 20-30 years, could drop to a threshold level below which recovery is impossible unless habitat is increased substantially within 1 to 2 generations (Kenney *et al.*, 2014). The only additional potential habitat for this small population exists on the Chinese side (Hebblewhite *et al.*, 2011), because on the southeast side of the southwest Primorskii Krai is the Sea of Japan. To improve habitat conditions, the Chinese Government has created a national park in Jilin and Heilongjiang Provinces on the border close to the Land of the Leopard National Park, with an area of 1.5 million ha (http://programmes.putin.kremlin.ru/en/tiger/news/25404).

After obtaining a clearer picture of the tiger population distribution and structure, we want to know how they survive in this area. The acquirement of food is a fundamental component for every predator's existence, and prey selection is critical for understanding the life history strategies of any carnivore (Miquelle *et al.*, 1996). Kerley *et al.* (2015) and Sugimoto *et al.* (2016a) report on the food habits and prey preference of Amur tigers in the south-west Primorskii Krai on the Russia side in, 2001–2003 and 2008–2012. However, because many tigers live on the territory of both countries, it is valuable to examine the situation on the Chinese side to understand the Amur tiger habitat expansion to China.

We conducted our study in Hunchun, which is connected to the Land of the Leopard National Park. We investigated the food habits and winter prey preferences of Amur tigers in this area, then compared our results with those from the Russian side to better understand this small Amur tiger population's food acquirements and to provide a reference for habitat recovery in China.

5.2 Materials and methods

5.2.1 Study area

Hunchun municipality has an area of 4938km^2, is located in Jilin, northeast China and contains the Hunchun Amur Tiger National Reserve (HNR), covering 1087km^2 (Fig. 5.1). The municipality borders Russia to the east and North Korea to the southwest and is connected to the Land of the Leopard National Park in Russia. This area is a key corridor for movement of Amur tigers and Amur leopards among China, Russia and North Korea. As part of the Changbai Mountains, Hunchun is in a temperate zone and has an average rainfall of 661mm, concentrated between July and September (50% of yearly precipitation). The main forms of vegetation here are mixed broad-leaved forest and secondary Mongolian oak (*Quercus mongolica*.) forest and the main animals include: Amur tigers, Amur leopards (*Panthera pardus orientalis*), red deer (*Cervus elaphus*), sika deer (*Cervus nippon*), wild boar (*Sus scrofa*), roe deer (*Capreolus pygargus*) and musk deer (*Moschus moschiferus*).

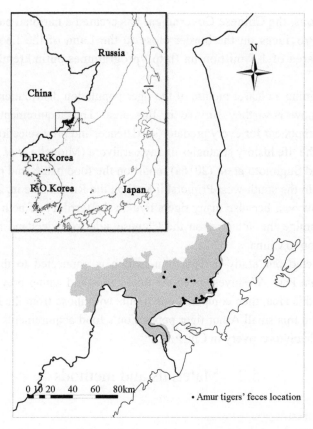

Fig. 5.1 The study area and Amur tigers' scat sample locations. Please scan the QR code on the back cover to see the color image.

5.2.2 Field research methods

We collected fecal samples from 2011 to 2016. Sampling areas covered most of the HNR and some areas outside the HNR (Fig. 5.1). We collected scats opportunistically along roads where tigers commonly deposit scats (Sunquist 1981; Karanth and Sunquist 1995) and while snow-tracking individual tigers (Yudakov et al., 1988). We also collected scats near kills that were found either by snow-tracking tigers or from information given by local citizens. Scats were placed in plastic bags and stored in freezers until analysis. Because the other sympatric big cats, Amur leopards, live here and their feces are similar to those of the Amur tiger, fecal samples were identified as belonging to either tigers or leopards using a fecal DNA-based method (Sugimoto et al., 2006a).

The relative prey abundance was estimated from the snow-track survey during the winter of 2015, during which standard survey routes were traversed after sufficient snow and all fresh tracks were recorded by species (Stephens et al., 2006). The survey assessed ungulate abundance on the basis of track encounters and the ungulates' daily movement distance, and was conducted over 18 routes (total length is approximately 90 km), covering most of the tiger habitats in Hunchun.

5.2.3 Scat analysis

The hair of prey is relatively undamaged in carnivore scat and thus can be used to identify the prey species eaten. We recorded the presence of a species in any one scat as a single occurrence. Collected Amur tigers' scats were washed in water using a 1.5-mm sieve to separate the hair from other organic matter and remains such as hair, bones, hooves, quills and teeth of the prey consumed were separated for species identification (Sunquist 1981; Karanth and Sunquist 1995). Among our samples, bones, hooves, quills and teeth of prey were not found, so prey species identification depended mainly on their hair characteristics. Separated hair was washed in hot water to remove surface oil. Each scat sample was washed separately in acetone and dehydrated in 100% ethanol. We selected at least 10 compete hairs from each scat sample. Slides were examined at 400× using an Olympus microscope. Species identification was based on the general appearance of the hair, color, length, width, medullary structure, medullary width/hair width ratio and cuticle pattern (Moore *et al.*, 1974; Mukherjee *et al.*, 1994), according to the database from the laboratory of the fur herbarium of Northeast Forestry University, Harbin, China.

Sometimes, it is hard to identify cervid hairs in scats to species by using the method above (Rozhnov *et al.*, 2011), so we used a DNA test for the cervid species identification. These hair samples were subjected to DNA extraction by TIAN amp Micro DNA Kit (Beijing), following the instructions to extract DNA. Standard PCRs were performed on 10μg of DNA extracted from scat in a 20μL volume containing 0.6μL of each primer, 10μL buffer, 4μL dNTPs, 2.5μL ddH$_2$O and 0.3μL KOD FX Neo DNA polymerase (TOYOBO). Thermal cycling conditions were as follows: 94℃ for 2min, then 35 cycles (94℃, 15s/55℃, 40s/68℃, 30s) followed by 68℃ for 20min. Sequences from the mitochondrial DNA gene of the prey species were obtained through direct sequencing of PCR products amplified using the primers mcb398 and mcb869; others primers were L14841 and H15149 (Kocher *et al.*, 1989; Ficetola *et al.*, 2010).

5.2.4 Data analysis

We estimated the contribution of each prey species to Amur tiger diet as percent frequency occurrence and biomass contribution. We used the regression equation developed for cougars (Puma concolor Linnaeus, 1771) by Ackerman *et al.* (1984) to

estimate biomass contribution to the diet:
$$y = 1.98 + 0.035x$$
where y is the weight of prey consumed/scat produced and x is the live weight of the prey, to relate the number of scats containing a prey species to the prey's biomass (Ackerman, 1984). We calculated y for each species and multiplied it by the number of occurrences of the species to estimate the relative biomass of each prey type consumed. We used the estimated live weights of prey consumed by tigers in the Russian Far East (Bromley and Kucherenko 1983; Dalnikin 1999; Prikhodko 2003). We then estimated percent biomass contribution (biomass of each prey type consumed/total biomass consumed × 100).

We estimated relative prey abundance from the winter track survey conducted in the study area, in which standard survey routes were traversed after sufficient snow and all fresh tracks were recorded by species in 2015 (Stephens et al., 2006). We used the Formozov-Malyshev-Pereleshin (FMP) formula, which considers ungulates' daily movement distance to analyze the population density with 95% confidence intervals (Stephens et al., 2006), and then changed the density into proportion data (relative abundance) for obtaining Jacobs's index (Jacobs 1974). This index ranges in value from +1 (strongly preferred) to −1 (strongly avoided). Because evidence of red deer in the tiger diet was limited to only a single observation in our winter samples, we restricted our analyses of preference to 3 species: wild boar, sika deer and roe deer during winter.

The feces sample were collected from 2011 to 2016 (77% of samples were from 2013 to 2015). During this period, there were 3 winter ungulate surveys in Hunchun in 2012, 2014 and 2015. The surveys in 2012 and 2014 used the sample plot method, which is different from the surveys used in 2015 and those used on the Russian side, so these results are not shown here. However, similar results about the ungulate species diversity and density of the dominant species were released from these 2 surveys, which taken together with the 2015 survey, means the ungulate relative abundance (as a percentage) order is stable in this area while the ungulate density may be different each year. Therefore, to obtain comparable results for the Russian side (using the same field methods), we used the 2015 ungulate data to extrapolate the whole period.

The results of the Amur tiger's biomass contribution, prey abundance and food preference were compared with the results of Kerley et al. (2015) and Sugimoto et al. (2016a) to enable better understanding of the small Amur tiger population.

5.3 Results

We collected 68 tiger scats in Hunchun from 2011 to 2016, among which 13 contained unidentifiable prey remains; 8 were multiple samples collected in the same tiger's trail or near the same killing sites. These 2 sample types were excluded from further analyses to reduce potential bias. Thus, 35 samples from the winter season (November to April) and 12 from the summer (May to October) were used to analyze the food habits and prey preferences of the Amur tiger in Hunchun (Hojnowski et al., 2012). We identified a total of 53 prey items (winter 38, summer 15) in the remaining 47 scats (Table 5.1). Because of limited samples in summer, seasonal relative biomass contribution was not compared.

Table 5.1 Prey species identified in scats of Amur tigers (2011–2016)

Species	Occurrence number		Total items
	Winter ($n = 35$)	Summer ($n = 12$)	
Sika deer	4	3	7
Roe deer	8	—	8
Wild boar	15	5	20
Red deer	1	—	1
Cow	3	2	5
Domestic dog	2	—	2
Horse	1	3	4
Leopard cat	1	—	1
Black bear	1	—	1
European badger	1	—	1
Sheep	1	—	1
Lynx	—	2	2
Total items	38	15	53

We identified a total of 11 species in winter tiger scats in Hunchun (Table 5.2), among which red deer, cow, horse and sheep were not recorded on the Russian side. There were a total of 16 winter and summer samples containing deer's hair (Table 5.1). For all samples, we successfully extracted DNA of hair bulbs for species identification and then we found 3 mistaken identifications based on the appearance observation of hair. Therefore, in our research, the accuracy rate of species identification by hair appearance observation was 81.25% (13/16), based on DNA analysis results. The top 3 prey species with the highest occurrence frequency in tiger scats were wild boar, roe deer and sika deer. Because domestic cows are much heavier than wild prey species, cow biomass contribution was in the top 4 with wild boar, roe deer and sika deer, which was the other difference compared with the Russian side. Total biomass contribution from 4 domestic animals was 33.85%, while on the Russian side was just 0.44 and 6.3 (Table 5.2).

Table 5.2 Percent occurrence frequency and biomass contribution of prey species to Amur tiger diet as determined by scat analysis on the Chinese side and the Russian side in winter

Prey species	Prey weight (kg)	Occurrence frequency (%)	Biomass contribution		
			China side (2011–2016) $n = 35$	Russia side[†](2008–2012) $n = 152$	Russia side[‡](2001–2003) $n = 63$
Sika deer	95	11.43	9.28	25.09 (18.12–32.06)	23.86
Roe deer	37	22.86	11.45	8.31 (4.89–12.22)	41.3
Wild boar	103	42.86	36.63	58.60 (49.47–67.74)	54.0
Red deer	187	2.86	4.19	—	—
Cow	418	8.57	21.79	—	—
Domestic dog	31	5.71	2.68	0.44 (0.00–1.31)	6.30
Horse	450	2.78	7.75	—	—
Leopard cat	4	2.86	0.93	—	1.60
Black bear	150	2.86	3.16	3.11 (0.00–7.25)	—
European badger	6	2.86	0.96	0.34 (0.00–1.01)	6.30
Sheep	50	2.86	1.63	—	—
Red fox	5	—	—	—	3.20
Hare	1.8	—	—	—	3.20
Musk deer	—	—	—	0.37 (0.00–1.10)	—
Weasel	—	—	—	0.94 (0.00–2.19)	—
Amur tiger				2.81 (0.70–5.62)	

†Cited from Kerley et al. (2015) Table S1, with the results from 2008 to 2012; "$n = 152$" was the total winter and summer scats in south-west Primorskii Krai of Russia; Kerley et al. (2015) did not mention total winter scat number.
‡Cited from Sugimoto et al. (2016a), showing the results from 2001 to 2003.

There were 2 key differences between the Chinese and Russian sides in this region (Fig. 5.2). First, Kerley et al. (2015) and Sugimoto et al. (2016a) did not find red deer footprints in their snow-track surveys, but on the Chinese side we found red deer footprints in 5 survey lines. Second, the relative abundance order of 4 prey species in Kerley et al. (2015) and Sugimoto et al. (2016a) was the same: sika deer > roe deer > wild boar (red deer = 0), but our results were: roe deer > sika deer > wild boar > red deer, which showed different ungulate population structures on the 2 sides.

Jacobs index values indicated that in our study Amur tigers showed a preference for wild boar (+0.849), which was similar with the Russian side, but a strong avoidance of roe deer (−0.693), while the Russia data showed a strong avoidance of sika deer (−0.698, −0.717). Besides roe deer, Amur tiger also avoided sika deer (−0.495) in proportion to their availability (Table 5.3).

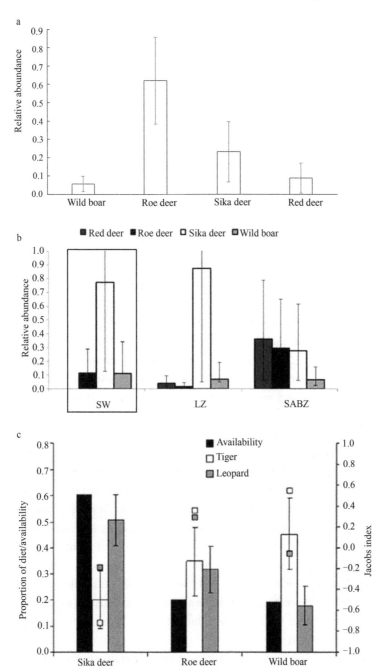

Fig. 5.2 Relative abundance of wild boar, sika, red and roe deer as estimated by winter track counts in Hunchun: (a) shows our results from the winter survey of 2015; (b) was cited from Kerley et al. (2015) in which SW represented the southwest Primorskii Krai of Russia and the results were from 2008–2012; (c) was cited from Sugimoto et al. (2016a) showing the results from 2001–2003.

Table 5.3 Jacobs' index scores with 95% confidence intervals measuring tiger preference for or avoidance of 4 ungulate species on the Chinese side and the Russian side

Species	Jacobs' index		
	Chinese side 2011–2016	Russian side[†] 2008–2012	Russian side[‡] 2001–2003
Wild boar	0.849 (0.747 to 0.957)	0.790 (0.618 to 0.962)	0.547
Red deer	—	−0.326 (−0.791 to 0.139)	—
Sika deer	−0.495 (−0.672 to 0.272)	−0.698 (−0.917 to −0.268)	−0.717
Roe deer	−0.693 (−0.906 to −0.355)	−0.368 (−1.000 to 0.275)	0.353

[†]Cited from Kerley et al. (2015), but the results were from 3 sites, not only SW, which represents the south-west Primorskii Krai of Russia. [‡]Cited from Sugimoto et al. (2016a).

5.4 Discussion

Considering the predation habits in southwest Primorskii Krai, Russia and Hunchun, China, the biggest difference in Amur tigers' diet and prey preferences between the two countries was caused by domestic animal consumption. With fewer residents in the forest on the Russian side, only domestic dogs became Amur tigers' prey in winter, while on the Chinese side, in addition to domestic dog consumption in winter, cows, horses and sheep were recorded throughout the year. There are 98 villages and 4 towns in our research area (Yi et al., 2014). Based on a questionnaire survey, 63 pastures are within the area and 3066 cows were recorded, among which 42 pastures had an area of 355km^2(Li et al., 2016). Li et al. (2016) found that grazing activities mainly affected tree density and shoot numbers, which is preferred and browsed upon by ungulates. For example, grazing activities had negative effects on the number of shoots, with a contribution of 30.9%. In winter, most domestic animals were brought back to the villages near people, however their damage effect on vegetation were still there which reduced the ungulate prey population. Even so, the preferred food of Amur tigers is still wild boar and the majority of a tigers' diet consists of medium to large ungulates, mirroring the results of previous studies on tiger diet (Karanth and Sunquist 1995; Biswas and Sankar 2002; Miquelle et al., 2007, 2010; Kapfer et al., 2011). Therefore, with the lower ungulate density (especially wild boar), in order to survive, Amur tigers on the Chinese side did not avoid eating sika deer as strongly as on the Russian side.

In our study, the biomass contribution of cows and horses may be overestimated, because we found that at the kill sites, tigers often eat only part of a cow or a horse, unlike wild boar, which are almost entirely eaten. In addition, tigers appear to attack calves and not adult cows. Domestic animal loss to tigers may also be overestimated, because if tigers eat a domestic animal, the owner can apply for compensation after submitting information to local government; however, local people will not report every incident of a tiger eating a wild animal. During the winter survey, we did not find cow or horse footprints in the forest, but it is not hard for tigers to find these domestic animals near villages on the Chinese side when they are seeking wild prey.

In regards to Amur tigers' prey preferences, a strong avoidance of roe deer (−0.693) was also shown in our research, while a strong avoidance for sika deer was shown on the Russian side (−0.698, −0.717). Prey selection analysis revealed that Amur tigers had a notable preference for or avoidance of particular ungulates compared with Amur leopards: the most frequently consumed prey were sika and roe deer for leopards and wild boar for tigers (Miquelle *et al.*, 1996; Sugimoto *et al.*, 2016a). Perhaps that is why Amur leopards can coexist with Amur tigers in this region.

The other issue worthy of concern is red deer, which are preferred in Amur tigers' diet (Zhivotchenko 1981; Miquelle *et al.*, 2010). In the southwest Primorskii Krai, Russia–Hunchun, China area, according to the results from Kerley *et al.* (2015) and Sugimoto *et al.* (2016a), red deer were not eaten by Amur tiger on the Russian side. However, on the Chinese side, we found through hair DNA analysis that tigers did eat red deer at least once (Table 5.1 and Fig. 5.2a). From 2012 to 2013, WWF-China released 67 red deer and sika deer into the wild in Wangqing, which is in the north of Hunchun, approximately 50-60km from the Sino–Russia border, to aide in the recovery of Amur tiger and leopard prey resources (http://www.wwfchina.org/pressdetail.php?id=1485). It is not known whether there is a wild red deer population surviving in this region or if the released newcomers have broken into the wild network of ungulates. This requires further tracking research.

On 24 May and 18 October 2013, we collected 2 Amur tiger scat samples that revealed lynx hair (Table 5.1), earlier than when Petrunenko *et al.* (2016) announced that they had found the first recorded case of a tiger killing a Eurasian lynx. Petrunenko *et al.* (2016) report that on 4 March 2014 in Bastak Nature Reserve (48°56′37 N, 133°07′13 E), located in the Jewish Autonomous Region of the Russian Far East, the team found the remains of a lynx surrounded by tiger tracks. In Hunchun, we camera-trapped lynx and showed that Amur tigers, Amur leopards and lynx co-exist there. It is not clear whether tigers depress the numbers of these smaller cats, as has been suggested elsewhere (Harihar *et al.*, 2011). Even less is known of the relationships between tigers and lynx. While documentation of this episode of direct competition between tigers and lynx is compelling, there is still much to be learned about the complex interrelationships of large carnivores in this ecosystem.

In regards to scat analysis, except for multiple samples of diet from the same individual, the majority of pseudo replicates were for deer species (Kerley *et al.*, 2015; Sugimoto *et al.*, 2016a). Using a compound microscope to examine the shape and thickness of hair medulla and hair cuticle patterns did not work every time for deer species identification. The simple removal of the pseudo replicates reduced the percentage of biomass for the combined deer category, and increased the percent biomass for other prey species. We strongly recommend that DNA testing be used in scat analysis to avoid pseudo replicates. Our research is a good example: if we had not used DNA testing, we would not have found a case where a tiger ate a red deer in this area.

Our results showed that Amur tigers in Hunchun, China had different food habits and prey preferences compared with those in the southwest Primorskii Krai of Russia, likely due to heavy human disturbance, such as grazing, which affects local vegetation and so reduces the number of prey species that survive on this vegetation. With the clear trend of the expansion of Amur tigers to the Chinese side, reducing or eliminating human disturbance is essential to the recovery of tiger prey, as well as the habitat in this region. The Sino-Russian joint ungulate annual survey is also indispensable for prey estimates of this isolated, small Amur tiger population.

5.5 Summary

We used scat analysis combined with data on the abundance of 4 prey species to examine Amur tiger diet and prey preferences in Hunchun. We examined 53 tiger scat samples from 2011 to 2016 and found that tigers preyed on 12 species (11 species in winter), 4 of which were domestic animals with 33.58% biomass contribution; this was the first record of Amur tigers eating lynx in this area. Tigers showed a strong preference for wild boar (Jacobs index: +0.849), which were also the most frequently consumed prey, and a strong avoidance of roe deer (Jacobs index: −0.693). On the Russian side, domestic animals (just dog) were rarely found in tiger scat, and tigers did not show strong avoidance of roe deer, but of sika deer. We also found red deer footprints during winter surveys and that tigers ate red deer on the Chinese side, while there was no record of red deer feeding on the Russian side. Reducing or eliminating human disturbance, such as grazing, is essential to recovering tiger prey and habitat in this area and the Sino–Russian joint ungulate annual survey is indispensable for prey estimates of this small, isolated Amur tiger population.

References

Ackerman B B, Lindzey F G, Hemker T P. 1984. Cougar food habits in southern Utah. The Journal of Wildlife Management, 48(1): 147.

Biswas S and Sankar K. 2002. Prey abundance and food habit of tigers (panthera tigris tigris) in pench national park, madhya pradesh, india. Proceedings of the Zoological Society of London, 256(3): 411-420.

Bromley G F, Kucherenko S P. 1983. Ungulates of the Southern Far East USSR. Moscow: Nauka Press.

Dalnikin A A. 1999. Mammals of Russia and Adjacent Regions: Deer(Cervidae). Moscow: GEOC.

Ficetola G F, Coissac E, Zundel S, *et al.* 2010. An in silico approach for the evaluation of dna barcodes. Bmc Genomics, 11(1): 434.

Harihar A, Pandav B, Goyal S P. 2011. Responses of leopard *Panthera pardus* to the recovery of a tiger *Panthera tigris* population. Journal of Applied Ecology, 48: 806-814.

Hebblewhite M, Miquelle D G, Murzin A A, *et al.* 2011. Predicting potential habitat and population size for reintroduction of the Far Eastern leopards in the Russian Far East. Biological Conservation, 144: 2403-2413.

Hebblewhite M, Miquelle D G, Robinson H, *et al.* 2014. Including biotic interactions with ungulate prey and humans improves habitat conservation modeling for endangered Amur tigers in the Russian Far East. Biological Conservation, 178: 50-64.

Henry P, Miquelle D, Sugimoto T, *et al.* 2009 . *In situ* population structure and *ex situ* representation of the endangered Amur tiger. Molecular Ecology, 18: 3173-3184.

Hojnowski C E, Miquelle D G, Myslenkov A I, *et al.* 2012. Why do Amur tigers maintain exclusive home ranges? Relating ungulate seasonal movements to tiger spatial organization in the Russian Far East. Journal of Zoology, 287: 276-282.

Jacobs J. 1974. International Association for Ecology Quantitative Measurement of Food Selection: A modification of the forage ratio and Ivlev's electivity index. Oecologia, 14: 413-417.

Jiang G, Sun H, Lang J, *et al.* 2014. Effects of environmental and anthropogenic drivers on Amur tiger distribution in northeastern China. Ecological Research, 29: 801-813.

Kapfer P M, Streby H M, Gurung B, *et al.* 2011. Fine-scale spatio-temporal variation in tiger *Panthera tigris* diet: Effect of study duration and extent on estimates of tiger diet in Chitwan National Park, Nepal. Wildlife Biology, 17: 277-285.

Karanth K U, Sunquist M E. 1995. Prey selection by tiger, leopard and dhole in tropical forests. Journal of Animal Ecology, 64: 439.

Kenney J, Allendorf F W, McDougal C, *et al.* 2014. How much gene flow is needed to avoid inbreeding depression in wild tiger populations? Proceedings of the Royal Society B Biological Sciences, 281 (1789): 20133337.

Kerley L L, Mukhacheva A S, Matyukhina D S, *et al.* 2015. A comparison of food habits and prey preference of Amur tiger (*Panthera tigris altaica* Timminck, 1884) at three sites in the Russian Far East. Integrative Zoology, 10: 354-364.

Kocher T D, Thomas W K, Meyer A, *et al.* 1989. Dynamics of mitochondrial DNA evolution in animals: Amplification and sequencing with conserved primers. Proceedings of the National Academy of Sciences of the United States of America, 86: 6196-6200.

Li Z, Kang A, Gu J, *et al.* 2016. Effects of human disturbance on vegetation, prey and Amur tigers in Hunchun Nature Reserve, China. Ecological Modelling, 353: 28-36.

Melville H I A S. 2004. Behavioural ecology of the caracal in the Kgalagadi Transfrontier Park, and its impact on adjacent small stock production units (MSc dissertation). Pretoria: University of Pretoria.

Miquelle D G, Goodrich J M, Smirnov E N, *et al.* 2010. Amur tiger: A case study of living on the edge. In: Macdonald D W, Loveridge A J, eds. Biology and Conservation of Wild Felids. Oxford: University of Oxford Press.

Miquelle D G, Pikunov D G, Dunishenko Y M, *et al.* 2007. 2005 Amur Tiger Census. Cat News, 46: 14-16.

Miquelle D G, Smirnov E N, Quigley H G, *et al.* 1996. Food habits of Amur tigers in Sikhote Alin Zapovednik and the Russian Far East, and implications for conservation. Journal of Wildlife

Research, 1: 138-147.

Moore T M, Spence L E, Dugnolle C E, et al. 1974. Identification of the dorsal hairs of some mammals of Wyoming. Cheyenne: Issue 14 of Bulletin, Wyoming Game and Fish Department.

Mukherjee S, Goyal S P, Chellam R. 1994. Standardisation of scat analysis techniques for leopard (*Panthera pardus*) in Gir National Park, Western India. Mammalia, 58: 139-143.

Petrunenko Y K, Polkovnikov I L, Gilbert M, et al. 2016. First recorded case of tiger killing Eurasian lynx. European Journal of Wildlife Research, 62: 373-375.

Prikhodko V I. 2003. Musk Deer: Distribution, Systematics, Ecology, Behaviour, and Communication. Moscow: GEOS.

Rozhnov V V, Chernova O F, Perfilova T B. 2011. *A Guide to Deer Species in the Diet of Amur Tigers* (*Microstructure of Deer Species Guard Hairs Found in Amur Tiger Excrement*). Moscow: Scientific Press. (in Russian)

Stephens P A, Zaumyslova O Y, Miquelle G D, et al. 2006. Estimating population density from indirect sign: track counts and he Formozov-Malyshev-Pereleshin formula. Animal Conservation, 9: 339-348.

Sugimoto T, Aramilev V V, Kerley L L, et al. 2014. Noninvasive genetic analyses for estimating population size and genetic diversity of the remaining Far Eastern leopard (*Panthera pardus orientalis*) population. Conservation Genetics, 15: 521-532.

Sugimoto T, Aramilev V V, Nagata J, et al. 2016a. Winter food habits of sympatric carnivores, Amur tigers and Far Eastern leopards, in the Russian Far East. Mammalian Biology–Zeitschrift für Säugetierkunde, 81(2): 214-218.

Sugimoto T, Nagata J, Aramilev V V, et al. 2006. Species and sex identification from faecal samples of sympatric carnivores, Amur leopard and Siberian tiger, in the Russian Far East. Conservation Genetics, 7: 799-802.

Sugimoto T, Nagata J, Aramilev V V, et al. 2016b. Population size estimation of Amur tigers in Russian Far East using noninvasive genetic samples. Journal of Mammalogy, 93: 93-101.

Sunquist M E. 1981. The Social Organization of Tigers (*Panthera Tigris*) in Royal Chitawan National Park, Nepal. Journal of Endocrinology, 336: 345-359.

The Amur Tiger Program. 2016. Land of the leopard, Chinese nature reserves exchange tiger and leopard data. http://programmes.putin.kremlin.ru/ en/tiger/news/25404[2016-10-20].

Tian Y, Wu J G, Kou X J, et al. 2009. Spatiotemporal pattern and major causes of the Amur tiger population dynamics. Biodiversity Science, 17: 211.

Uphyrkina O, Miquelle D, Quigley H, et al. 2002. Conservation Genetics of the Far Eastern Leopard (*Panthera pardus orientalis*). Journal of Heredity, 93: 303.

WWF-China. 2016. WWF released deer to the wild again. http://www.wwfchina.org/pressdetail. php?id=1485[2016-10-26].

WWF-Russia. 2016. Leopards and tigers freely cross the border of Russia and China. http://www.wwf.ru/resources/news/article/eng/14752[2016-12-5].

Yi H S, Carrasco L R, Miquelle D G, et al. 2014. Spatial correlates of livestock depredation by Amur tigers in Hunchun, China: Relevance of prey density and implications for protected area management. *Biological Conservation*, 169: 117-127.

Yudakov A G, Nikolayev E G, Olson E C. 1988. Ecology of the Amur Tiger. The Quarterly Review of Biology, 63(4): 472.

Zhivotchenko V E. 1981. Food habits of the Amur tiger. In: *Predatory Mammals*. Moscow: TSNIL Glavokhoty RSFSR: 64-75. (in Russian)

Chapter 6 A comparison of reproductive parameters of female Amur tiger in the wild and captivity

6.1 Introduction

The Amur tiger (*Panthera tigris altaica* Temminck, 1844) is the most northernmost tiger subspecies. At present, the majority of this wild population resides in Russia, where an estimated 430–502 individuals occur(Miquelle *et al.*, 2007), following a long-term recovery from 20–30 tigers in the 1940s (Kaplanov, 1948). There are an additional 20 or so individuals surviving in China(Zhang *et al.*, 2013) and few or none in Korea (Miquelle and Pikunov, 2003; Miquelle *et al.*, 2005).

In Russia, research and conservation efforts are focused on recovery of the wild population. Kerley *et al.* (2003) studied the reproductive parameters of 11 wild female Amur tigers and found that only 0.7 cubs/year survived up to 12 months of age, with these results plus others (Miquelle *et al.*, 2015) suggesting that tiger populations may not recover as quickly as others had hoped. Russello *et al.* (2004) reveal extremely low levels of haplotype diversity from samples collected from 27 tigers in Russia, and with a larger sample ($n = 95$). Henry *et al.* (2009) estimate an effective population size of 27–35 individuals. Loss of genetic diversity can lead to inbreeding, and can reduce reproductive fitness(Roelke *et al.*, 1993; Reed and Frankham, 2003). Given that the reproductive parameters of female tigers strongly influence the dynamics of a population, it is worthwhile comparing such parameters to look for variation between captive and wild populations that may be potential indicators of underlying genetic pitfalls.

The Amur tiger has been successfully bred in captivity in China for more than 30 years, with a captive population in Hengdaohezi Felid Breeding Center having a similar number of founders as the Russian wild population: as of 2010 a population of more than 350 captive Amur tigers were derived from 27 founders, with studbook records clearly demonstrating a purely Amur tiger lineage (Zheng *et al.*, 2005) collected from zoos in China and USA since 1986. Given their genetic status, the reproductive parameters of this captive closed population may provide a useful reference point for comparison to the wild population.

Ex-situ management involving sufficient numbers of individuals could potentially assist in the recovery and maintenance of wild tigers in China as well as enhance their genetic diversity (Russello *et al.*, 2004). Whether captive-born tigers are capable of successfully adapting to life in the wild without representing a threat to humans is still unknown, but recent attempts in Russia to reintroduce orphaned tiger cubs into the wild (Shaer, 2015) suggest that such an approach may be viable. However, after several decades of captive breeding, changes might have occurred in reproductive parameters to make these animals unsuitable for release.

We sought to compare the reproductive characteristics of captive female Amur tigers in Hengdaohezi Felid Breeding Center (HFBC) to the same parameters of wild female Amur tigers as reported in Kerley *et al.* (2003). One wild female gave birth 7 months after she lost her 1-week-old litter due to unknown causes (Kerley *et al.*, 2003), so we sought to compare normal birthing cycles to situations when a female prematurely lost a litter. Finally, we sought to determine whether timing of birthing might be shifting in this captive population by comparing birth dates across 16 years of data.

6.2 Materials and methods

6.2.1 Study population

Hengdaohezi Felid Breeding Center is situated in Heilongjiang Province of China within the historical range of Amur tigers (44°47′38″N, 129°06′39″E) (Yu *et al.*, 2009). Records of breeding females have been kept since 1986, resulting in a database that includes more than 350 individuals. A separate studbook, recording ancestry, mates, and offspring is kept for each tiger. In 2003, in cooperation with the State Forestry Administration Detection Center for Wildlife, the center began to reconstruct the precise molecular genealogy of this population, and developed a management system to minimize the loss of genetic diversity (Siberia Tiger Park, 2003). All tigers at HFBC are managed in accordance with *the Technical Code of Feeding and Management for Wild Animals—Siberian Tiger* (LY/T 2199—2013) (State Forestry Administration, 2013). Tigers are fed 6 days a week; males consume 6–8kg/day and females consume 5–7kg/day of meat. The diet of the tigers is comprised of locally-sourced beef, chicken and pig (with bones). Young tigers are kept together in outdoor enclosures until the first estrus of females. At that point, females are moved to separate enclosures with outdoor and indoor compartments, with males introduced to these pens during estrus periods and according to breeding plans. The indoor area is heated during winter. The pens are cleaned using water from hoses and some flame sterilization is conducted. Most cubs are separated from their mother at approximately 6 months old (occasionally up to 1 year old), to induce estrus of females and a shorter reproductive cycle.

We measured 7 reproductive parameters, including the age of first parturition, mating date, the duration of the last mating event (the time between the first and last day of the most recent mating event that lead to conception), birth date, length of gestation (derived from conception date to birth date), litter size and interval between litters. The date of conception was defined as the first copulation of a mating event that led to a pregnancy.

We measured neonatal mortality rates (from birth to 7 days) as well as cub mortality rate for the period from birth to 6 months from 2004 to 2010. We classified neonatal mortalities into 4 categories: premature birth, stillbirth, congenital abnormalities and abandonment (or inadequate care by mother). Congenital abnormalities were defined as externally-visible abnormalities.

6.2.2 Data analysis

We calculated the mean and standard deviation for age of first parturition, interval between litters, gestation length, litter size and cub mortality rate, and compared the results with those of wild tigers. We compared the mean litter size for captive and wild populations using a 2-sample Kolmogorov-Smirnov Test(Lilliefors, 1969).

We estimated monthly frequencies of mating and birth dates for captive females and then grouped data into 4 seasons: early winter (November–January of the following year), late winter (February–April), early summer (May–July) and late summer (August–October) after Kerley et al. (2003). Because the data were not normally distributed, we used a Kruskal-Wallis test to compare mating and birth peaks of captive and wild tigers.

To explore how timing of reproduction changed over time in captivity, we regressed the day of birth (from the beginning of the year) with the year (1995 to 2010) to look for any trends over the 16 years of data.

Lastly, we compared the characteristics of a subsequent pregnancy and birthing after losing cubs prematurely with the normal cycles of birthing to determine whether reproductive parameters changed after losing cubs.

6.3 Results

From 1995 to 2010, 68 female Amur tigers at HFBC produced 252 litters, totaling 724 cubs. Of those 252 litters, only 123 mating events were recorded that led to conception.

6.3.1 Reproductive parameters of captive and wild Amur tigers population

Age of first parturition (4.10 ± 1.12 years) was very similar to estimates derived from the few samples in the wild (Table 6.1). Gestation length was estimated to be (108.27 ± 3.64) days ($n = 126$), similar to the findings of previous studies in captivity (103 days in (Kitchener, 1991; Sankhala, 1978)). The average duration of a mating event prior to conception was 6.56 ± 2.85 days (a measurement with no comparison from the wild). The mean litter size at birth of captive Amur tigers (2.91 ± 0.97 cubs) was significantly higher ($P = 0.034$) than the mean litter size of wild Amur tigers when first documented (Table 6.1). The interval between litters was much shorter in captivity (384.96 ± 198.84 days) than in the wild (642 ±132 days).

Table 6.1 Reproductive parameters compared between captive and wild Amur tigers (mean ± SD, if available)

Mean	Captive	n	Wild	n	P
Age of first parturition	4.10 years ± 1.12 years	68	4 years± 0.4 years	2	—
Interval between litters	384.96 days ± 198.84 days	184	21.4months ± 4.4 months (approximately 642 day ±132 day)	7	—
Duration of mating period prior to conception	6.56 days ± 2.85days	87	—	—	—
Gestation length	108.27 days ± 3.64 days†	126	103	—	—
Cub mortality	41.99% ± 40.96%	252	41%-47%	19	—
Litter size	2.91 cubs ± 0.97cubs	252	2.4 cubs ± 0.6 cubs	16	0.034

†Gestation length is derived as the date from first copulation of the most recent mating event to birthing. Wild tiger data are from Kerley *et al.* (2003) except gestation length, which is from Kitchener (1991).

6.3.2 Cub mortality

Within each litter, 41.99% ± 40.96% of the cubs died, within the same range as the estimate derived from the wild population (Table 6.1). We also recorded a 13% neonatal mortality rate from 2004 to 2010. A total of 76 neonatal mortalities were recorded (Table 6.2), with still births and congenital abnormalities the 2 most important causes of cub mortality. Inexperienced and some older females lost cubs due to inadequate care or abandonment, especially in 2008. The neonatal losses were usually low in 2004 and 2010.

Table 6.2 The frequency of four types of neonatal mortality

	2004	2005	2006	2007	2008	2009	2010	Total
Premature birth					1	4		5
Stillbirth	1	5	3	9	7	5	1	31
Congenital abnormalities		1	7	4	1	6	2	21
Abandonment/inadequate care	3	4	2	2	7	1		19
n	4	10	12	15	16	16	3	76
n/total cubs	0.07	0.17	0.12	0.14	0.19	0.13	0.04	0.13

6.3.3 Seasonality of reproduction

Captive tigers at HFBC conceived in all but 2 months (September and October) of the year, but most commonly in late winter (February-April) (x^2 = 8.996, df = 3, P = 0.004) (Fig. 6.1a). Wild tigers also conceived in all but 3 months of the year, but conceptions were most frequent from March to May (Fig. 6.1a). Captive tigers at

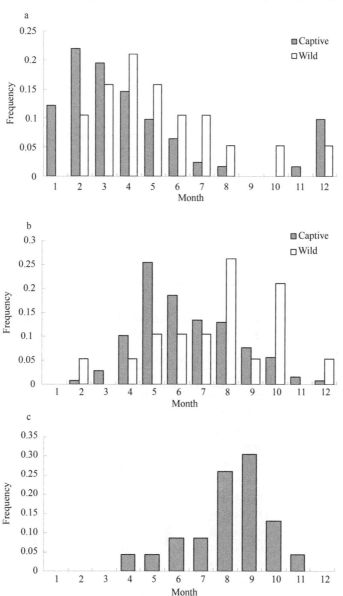

Fig. 6.1 Monthly frequencies of reproductive parameters of captive tigers at Hengdaohezi Felid Breeding Center and wild Amur tigers in Russia (from Kerley *et al.* 2003): a. date of breeding (in captivity, first copulation observed during a mating event); b. birth date; c. birth date of litter subsequent to early loss of the previous litter (captive population only).

HFBC gave birth in all but 1 month (January) of the year, but birthing peaked in early summer (May–July) (x^2 = 8.728, df = 3, P = 0.006), again earlier than wild tigers (August–October).

The mean date of birth appeared to decrease (i.e., births coming earlier in the year) over the first 6 years of observation (1995–2000) and then increased from 2001 to 2012. The average birth date was significantly correlated with year only for this second period (P = 0.01, r = 0.764), suggesting that females were giving birth later in the year (Fig. 6.2). During this same period there was no significant change in the age of females in each year (P = 0.68) (Table 6.3).

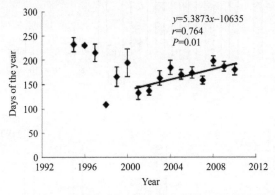

Fig. 6.2 Changes in mean date of birth (Julian dates) for tigers held at Hengdaohezi Felid Breeding Center across 16 years, 1995–2010 (significant increase in birth date for the period from 2001 to 2010)

Table 6.3 The annual breeding records (mean ± SD)

Year	Litters	Days of the year	Mean age	Minimum age	Maximum age
1995	2	232.50 ± 20.50	4.94 ± 2.55	3.1	6.7
1996	3	230.67 ± 2.51	5.22 ± 2.15	4.0	7.7
1997	4	215.50 ± 35.38	4.77 ± 0.40	4.2	5.0
1998	1	109.00	9.37	9.4	9.4
1999	4	166.00 ± 40.17	7.06 ± 2.50	4.9	10.6
2000	7	195.00 ± 75.74	6.52 ± 3.04	2.9	11.4
2001	7	132.86 ± 35.67	7.09 ± 3.12	3.0	12.3
2002	7	137.14 ± 23.98	4.34 ± 2.54	2.8	9.8
2003	13	163.15 ± 55.03	4.99 ± 2.09	2.7	9.9
2004	20	184.95 ± 64.54	5.21 ± 2.61	2.9	11.8
2005	19	170.79 ± 43.03	4.97 ± 1.85	3.0	10.9
2006	31	174.16 ± 65.35	5.55 ± 1.91	3.0	13.0
2007	32	158.97 ± 46.12	6.33 ± 2.42	3.7	14.0
2008	31	198.71 ± 53.59	5.93 ± 1.86	3.3	10.3
2009	43	187.40 ± 68.25	5.86 ± 1.58	3.3	9.9
2010	28	181.07 ± 57.78	6.84 ± 2.03	3.9	11.3
Total	252	178.12 ± 58.42	5.85 ± 2.16	2.7	14.0

6.3.4 Reproductive characteristics of females after losing cubs

Twenty-one females from the HFBC population lost 26 entire litters over the study period. The length of gestation of the pregnancy after the lost litter (108.89 days ± 2.02 days) did not vary from pregnancies after normal birthing cycles (108.17 d ± 3.83 d, $P = 0.21$). Mean litter size after losing a litter was 2.73 cubs ± 1.15 cubs; not significantly different than in normal cycles (2.93 cubs ± 0.94 cubs, $P = 0.37$). Mean cub mortality was 47.35% ± 42.72%, not statistically different from normal cycles (41.33% ± 40.69%, $P = 0.12$). However, the timing of birthing did vary: females gave birth to a second litter usually between April and November, with the peak in late summer (August–October) ($x^2 = 9.17$, df = 3, $P = 0.01$) (Fig. 6.1c).

6.4 Discussion

Litter sizes of the captive population at HFBC were higher than estimates derived from wild Amur tiger population. However, these differences are likely due to the timing of when litter size was measured. We were able to determine litter size at birth, while Kerley et al. (2003), on average, were only able to determine litter size at nearly 4 months of age, making it impossible to estimate neonatal mortality and consequently clearly underestimating initial litter size. Similar to our findings, (Christie and Walter, 2000) report mortality rates of nearly 40% during the first 2 months for captive Amur tigers. Kerley et al. (2003) estimate mortality rates over an older age class, but collectively it appears that cub mortality is high in both captive and wild settings.

Our observations suggest that in HFBC captive tigers were "weakly" seasonal breeders, with a peak in early summer, but with birthing occurring throughout the year. However, we did note a significant shift over years, with the mean date of birthing getting later from 2001 to 2010 (Fig. 6.2). With no significant difference in ages of females during this period (Table 6.3), it is likely that changes in management schemes were responsible for the shift. However, this phenomenon needs continued observation and research.

The main difference in reproduction parameters between the wild and HFBC populations is the timing of mating/birthing, and the interval between litters. Our results are consistent with the majority of the literature on wild Amur tiger populations that report peak conception rates in January and February (Baikov, 1925; Kucherenko, 1985; Dunishenko and Kulikov, 1999), suggesting a spring (April–May) birth peak. However, data for all these references were dependent on tracking tigers in snow, which will clearly bias results towards those months when snow is on the ground. Nonetheless, (Seal, 1987) also report a distinct spring birth peak in the captive North American population. Kerley et al. (2003) were dependent on a small sample size to estimate birthing peaks, which may have resulted in a skewed estimate. However, the changing birth dates we documented over 16 years at HFBC also suggest that management activities there clearly influenced timing of births (Fig. 6.2). Further data from the wild and captive populations will be needed to resolve these differences.

The normal interval between births in the wild is nearly 2 years, while at HFBC estrus was induced over a shorter time interval (about 13 months) by removing cubs at 6 months to one year of age. Cases in which females prematurely lose litters in the wild are rarely observed, making our observations of a captive population a valuable opportunity to better understand the dynamics of this process, and to better define the interval between births after premature loss of a litter. Our results suggest that reproductive parameters after losing a litter are quite similar to normal cycles, with only the birth date shifting in the captive population. How a shift in the birthing date may influence survival rates of such litters in the wild is still unknown.

In general, except for the timing of birthing and the shorter interval between litters, most of the reproductive parameters of the HFBC captive population and the wild population are similar. These results provide no indication of major problems in the use of captive females for a breeding program for release of cubs into the wild, but additional information is still needed to assess their suitability. We suggest that further research in the following areas is still needed. First, more information is required on cub survival and the causes of mortality to assess the potential impact of inbreeding depression and environmental factors on cub survival. Second, further examination of the cause of the high neonatal losses both in captive and wild Amur tiger population is necessary. Finally, reasons for the shifting birth date of the captive Amur tiger population should be investigated.

6.5 Summary

A healthy population of captive Amur tigers might assist recovery of the wild population in Northeast China if individuals were properly prepared and considered suitable for release in the wild. We analyzed the breeding records of 68 female Amur tigers from 1995 to 2010 in the Hengdaohezi Felid Breeding Center of China and compared the reproductive parameters of this population to wild female Amur tigers. We found that the reproductive parameters of the captive population (the age of first parturition, mating date, length of gestation and litter survival rate) were not significantly different from those of wild Amur tigers. Differences in birth date and litter size between wild and captive populations may be caused by management protocols for the captive population or insufficient field data from the wild population. Reproductive parameters of females giving birth after losing a litter were similar to parameters of females that did not lose a litter, except for birth date. These results provide no indication of major problems in using captive females for a breeding program for release of cubs into the wild, but additional information is still needed to assess their suitability.

References

Baikov N. 1925. The Manchurian tiger. Harbin: Obshchestvolzuchenia Manchurskogo Kraya. (in Russian)

Christie S, Walter O. 2000. European and Australasian Studbook for Tigers (*Panthera tigris*). London: Zoological Society of London.

Dunishenko Y, Kulikov A. 1999. Amur tiger. Khabarovsk, Russia: Khabarovski Izadatilstva. (in Russian)

Kaplanov L G. 1948 Tigers in Sikhote-Alin. Tiger, red deer, and moose. Moscow: Obschestva Ispytateley Prirody. (in Russian)

Kerley L L, Goodrich J M, Miquelle D G, et al. 2003. Reproductive parameter of wild female Amur tigers(*panthera tigris altaical*). Journal of Mammalogy, 84(1): 288-298.

Kitchener A. 1991. The Natural History of the Wild Cats. New York: Comstock Publishing. Associates.

Kucherenko S. 1985. Tiger. Moscow, Russia: Agropromizdatilstva. (in Russian)

Lilliefors H W. 1969. On the Kolmogorov-Smirnov test for the exponential distribution with mean unknown. Journal of the American Statistical Association, 64(325): 387-389.

Miquelle D G, Pikunov D G, Dunishenko Y M, et al. 2005. Amur tiger census. Cat News, 46: 11-14.

Miquelle D G, Pikunov D G. 2003. Status of the Amur tiger and Far Eastern leopard // The Russian Far East: A reference guide for conservation and development. McKinleyville, California: Daniel and Daniel Publishers: 106-109.

Miquelle D G, Smirnov E N, Zaumyslova O Y, et al. 2015. Population dynamics of Amur tigers (*Panthera tigris altaica*) in Sikhote-Alin Biosphere Zapovednik: 1966-2012. Integrative Zoology, 10(4): 315-328.

Miquelle D, Pikunov D, Dunishenko Y. 2007. 2005 Amur tiger census. CAT News, (46): 14-16.

Reed D H, Frankham R. 2003. Correlation between fitness and genetic diversity. Conservation Biology, 17(1): 230-237.

Roelke M E, Martenson J S, Obrien S J. 1993. The Consequences of Demographic Reduction and Genetic Depletion in the Endangered Florida Panther. Current Biology, 3(6): 340-350.

Russello M A, Gladyshev E, Miquelle D, et al. 2004. Potential genetic consequences of a recent bottleneck in the Amur tiger of the Russian Far East. Conservation Genetics, 5(5): 707-713.

Sankhala K. 1978 Tiger!: The Story of the Indian Tiger. London: HarperCollins.

Seal U S. 1987. Behavioral indicators and endocrine correlates of estrus and anestrus in Siberian tiger. Tigers of the World: 244-254.

Shaer M. 2015. Cinderella story. Smithsonian, (46): 30-41.

Siberia Tiger Park (2003). Siberia tiger park timeline. http://www.dongbeihu.net.cn/news/view.asp?id=585[2003-9-29].

State Forestry Administration. 2013. Technical Code of Feeding and Management for Wild Animals-Siberian Tiger. http://www.forestry.gov.cn/uploadfile/mama/2013-10/file/2013-10-24-2283c308377341588ba7649f025c232e.pdf[2013-10-24].

Yu T, Jianguo W, Xiaojun K, et al. 2009. Spatiotemporal pattern and major causes of the Amur tiger population dynamics. Biodiversity Science, 17(3): 211-225.

Zhang C, Zhang M, Stott P. 2013. Does prey density limit Amur tiger Panthera tigris altaica recovery in northeastern China. Wildlife Biology, 19(4): 452-461.

Zheng D, Liu X, Ma J. 2005. Patterns of genetic variation within a captive population of Amur tiger Panthera tigris altaica. Acta Theriologica, 50(1): 23-30.

Chapter 7   Innate preference for native prey and personality implications in captive Amur tigers

7.1   Introduction

Recognition of prey, a prerequisite step for predation, is vital for the survival of carnivores, because predation behaviour is costly in energy and is sometimes risky (Benoit-Bird, 2004; Liznarova and Pekar, 2013). To be successful, carnivores trade off the costs and gains of predation. Effective prey recognition increases both predation efficiency and net energy gain (McNamara and Houston, 1997; Van Gils et al., 2005). It is commonly assumed that carnivores recognize prey by instinct and/or learning, which have been termed the instinct and learning hypotheses. The learning hypothesis has been strongly supported by previous studies (Darmaillacq et al., 2004; Fabregas et al., 2015; Lillywhite et al., 2015). However, evidence for the instinct hypothesis is limited (Dolev and Nelson, 2014). So, the validity and generalization of the instinct hypothesis needs more testing.

The Amur (or Siberian) tiger (*Panthera tigris altaica*) is one of the most seriously endangered felid species (Miquelle et al., 2015). The number of Amur tigers living in the wild today is about 500 individuals (Miquelle et al., 2006). The majority of wild Amur tigers reside in the Russian Far East (Carroll and Miquelle, 2006). Additionally, there are about 20 wild tigers in the northeast of China (Jiang et al., 2014; Dou et al., 2016). In the wild, Amur tigers mainly prey on native medium and large ungulates (Miller et al., 2013; Kerley et al., 2015), such as wild boar (*Sus scrofa*), roe deer (*Capreolus pygargus*), sika deer (*Cervus nippon*) and red deer (*Cervus elaphus*). However, it remains unknown if Amur tigers instinctually recognize native prey. Elucidating this question will not only help us know more about the biological character of Amur tigers, it will also provide a good opportunity to test the validity of the instinct hypothesis for prey recognition.

Rewilding captive individuals is a useful manner to conserve endangered species (Fabregas et al., 2015). In order to conserve Amur tigers, the Siberia Tiger Park in Heilongjiang, China, has successfully bred about 1000 individuals and plans to train captive individuals for reintroduction into the wild, which would facilitate the restoration of the wild Amur tiger population. To enhance the survival rate of reintroduced Amur tigers, it is vital to screen and train suitable candidates that retain good behavioural and cognitive characteristics.

It has been shown that animals have various personalities, revealed by individual peculiarities in behaviour that consistently differ from one another (Briffa and Weiss, 2010; Stamps and Groothuis, 2010). In light of the relationships that exist between behaviour, physiology and cognition (Dolan, 2002; Killen et al., 2013), studies of the personalities of endangered species have been used for estimating health status and improving conservation management (Kenneth and Terry, 1994; Nadja, 1999; Gartner and Weiss, 2013a). For example, fecundity was predicted by the 'tense-fearful' trait in captive cheetahs, which offered new insights into solving breeding problems (Nadja, 1999). For Amur tigers, if this species recognizes native prey instinctually, we speculate that there will be a correlation between prey recognition instinct and personality, which would provide essential insights for the selection of rewilding candidates with keen prey recognition instinct to promote survival.

In this study, we hypothesized that the response of captive Amur tigers, isolated from almost every other animal species since birth, to the native prey stimuli would be stronger than the response to stimuli given by non-native animals. This would indicate that Amur tigers have prey recognition instincts. Further, we explored if prey recognition instincts were correlated with the personalities of individual Amur tigers.

7.2 Materials and methods

7.2.1 Experimental animals

Amur tigers used in this study were bred and reared in the Siberia Tiger Park (Harbin, Heilongjiang, China), which was established in 1996 to protect the Amur tiger population. All the Amur tigers in the park were born in captivity and have been geographically isolated from native prey and almost all other animals. In total, 45 tigers, aging from 2 to 15 years, were used for our experiments. Each tiger was housed individually, each in their own cage. Experiments were conducted between 9: 00 h and 15: 00 h. Water was available *ad libitum*. Food was offered after the daily experiment. All experiments complied with the Wild Animals Protection Law of the People's Republic of China, and adhered to the principles and rules of the Siberia Tiger Park.

7.2.2 Experimental apparatus

The behaviour of the Amur tigers was recorded by a digital video monitoring system (Hivision, China; video frame rate of 30 frames per second), installed on a scaffold 3 m in front of the tiger cage. A loudspeaker was also installed on the scaffold. The scaffold and attached apparatus were constructed 3 days before performing experiments, for habituation.

7.2.3 Stimuli

The images, sounds and faeces of four native prey, three non-native animal species and a control group of non-animal stimuli were used as visual, auditory and olfactory stimuli. We collected the images, sounds and faeces of three native prey (wild boar, sika deer and red deer) and three non-native animals (wildebeest (*Connochaetes taurinus*), alpaca (*Vicugna pacos*) and hippopotamus (*Hippopotamus amphibius*)) from the Harbin Northern Forest Zoo (Heilongjiang). The image, sound and faeces of roe deer (a native prey) were collected from a private farm located in the Jixi region (Heilongjiang). Additionally, the irregular image, white noise and dry grass were used as stimuli for the control group.

The images of animals were all photographed with the same posture (i.e., lateral view of the body, head turned to face the observer). The irregular image of control group was made of seven randomly arranged squares, tailored from the images of seven animals. Every image was digitally transformed to the same size (0.66m^2 surface area, similar to the actual lateral body area of each animal), and then printed on a piece of waterproof paper (1.5m × 1.5m) with a white background.

The animal sounds, including the squeal of a wild boar, bark of a roe deer, bleat of a sika deer, bellow of a red deer, bellow of a wildebeest, bleat of an alpaca and the roar of a hippopotamus, were recorded by a digital voice recorder (Aigo, China). White noise was generated by audio software (Adobe Audition, USA). All sounds were adjusted to the same intensity (65dB measured 1.0m from the loudspeaker).

All faeces was freshly collected and frozen at $-20°C$ until use.

7.2.4 Prey recognition test

To test if captive Amur tigers could innately recognize the stimuli of native prey, we compared the responses of Amur tigers to the stimuli of native prey with responses to the stimuli of non-native animals and a control.

For each Amur tiger, one visual, one auditory and one olfactory stimulus were presented sequentially in daily sessions. The inter-trial interval was 30 min. These three kinds of stimuli were independently selected in a random order, and each stimulus was presented only once for each tiger.

7.2.4.1 Image presentation trial

The image was hung on the scaffold and covered with a cloth cover. At the end of a 10 min baseline period, the cloth cover was torn remotely off using a piece of string. Then, the image was presented for 10 min (stimulus period). To prevent Amur tigers from being afraid, each Amur tiger was habituated to the action of tearing off the cloth cover, placed over a blank waterproof paper, 20 times before the experiment. We empirically found that Amur tigers almost habituated to tearing off the cloth cover in 10 repetitions.

7.2.4.2 Sound playback trial

Following a 10 min baseline period, sound was played via the loudspeaker using a pulsed mode consisting of 10 sec/min for 10 min (stimulus period).

7.2.4.3 Faeces presentation trial

Thawed faeces or dry grass was put in an iron box (8cm× 8cm× 2cm, with small holes distributed on the lid). Then, the iron box was presented to the Amur tiger for 10 min (stimulus period). Each Amur tiger was habituated to an empty ion box for 24 hours before the experiment. In addition, the iron box was rinsed with water to clear any scent and dried before the next usage.

7.2.5 Personality assessment

Like other studies, the methods we used to assess the personality of Amur tigers included personality rating, behavioural observation and behavioural testing (Bergvall et al., 2011; Gartner and Powell, 2012).

Initially, a personality rating was used to assess the personality structure of each Amur tiger. A personality survey, consisting of 25 personality traits (Table 7.1), was designed based on previous personality studies (Gartner and Weiss, 2013b; Gartner et al., 2014). Three keepers were required to individually assess the personality of each Amur tiger and rate these 25 personality traits on a seven-point scale (one meant 'not at all' and seven meant 'very much so'). The reliability of personality ratings was preliminarily assessed by intraclass correlation coefficient (ICC) analysis, by which 16 reliable personality traits were generated (F-test: $n = 45$, $P < 0.05$; Table 7.1).

Table 7.1 Reliability assessment of personality ratings by intraclass correlation coefficient (ICC)

Personality trait	ICC(3, 1)	ICC(3, k)	P-value
Active	0.308	0.572	**0.000**
Aggressive to people	0.245	0.494	**0.002**
Anxious	0.017	0.049	0.406
Calm	0.161	0.365	**0.036**
Cooperative	0.247	0.496	**0.004**
Curious	0.423	0.688	**0.000**
Depressed	0.101	0.252	0.113
Distractible	0.175	0.389	**0.007**
Eccentric	−0.028	−0.090	0.677
Excitable	0.373	0.641	**0.000**

Continuation Table

Personality trait	ICC(3, 1)	ICC(3, k)	P-value
Fearful of people	0.148	0.343	**0.044**
Friendly to people	0.207	0.439	**0.006**
Impulsive	0.049	0.134	0.259
Insecure	0.055	0.148	0.253
Irritable	0.156	0.357	**0.020**
Playful	0.403	0.670	**0.000**
Reckless	0.144	0.335	**0.049**
Self-assured	0.098	0.246	0.133
Smart	0.063	0.168	0.212
Suspicious	0.236	0.480	**0.005**
Tense	0.122	0.295	0.058
Timid	0.308	0.572	**0.000**
Trusting	0.295	0.557	**0.001**
Vigilant	0.080	0.208	0.153
Vocal	0.429	0.692	**0.000**

Note: ICC(3, 1) indicates reliability of individual ratings. ICC(3, k) indicates the reliability of mean ratings. Significant P value is in boldface (F-test, two tailed: $n = 45$, $P < 0.05$). The trait with $P < 0.05$ is significantly reliable and is used in further analysis.

Then, to assure the rating reliability of personality traits, we performed behavioural observations. All 10 min baseline periods of image/sound presentation trials were used to observe two kinds of behaviours, namely moving and vocalizing. 'Move frequency', reflecting the 'active' trait, was calculated as the average moving time during each 10 min baseline period. 'Vocalize frequency', reflecting the 'vocal' trait, was calculated as the average vocalizing time during each 10 min baseline period.

Finally, behavioural testing use 'novel object test' to assure the rating reliability of some personality traits, such as 'curious', 'excitable', 'playful', 'reckless', 'suspicious', 'timid' and 'trusting'. In this test, one novel object was placed in the tiger cage. Then, the tiger was introduced into the cage and monitored for 20 min (stimulus period). In total, two novel objects, an iron chair leg and a folded iron pipe, were presented once each in a random sequence. The inter-trial interval was 1 hour. Objects were rinsed with water to clear scent and dried for the next usage. Contact latency was calculated as the average time from stepping into the cage to sniffing, licking or touching the novel objects for the first time. Contact time was calculated as the average time spent sniffing, licking or touching the novel objects.

7.2.6 Data analysis

Behavioural responses were replayed and counted by one colleague who was blind to the experiments. Unless noted, data analysis was all conducted by SPSS software (SPSS, Inc., USA). Data in the text and figures were expressed as means ± SEM (standard error of mean). Significance level was set at 0.05 ($*P < 0.05$; $**P < 0.01$; $***P < 0.001$; NS = not significant).

To analyze the data from the prey recognition test, because the distributions of behavioural response were significantly non-normal (Kolmogorov-Smirnov test: $P < 0.05$), we used a Friedman test to test for differences in behavioural responses to different stimuli. If the difference was significant (Friedman test: $P < 0.05$), a Wilcoxon test was used for multiple comparisons between any two stimuli.

In addition, we used the preference index to quantify the strength of preference for native prey (Dix and Aggleton, 1999). Preference index (PI) was calculated using the following formula:

$$\text{PI} = \frac{(T_{\text{wild boar}} + T_{\text{roe deer}} + T_{\text{sika deer}} + T_{\text{red deer}}) - (T_{\text{wildebeest}} + T_{\text{alpaca}} + T_{\text{hippopotamus}} + T_{\text{control}})}{(T_{\text{wild boar}} + T_{\text{roe deer}} + T_{\text{sika deer}} + T_{\text{red deer}}) + (T_{\text{wildebeest}} + T_{\text{alpaca}} + T_{\text{hippopotamus}} + T_{\text{control}})}$$

In this formula, 'T' was the time spent on the behavioural response (staring or sniffing). The subscript of 'T' indicates to which stimulus Amur tiger responded. In theory, PI ranges from +1.0 (complete preference for native prey) to −1.0 (complete preference for non-native animals). '0' indicates no discrimination between native prey and non-native animals.

Because there was no significant difference between the distributions of preference index and the normal distribution (Kolmogorov-Smirnov test: $P > 0.05$), we used linear regression to analyze the relationship between tiger age and preference index, and used t-tests (two tailed) to test for the differences between the preference indexes of any two groups. Simulated preference index was generated by randomly shuffling 'T' in the preference index formula 100 times.

For the personality assessment tests, we initially used an intraclass correlation coefficient (ICC) analysis to test the reliability of personality ratings. The significantly reliable traits (F-test, two tailed: $P < 0.05$) were averaged across three raters for further analysis. Then, to further assure the reliability of personality ratings, we analyzed Spearman's correlations between personality ratings and response parameters of behavioural observation/test (Spearman correlation analysis: $n = 45$, P-values were Bonferroni corrected; Table 7.2).

Additionally, because there was a significant multicollinearity among personality traits (Bartlett's test of sphericity: $\chi^2_{120} = 458.951$, $P < 0.05$), we used principal component analysis to reduce the dimension of personality structure (R software version 3.4.1, princomp function; R Development Core Team, 2017). The number of principal components was determined via parallel analysis and a scree plot (R software, fa.parallel function; Fig. 7.1). Like a previous study (Gartner et al., 2014), we defined factor loading $\geq |0.4|$ as salient for the principal component analysis (Table 7.3). Finally, we used linear regression analysis to test the relationship between preference index and principal components.

Table 7.2 Correlation between personality ratings and behavioural observation/test

Personality trait	Behavioural observation		Novel object test	
	Move frequency	Vocalize frequency	Contact latency	Contact time
Active	**0.811**	0.197	−0.031	−0.022
Aggressive to people	−0.011	−0.205	0.423	−0.279
Calm	−0.114	0.141	−0.292	0.266
Cooperative	−0.020	0.327	**−0.455**	0.269
Curious	−0.007	−0.037	**−0.696**	**0.806**
Distractible	−0.009	−0.211	0.028	0.193
Excitable	−0.015	0.195	**−0.514**	**0.597**
Fearful of people	−0.002	0.014	0.166	0.002
Friendly to people	0.240	0.331	−0.399	0.254
Irritable	−0.078	−0.405	**0.443**	−0.307
Playful	0.044	−0.017	−0.424	**0.626**
Reckless	0.133	−0.016	**−0.630**	**0.499**
Suspicious	0.000	**−0.442**	**0.548**	−0.357
Timid	0.077	0.025	**0.698**	**−0.677**
Trusting	−0.013	0.345	**−0.517**	0.248
Vocal	0.133	**0.753**	−0.126	−0.118

Note: Significant correlation coefficients are in boldface (Spearman correlation test, Bonferroni correction is performed).

Table 7.3 The principal components of personality structure

Personality trait	Principal components	
	PC 1	PC 2
Active	−0.08	0.26
Aggressive to people	−0.38	**−0.72**
Calm	0.30	0.16
Cooperative	0.36	**0.65**
Curious	**0.91**	0.01
Distractible	0.14	**−0.48**
Excitable	**0.82**	0.18
Fearful of people	−0.28	**−0.62**
Friendly to people	0.33	**0.78**
Irritable	**−0.47**	**−0.69**
Playful	**0.87**	0.10
Reckless	**0.63**	0.03
Suspicious	−0.40	**−0.64**
Timid	**−0.84**	−0.11
Trusting	0.29	**0.64**
Vocal	−0.27	**0.59**

Note: Salient loadings are in boldface.

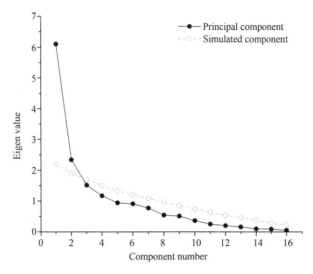

Fig. 7.1 Scree plot and parallel analysis. The first two principal components located above simulated components are significant and used for further analysis.

7.3 Results

7.3.1 Preferential response to native prey visual stimuli

As shown in Fig.7.2a, Amur tigers showed an obvious period when they stared at images of native prey(wild boar, roe deer, sika deer and red deer), non-native prey (wildebeest, alpaca and hippopotamus) and a non-animal control (irregular image) during the stimulus presentation period. Amur tigers stared at the wild boar image for a longer time than all other image kinds during the first presentation minute (Wilcoxon test: $n = 45$, $P < 0.05$; Fig. 7.2a). The sum of time spent staring at the wild boar image was significantly longer than the images of non-native prey or non-animal control (Wilcoxon test: $n = 45$, $P < 0.05$). Tigers also tended to starelonger at the wild boar than the images of other native prey, although this difference was not significant (Fig.7.2b). In addition, the averaged sum of time spent on staring at the images of native prey was significantly longer than for non-native prey and non-animal control images (Wilcoxon test: $n = 45$, $P < 0.001$ for both comparisons; Fig. 7.2c).

Because aging is a potential factor impairing cognition (Burke and Barnes, 2006; Bishop et al., 2010), we explored if aging affected innate visual recognition capabilities of native prey in Amur tigers. We found that there was no significant linear relationship between tiger age and image preference index which was used to

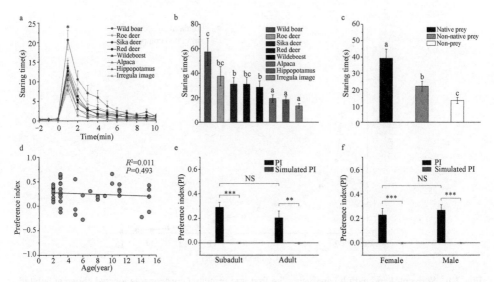

Fig. 7.2 Preference for the images of native prey. (a) The time course of staring visual stimuli. The time '0' indicates the initiation of presenting image to a tiger. Asterisk (*) marks that the time spent staring at the wild boar image was significantly longer than all other images during the first minute of stimulus period (Wilcoxon test: $n = 45$, $P < 0.05$). Native prey is in boldface in both (a) and (b). (b) The sum of staring time at visual stimuli during the 10 min stimulus period. Superscript letters (a, b or c) indicate significant difference between any two stimuli (Wilcoxon test: $n = 45$, $P < 0.05$). (c) The averaged sum of time spent staring at the images of native prey, non-native prey and non-animal control. Superscript letters (a, b or c) indicate significant difference between any two group (Wilcoxon test: $n = 45$, $P < 0.05$). (d) Linear regression analysis between tiger age and image preference index. (e and f) Comparison of image preference index between subadult and adult groups ($n = 24$ and 21, respectively), or between female and male groups ($n = 19$ and 26, respectively) with unpaired t-tests. The significance of image preference index of these four groups was tested with one sample t-test by comparison with the mean of corresponding simulated preference index, respectively. Data are means ± SEM; *$P < 0.05$; **$P < 0.01$; ***$P < 0.001$; NS = not significant.

quantify preference strength for the images of native prey (linear regression analysis: $R^2 = 0.011$, $n = 45$, $P = 0.493$; Fig.7.2d). Furthermore, the image preference indexes of both subadult (age 2-3) and adult (age 4-15) groups were significantly bigger than simulated preference indexes (one sample t-test: $t_{23} = 7.218$, $P < 0.001$; $t_{20} = 3.727$, $P < 0.01$, respectively; Fig.7.2e). Meanwhile, there was no significant difference between image preference indexes of the two groups (unpaired t-test: $t_{43} = 1.280$, $P = 0.208$; Fig.7.2e).

Additionally, the image preference indexes of both female and male tigers were significantly larger than simulated preference indexes (one sample t-test: $t_{18} = 4.319$, $P < 0.001$; $t_{25} = 6.002$, $P < 0.001$, respectively; Fig.7.2f), and there were also no significant differences between the image preference indexes of the two groups (unpaired t-test: $t_{43} = -0.570$, $P = 0.572$; Fig.7.2f).

7.3.2 Preferential response to native prey auditory stimuli

Similar to the response to visual stimuli, Amur tigers stared in the direction of the wild boar sound for much longer than all other sounds during the first five minutes (Wilcoxon test: $n = 45$, $P < 0.05$; Fig. 7.3a). The sum of time spent staring in the direction of the wild boar sound was significantly longer than all other sounds (Wilcoxon test: $n = 45$, $P < 0.05$; Fig. 7.3b). Notably, the averaged sum of time spent paying attention to the sounds of native prey was significantly longer than those of non-native prey and non-animal control (Wilcoxon test: $n = 45$, $P < 0.001$ both; Fig. 7.3c).

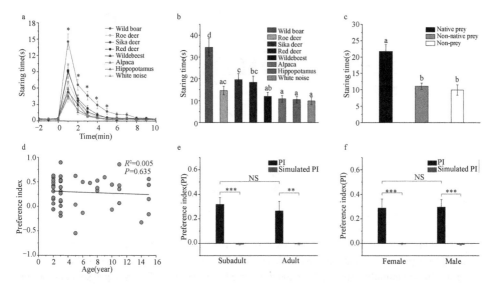

Fig. 7.3 Preference for the sounds of native prey. (a) The time course of staring auditory stimuli. Asterisk (*) marks that the time spent staring in the direction of the wild boar sound was significantly longer than all other sounds during the first 5 min of the stimulus period (Wilcoxon test: $n = 45$, $P < 0.05$). Native prey is in boldface in both (a) and (b). (b) The sum of staring time towards the auditory stimuli during the 10 min stimulus period (Wilcoxon test: $n = 45$, $P < 0.05$). (c) The averaged sum of time spent noticing the sounds of native prey, non-native prey and non-animal control. Superscript letters (a, b or c) indicate significant difference between any two group (Wilcoxon test: $n = 45$, $P < 0.05$). (d) Linear regression analysis between tiger age and sound preference index. (e and f) Comparison of sound preference index between subadult and adult groups ($n = 24$ and 21, respectively), and between female and male groups ($n = 19$ and 26, respectively). The convention is the same as Fig. 7.2.

We found no significant linear relationship between the age and sound preference index (linear regression analysis: $R^2 = 0.005$, $n = 45$, $P = 0.635$; Fig. 7.3d). Both subadult and adult individuals, female and male individuals, showed significant preference for the sounds of native prey (Fig. 7.3c, f).

7.3.3 Preferential response to native prey olfactory stimuli

Though there was no obvious difference between the time spent sniffing the faeces of individual species (Fig. 7.4a and b), the averaged sum of time spent sniffing the olfactory stimuli of native prey was significantly longer than those of non-native prey and non-animal control (Wilcoxon test: $n = 31$, $P < 0.05$ and $P < 0.001$ respectively; Fig. 7.4c).

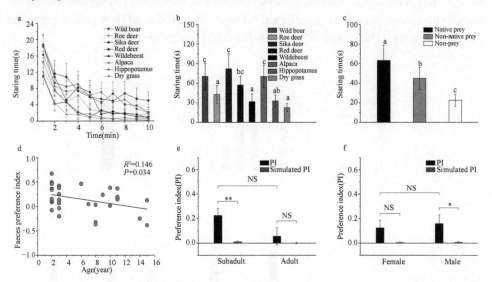

Fig. 7.4 Preference for the smell of faeces of native prey. (a) The time course of sniffing olfactory stimuli. Native prey is in boldface in both (a) and (b). (b) The sum of sniffing time to olfactory stimuli during the 10 min stimulus period (Wilcoxon test: $n = 31$, $P < 0.05$). (c) The averaged sum of time spent sniffing the olfactory stimuli of native prey, non-native prey and non-animal control. Superscript letters (a, b or c) indicate significant difference between any two group (Wilcoxon test: $n = 31$, $P < 0.05$). (d) Linear regression analysis between tiger age and faeces preference index. (e and f) Comparison of faeces preference index between subadult and adult groups ($n = 16$ and 15, respectively) or between female and male groups ($n = 14$ and 17, respectively). The convention is the same as Fig. 7.2.

The faeces preference index had a significantly negative correlation with age (linear regression analysis: $R^2 = 0.146$, $n = 31$, $P = 0.034$; Fig. 7.4d). Also, the faeces preference index of the subadult group, but not the adult group, was significantly larger than the simulated preference index (one sample t-test: $t_{15} = 3.632$, $P < 0.01$; $t_{14} = 0.783$, $P = 0.447$, respectively; Fig. 7.4e). The faeces preference index of the subadult group was relatively larger than that of the adult group, although no significant difference was found (unpaired t-test: $t_{29} = 1.854$, $P = 0.074$; Fig. 7.4e).

Additionally, there was no significant difference between the faeces preference indexes of the two gender groups (unpaired t-test: $t_{29} = -0.357$, $P = 0.724$; Fig. 7.4f), though the faeces preference index of the female group was relatively, but not significantly, larger than the simulated preference index (one sample t-test: $t_{13} = 1.911$, $P = 0.078$; Fig. 7.4f).

7.3.4 Correlation between innate preference for native prey and personality

To analyze the correlation between innate prey recognition capability as quantified by preference index and personality traits, we first reduced the dimensions of personality structure by principal component analysis. Two principal components were described in the personality structure (Fig. 7.1, Table 7.3). The first component had the highest loadings on the traits curious, playful, and excitable. The second component had the highest loadings on the traits friendly to people, cooperative, and trusting (Table 7.3). Both image and sound preference index were negatively correlated with principle component 1 (linear regression analysis: $R^2 = 0.113$, $n = 45$, $P < 0.05$; $R^2 = 0.205$, $n = 45$, $P < 0.01$, respectively; Fig. 7.5a and c), but had no significant correlation with principle component 2 (linear regression analysis: $R^2 = 0.009$, $P = 0.527$; $R^2 = 0.049$, $P = 0.144$, respectively; Fig. 7.5b and d). However, faeces preference index had no significant correlation with the two principle components (linear regression analysis: $R^2 = 0.020$, $n = 31$, $P = 0.452$; $R^2 = 0.014$, $n = 31$, $P = 0.528$, respectively; Fig. 7.5e and f).

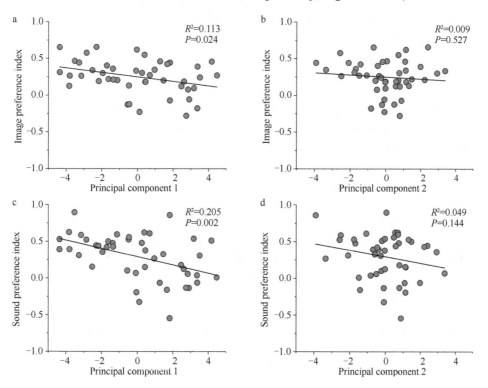

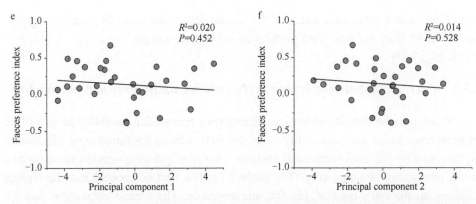

Fig. 7.5 Correlation between preference index and the principal components of personality structure (a-d) Image/Sound preference index has a significantly negative correlation with principal component 1 (linear regression analysis: $P < 0.05$, $P < 0.01$, respectively), but no significant correlation with principal component 2.(e and f) Faeces preference index had no significant correlation with both principal components.

7.4 Discussion

Here, we demonstrated that captive Amur tigers showed preferential response to stimuli of native prey over those of non-native prey and the non-animal control. Because these captive Amur tigers were isolated from native prey and non-native prey since birth, innate preference for native prey stimuli provided support for the hypothesis that Amur tigers instinctually recognize native prey. Given that predatory behaviour is costly in energy and is even hazardous (Benoit-Bird, 2004; Liznarova and Pekar, 2013), animals benefit from the capacity to distinguish between suitable and unsuitable prey. The risk of preying on unsuitable prey should be averted if prey recognition capability is innate and only suitable prey are attacked. So, it appears that selective pressure drove prey recognition capacities to evolve into a conservative instinct, by which the fitness of carnivores, likely both invertebrates and vertebrates, increased (Dolev and Nelson, 2014; Holding *et al.*, 2016). Though several exploratory studies have reported a neural or genetic basis for some types of instinctual behaviour (Musso *et al.*, 2003; Liedtke *et al.*, 2011; Young *et al.*, 2016), there is little research investigating the biological underpinnings of the prey recognition instinct, and further research about prey recognition could focus on this.

Our present study showed that captive Amur tigers exhibited the strongest preference for the image and sound of the wild boar. Because previous studies have provided strong evidence that Amur tigers preferentially preyed on boars in the wild (Kerley *et al.*, 2015; Sugimoto *et al.*, 2016), it is interesting that the innate preferential response to the image and sound of wild boar in captive Amur tigers coincided with the preferential predation on boar in their wild counterparts. In fact, similar phenomena have also been found in other studies (Darmaillacq et al., 2006; Portela et al., 2014; Pekar and Cardenas, 2015). For instance, octopuses showed innate preferential attack on their preferred prey (crabs) rather than non-preferred prey (shrimps), implying that octopuses innately recognized the preferred prey (Darmaillacq et al., 2006). Considering that preferred prey could provide more energy per ingested biomass or be preyed upon more easily (Bittar et al., 2012; Kerley et al., 2015), carnivores could gain greater benefits by innately recognizing preferred prey. So, it is probable that the innate preference for the image and sound of wild boar represents a cognitive adaptation, similar to physical adaptations in tigers (Meachen-Samuels and Van Valkenburgh, 2009a, 2009b).

It was also interesting that only olfactory, but not visual and auditory, innate prey recognition capability had a significantly negative correlation with age. A possible explanation is that olfactory prey recognition might be not crucially necessary and is prone to degrade in Amur tigers. Actually, it has been implied that olfaction was more helpful for detecting prey in nocturnal animals as compared with diurnal animals (Bicca-Marques and Garber, 2004; Piep et al., 2008), and tigers mainly forage at dawn or dusk, so are not considered a nocturnal species (Naha et al., 2016).

Although we primarily assessed the personality of Amur tigers by using subjective ratings, the reliability of personality ratings was confirmed by further objective analysis using behavioural observations and novel object tests. The methods of assessing personality used in the present study were the same as those used in other similar studies (Bergvall et al., 2011; Gartner and Powell, 2012). Also, the two principle components of Amur tiger personality in our study were comparable with those of other big cats in other studies (Gartner et al., 2014). However, due to the fact that the Amur tigers used in present study were housed individually, the social traits of personality were difficult to assess and were not considered.

The present study has potentially valuable implications for the conservation management of Amur tigers, both living in captivity and in the wild. In light of the heritability of instinct (Ricker and Hirsch, 1985), it is advisable to screen and breed the individuals that have keen prey recognition instincts. However, conventional processes of testing instinct are cumbersome; so, it may be feasible to preliminarily estimate instincts by personality assessment. In addition, personality is an important factor affecting the responses of animals to external stimuli (Stamps and Groothuis, 2010). Individuals with specific personality traits may have a much greater chance of surviving in the wild (Dingemanse et al., 2004; Boon et al., 2007). These results suggest that individual tigers with specific personality traits associated with increased prey recognition instinct should be prioritized for training and reintroduction into the wild.

7.5 Summary

Prey recognition is vital for predation and the survival of reintroduced Amur tigers. So it would be meaningful to screen suitable candidates which have good prey recognition capability. Our study indicated that Amur tigers recognized native prey instinctually. The innate prey recognition capability was significantly correlated with individual tiger personality traits. These results provide a support for the instinct hypothesis of prey recognition, and also provided a potential method to preliminarily screen tiger individuals with keen prey recognition instinct by personality assessment for breeding and wild training.

References

Benoit-Bird K J. 2004. Prey caloric value and predator energy needs: foraging predictions for wild spinner dolphins. Marine Biology, 145(3): 435-444.
Bergvall U A, Schapers A, Kjellander P, et al. 2011. Personality and foraging decisions in fallow deer, Dama dama. Animal Behaviour, 81(1): 101-112.
Bicca-Marques J C, Garber P A. 2004. Use of spatial, visual, and olfactory information during foraging in wild nocturnal and diurnal anthropoids: a field experiment comparing Aotus, Callicebus, and Saguinus. American Journal of Primatology, 62(3): 171-187.
Bishop N A, Lu T, Yankner B A. 2010. Neural mechanisms of ageing and cognitive decline. Nature, 464(7288): 529-535.
Bittar V T, Awabdi D R, Tonini W C T, et al. 2012. Feeding preference of adult females of ribbonfish Trichiurus lepturus through prey proximate-composition and caloric values. Neotropical Ichthyology, 10(1): 197-203.
Boon A K, Reale D, Boutin S. 2007. The interaction between personality, offspring fitness and food abundance in North American red squirrels. Ecology Letters, 10(11): 1094-1104.
Briffa M, Weiss A. 2010. Animal personality. Current Biology, 20(21): 912-914.
Burke S N, Barnes C A. 2006. Neural plasticity in the ageing brain. Nature Reviews Neuroscience, 7(1): 30-40.
Carroll C, Miquelle D G. 2006. Spatial viability analysis of Amur tiger *Panthera tigris altaica* in the Russian Far East: the role of protected areas and landscape matrix in population persistence. Journal of Applied Ecology, 43(6): 1056-1068.
Darmaillacq A S, Chichery R, Poirier R, et al. 2004. Effect of early feeding experience on subsequent prey preference by cuttlefish, *Sepia officinalis*. Developmental Psychobiology, 45(4): 239-244.
Darmaillacq A S, Chichery R, Shashar N, et al. 2006. Early familiarization overrides innate prey preference in newly hatched *Sepia officinalis* cuttlefish. Animal Behaviour, 71: 511-514.
Dingemanse N J, Both C, Drent P J, et al. 2004. Fitness consequences of avian personalities in a fluctuating environment. Proceedings of the Royal Society B-Biological Sciences, 271(1541): 847-852.
Dix S L, Aggleton J P. 1999. Extending the spontaneous preference test of recognition: evidence of object-location and object-context recognition. Behav Brain Res, 99(2): 191-200.

Dolan R J. 2002. Emotion, cognition, and behavior. Science, 298(5596): 1191-1194.
Dolev Y, Nelson X J. 2014. Innate pattern recognition and categorization in a jumping spider. PLoS ONE, 9(6): e97819.
Dou H L, Yang H T, Feng L M, et al. 2016. Estimating the population size and genetic diversity of Amur tigers in Northeast China. PLoS ONE, 11(4): e0154254.
Fabregas M C, Fosgate G T, Koehler G M. 2015. Hunting performance of captive-born South China tigers (*Panthera tigris amoyensis*) on free-ranging prey and implications for their reintroduction. Biological Conservation, 192: 57-64.
Gartner M C, Powell D M, Weiss A. 2014. personality structure in the domestic cat (*Felis silvestris catus*), scottish wildcat (*Felis silvestris grampia*), clouded leopard (*Neofelis nebulosa*), snow leopard (*Panthera uncia*), and African lion (*Panthera leo*): a comparative study. Journal of Comparative Psychology, 128(4): 414-426.
Gartner M C, Powell D. 2012. personality assessment in snow leopards (*Uncia uncia*). Zoo Biology, 31(2): 151-165.
Gartner M C, Weiss A. 2013a. Scottish wildcat (*Felis silvestris grampia*) personality and subjective well-being: implications for captive management. Applied Animal Behaviour Science, 147(3-4): 261-267.
Gartner M C, Weiss A. 2013b. Personality in felids: a review. Applied Animal Behaviour Science, 144(1-2): 1-13.
Holding M L, Kern E H, Denton R D, et al. 2016. Fixed prey cue preferences among dusky pigmy rattlesnakes (*Sistrurus miliarius barbouri*) raised on different long-term diets. Evolutionary Ecology, 30(1): 1-7.
Jiang G S, Sun H Y, Lang J M, et al. 2014. Effects of environmental and anthropogenic drivers on Amur tiger distribution in northeastern China. Ecological Research, 29(5): 801-813.
Kenneth C G, Terry L M. 1994. Personality assessment in the gorilla and its utility as a management tool. Zoo Biology, 13: 509-522.
Kerley L L, Mukhacheva A S, Matyukhina D S, et al. 2015. A comparison of food habits and prey preference of Amur tiger (*Panthera tigris altaica*) at three sites in the Russian Far East. Integrative Zoology, 10(4): 354-364.
Killen S S, Marras S, Metcalfe N B, et al. 2013. Environmental stressors alter relationships between physiology and behaviour. Trends in Ecology & Evolution, 28(11): 651-658.
Liedtke W B, Mckinley M J, Walker L L, et al. 2011. Relation of addiction genes to hypothalamic gene changes subserving genesis and gratification of a classic instinct, sodium appetite. Proceedings of the National Academy of Sciences of the United States of America, 108(30): 12509-12514.
Lillywhite H B, Pfaller J B, Sheehy C M. 2015. Feeding preferences and responses to prey in insular neonatal Florida cottonmouth snakes. Journal of Zoology, 297(2): 156-163.
Liznarova E, Pekar S. 2013. Dangerous prey is associated with a type 4 functional response in spiders. Animal Behaviour, 85(6): 1183-1190.
Mcnamara J M, Houston A I. 1997. Currencies for foraging based on energetic gain. American Naturalist, 150(5): 603-617.
Meachen-Samuels J, Van Valkenburgh B. 2009. Craniodental indicators of prey size preference in the Felidae. Biological Journal of the Linnean Society, 96(4): 784-799.
Meachen-Samuels J, Van Valkenburgh B. 2009. Forelimb indicators of prey-size preference in the Felidae. Journal of Morphology, 270(6): 729-744.
Miller C S, Hebblewhite M, Petrunenko Y K, et al. 2013. Estimating Amur tiger (*Panthera tigris*

altaica) kill rates and potential consumption rates using global positioning system collars. Journal of Mammalogy, 94(4): 845-855.

Miquelle D G, Pikunov D G, Dunishenko Y M, et al. 2006. A survey of Amur (Siberian) tigers in the Russian Far East, 2004-2005. Final Report to Save the Tiger Fund.

Miquelle D G, Smirnov E N, Zaumyslova O Y, et al. 2015. Population dynamics of Amur tigers (*Panthera tigris altaica*) in Sikhote-Alin Biosphere Zapovednik: 1966-2012. Integrative Zoology, 10(4): 315-328.

Musso M, Moro A, Glauche V, et al. 2003. Broca's area and the language instinct. Nature Neuroscience, 6(7): 774-781.

Nadja C W. 1999. Behavioral differences as predictors of breeding status in captive cheetahs. Zoo Biology, 18: 335-349.

Naha D, Jhala Y V, Qureshi Q, et al. 2016. Ranging, activity and habitat use by tigers in the mangrove forests of the Sundarban. PLoS ONE, 11(4): e0152119.

Pekar S, Cardenas M. 2015. Innate prey preference overridden by familiarisation with detrimental prey in a specialised myrmecophagous predator. Science of Nature, 102(1-2): 8.

Piep M, Radespiel U, Zimmermann E, et al. 2008. The sensory basis of prey detection in captive-born grey mouse lemurs, *Microcebus murinus*. Animal Behaviour, 75: 871-878.

Portela E, Simoes N, Rosas C, et al. 2014. Can preference for crabs in juvenile *Octopus maya* be modified through early experience with alternative prey? Behaviour, 151(11): 1597-1616.

R Development Core Team. 2017. R: a language and environment for statistical computing. Vienna, Austria: R Foundation for Statistical Computing.

Ricker J P, Hirsch J. 1985. Evolution of an instinct under long-term divergent selection for geotaxis in domesticated populations of *Drosophila melanogaster*. J Comp Psychol, 99(4): 380-390.

Stamps J, Groothuis T G G. 2010. The development of animal personality: relevance, concepts and perspectives. Biological Reviews, 85(2): 301-325.

Sugimoto T, Aramilev V V, Nagata J, et al. 2016. Winter food habits of sympatric carnivores, Amur tigers and Far Eastern leopards, in the Russian Far East. Mammalian Biology, 81(2): 214-218.

Van Gils J A, De Rooij S R, Van Belle J, et al. 2005. Digestive bottleneck affects foraging decisions in red knots *Calidris canutus*. I. Prey choice. Journal of Animal Ecology, 74(1): 105-119.

Young K S, Parsons C E, Elmholdt E M J, et al. 2016. Evidence for a caregiving instinct: rapid differentiation of infant from adult vocalizations using magnetoencephalography. Cerebral Cortex, 26(3): 1309-1321.

Chapter 8 Risks involved in fecal DNA-based genotyping of microsatellite loci in the Amur tiger a pilot study

8.1 Introduction

The emergence of molecular genetic marker-based technologies presents opportunities for mining information from feces and hair, materials that are both commonly collected in the field but made little use of in traditional monitoring methods. Non-invasive sampling of feces and hair and relevant genetic analyses have become popular in wildlife monitoring and management (Taberlet and Luikart, 1999), especially for endangered species (Woodruff, 1993). The quality and quantity of DNA extracted from fecal samples are more or less influenced by diet composition (Stenglein et al., 2010). Research shows that an interaction exists between the complexity of diet composition and the probability of false allele amplification in polymerase chain reaction (PCR), and compared with herbivores the quality and quantity of DNA extracted from carnivores is poorer because of complicated scat composition (Panasci et al., 2011).

As a rare and elusive carnivore species, it is difficult to collect samples like blood or tissue from Amur tigers in the field, but feces are easily obtained during wildlife surveys. Routine procedures for fecal analysis include isolation of genomic DNA and amplification of a panel of microsatellite loci and bioinformatics analyses. However, due to poor quality and small quantities of template DNA resulting from degradation or contamination (Bradley et al., 2000; Regnaut et al., 2006), microsatellite loci amplification and genotyping can greatly impact subsequent analyses and have wildlife management implications. Although some population genetic studies of Amur tiger have used feces as a DNA source (Rozhnov et al., 2009), their use has not been tested through scientific pilot studies that help to reduce genotyping errors associated with template DNA of poor quality (Arandjelovic et al., 2009).

Factors affecting microsatellite analysis of Amur tigers include efficiency of DNA isolation and PCR. The efficiency of DNA isolation largely depends on storage methods (Nsubuga et al., 2004), season of collection (Hájková et al., 2006), and age of the scat (Murphy et al., 2007; Santini et al., 2007). The efficiency of PCR mainly depends on the nature of primer sequences, amplification fragment length, secondary structure, melting temperature and nucleotide composition (Sambrook et al., 2001).

Software based estimation of genotyping errors is based on statistical models sensitive to sample size; however, the population size of wild Amur tigers is small, and the chance for sampling different tiger individuals is biased and software may not be effective. In this research, we improved the efficiency of DNA isolation and only studied the effect of microsatellite-based PCR protocols on the genetic analysis of Amur tigers. We compared the accordance rate of genotyping efficiency, individualization and inter-individual genetic relationships between blood and fecal samples for 10 captive Amur tigers (*Panthera tigris altaica*) with 12 microsatellite loci utilized in wild tiger population monitoring (Bhagavatula and Singh 2006; Rozhnov et al., 2009). Specifically, we quantified the variation in genotyping risk among the 12 microsatellite loci used for individualization and genetic relationship estimation, and further, whether among loci variation was influenced by repeated PCR when fecal DNA was used. Our pilot study and the results will provide a frame of reference for conservation programs designed on DNA analyses of wild Amur tigers.

8.2 Materials and methods

8.2.1 Samples

Feces and peripheral blood samples of 10 adult Amur tigers were collected in Heilongjiang Amur Tiger Park in November 2013. Blood samples were collected by veterinarians three months after birth for establishing genetic lineage. These blood samples were anticoagulated with sodium citrate (3.8%) and temporarily stored at 4℃ until DNA was extracted. Feces were collected from the same group of tigers using plastic bags within 12 h after defecation and temporarily stored in an ice bag and then transferred to a refrigerator (–20℃) before DNA extraction.

8.2.2 DNA extraction

DNA of blood samples was extracted using the standard phenol and chloroform method (Sambrook et al., 1989). DNA of fecal samples was extracted using the modified QIAamp DNA Stool Mini Kit (Qiagen, Hilden, Germany) as described by Zhang et al. (2009) and with optimized preparation work as follows. About 5g of feces was peeled off from the surface of a fecal pellet and deactivated in 100% ethanol with volume a ratio of 1: 1 at room temperature for 12h. Feces and ethanol were then vortexed at 2200 r/min for 3min. The mixture was then filtered through a piece of sterile gauze. The filtrate was collected and centrifuged at 3500r/min for 15min and pellets at the bottom were transferred to a new tube for DNA extraction. The quality and quantity of DNA extracted from blood and feces were evaluated using routine agarose (1.0%) electrophoresis with quantification molecular markers.

8.2.3 Microsatellite data analysis

A panel of 12 microsatellite loci was selected from former studies on tiger monitoring and individualization, including D10, FCA043, FCA304, E21B, E6 and E7 for wild tigers (Bhagavatula and Singh, 2006; Rozhnov et al., 2009), and FCA391, FCA441, FCA094, FCA152, Pti007 and Pti010 for captive tigers (Menotti-Raymond et al., 1999; Xu et al,. 2005; Zhang et al., 2009). Primer sequences, repeating motifs, annealing temperatures and expected allele size ranges of these microsatellite loci are shown in Table 8.1. The 50 end of each forward primer was labeled with fluorescent dye (e.g., 5-FAM, TAMRA and HEX).

Table 8.1 Characterizations and conditions of polymerase chain reaction for 12 microsatellite loci of Amur tiger

Microsatellite locus	Primers (5′→3′)	Repeat motif	Annealing temperature (℃)	PCR product size range (bp)
E6	CCTGGGGATAATAAAACTAGTA CATGAATGAATCTTTACACTGA	$(TAA)_{11}$	58	147–162
E21B	GCGATAAAGGCTGGCAGAGG CTTTGAGGGTCTGTTCTACTGTGA	$(CA)_{21}$	62	154–168
D10	CCCTCTCTGTCCCTCCCTTG GCCGTTTCCCTCATGCTACA	$(GT)_{14}$	62	134–150
E7	GCCCCAAAGCCCTAAAATAA GCATGTCGGACAGTAAAGCA	$(CA)_{11}CG(CA)_4$	58	136–156
FCA304	TCATTGGCTACCACAAAGTAGG CTGCATGCCATTGGGTAAC	$(GT)_{17}(GG)_1(GT)_6$	58	120–134
FCA043	GAGCCACCCTAGCACATATACC AGACGGGATTGCATGAAAAG		58	116–130
FCA391	GCCTTCTAACTTCCTTGCAGA TTTAGGTAGCCCATTTTCATCA	$(ATGG)_{10}(GATA)_{11}$ $(TAGA)_2TGA(TAGA)_1$	55	190–230
FCA152	TTTAGTCAGCTTAGGCTTCCA CTTCCCAGCTTCCAGAATTG	$(AC)_{21}$	58	129–147
Pti007	ATCAGGAGTTCTATCACC CATGATTAGGGAGTTGAG	$(AC)_{16}$	52	139–193
FCA441	ATCGGTAGGTAGGTAGATATAG GCTTGCTTCAAAATTTTCAC	$(ATAG)_9(GTAG)_1$ $(ATAG)_2AG(ATAG)_1$	58	130–168
FCA094	TCAAGCCCCATTTTACCTTC CACCTGAGCCAAAGGCTATC	$(GT)_{19}(AG)_{22}$	58	193–215
Pti010	GGGACAACTGAGAGAAGA CAAGATATGTTCTCAGACTG	$(AC)_8$	58	118–134

PCR was carried out in a 20μL system containing 19 PCR buffer containing 50mmol/L Tris-HCl (pH 8.0), 25mmol/L KCl, 0.1mmol/L EDTA, 1mmol/L DTT;

0.4mmol/L each of 4 dNTP (TOYOBO), 0.2μmol/L each of forward and reverse primer, 0.4U units of KOD FX Neo DNA polymerase (TOYOBO) and about 80ng of total DNA. PCR amplification was performed on a Model 9700 Thermocycler (Perkin-Elmer) using the following condition: 1 cycle of 2 min at 94℃, 35 cycles of 98℃ for 20s, annealing temperature (i.e., Tm) (52–62℃, Table 8.1) for 30 s, 68℃ for 20s, and 1 cycle of 68℃ for 20min. A positive and two negative controls were included for each set of amplification.

PCR products were analyzed with an ABI 3100 Automated DNA Sequencer (Applied Biosystems) and geno- Typing data collected using GeneScan3.1 and Geno-Type3.1 (Applied Biosystems). Amplification and genotyping ere repeated 7 times for fecal DNA, while for blood DNA, homozygotes whose signals were not perfect were reamplified twice to confirm the genotypes as described by Liu et al., (2013).

8.2.4 Evaluation of genotyping correctness at each locus

The blood genotype of each microsatellite locus for each tiger was regarded as 100% correct and used as the standard for our evaluation of genotyping correctness of fecal DNA amplifications. A genotype from a fecal sample at a locus was determined as 'correct genotyping' when it matched the genotype of blood DNA of the same individual, and as 'false genotyping' when it failed to match the blood DNA genotyping result. For each locus we calculated the accumulative matching rate of genotypes (R_m)between blood and fecal samples for each locus as the total number of correctly genotyped tigers divided by the total number of genotyping trials:

$$R_m = \frac{\sum_{i=1}^{m} N_i}{nm}$$

where, R_m is the cumulative matching rate of genotypes between blood and feces when PCR is repeated m times, n the number of tigers ($n = 10$), and N_i is the number of tigers with correct genotyping at the ith PCR. The genotyping risk of fecal samples for each locus is then expressed as $1-R_m$.

8.2.5 Evaluation of population genetic parameters using fecal samples

Population genetic parameters were computed for blood samples and fecal samples using POPGENE v1.32 (Yeh et al., 2001), including observed heterozygosity (Ho), expected heterozygosity (He), the number of alleles (A), effective number of alleles (Ne), allelic frequency, and polymorphism information content (PIC). Discrimination power (DP) and the exclusion probability of paternity (EPP) were computed using CERVUS v3.0 (Marshall et al., 1998). Pearson's bivariate correlation between genotyping risk ($1-R_m$) and absolute differences of these parameters between blood and feces groups were calculated using SPSS 13.0. Pairwise relatedness of individuals (r_R) was computed by Coancestry v1.0.12 based on genotyping data from blood and feces. Linear regression of r_R between blood and fecal samples was analyzed using SPSS 13.0 (SPSS, Inc., Chicago, USA).

8.3 Results

8.3.1 Evaluation of genotyping correctness at each locus

For each fecal sample and microsatellite marker, PCR was performed 7 times and the matching rate for genotypes varied from 30%–100%, averaging 71% across the 12 loci. According to the association between the cumulative matching rate of genotypes (R_m) and number of PCR repeats, for 8 loci R_m plateaued by the third PCR and for 11 loci, R_m plateaued by the fifth PCR (Fig. 8.1). Pti010 had a fluctuating R_m, indicating an unstable allele amplification efficiency for each PCR round.

R_m at the plateau (e.g. after the fifth PCR) varied significantly among loci (Fig. 8.1; Table 8.2). E6 had the greatest plateaued cumulative R_m at 0.871, while Pti007 had the lowest value at 0.357. The mean plateau cumulative R_m was 0.710 ± 0.139 for the 12 microsatellite loci.

8.3.2 Effects of genotyping error on estimation of population genetic parameters

For each fecal sample and microsatellite locus, PCR was performed 7 times. The fecal genotype of each locus was regarded as correct when it matched the genotype obtained from a blood sample of the same tiger 4 times or more. Characteristics of the 12 microsatellite loci and comparisons between fecal and blood samples are listed in Table 8.2.

The number of alleles (A) observed per locus ranged from 3 to 6 for blood samples ($\bar{X}=4.25$), and from 3 to 5 for fecal samples ($\bar{X}=4.08$). Allelic frequency was different between blood and feces for all loci except E6. This influenced population genetic parameters based on allelic frequencies. Parameter values calculated from blood samples and fecal samples were differed insignificantly (t test, all $P>0.05$) for mean effective number of alleles (mean Ne = 2.629 for blood; mean Ne = 2.439 for feces), mean expected heterozygosity (mean He = 0.616 for blood; mean He = 0.597 for feces), PIC (mean PIC = 0.543 for blood; mean PIC = 0.518 for feces), DP (mean DP = 0.228 for blood; mean DP = 0.245 for feces), and EPP (mean EPP = 0.470 for blood; mean EPP = 0.502 for feces). Pearson's bivariate correlation test showed that genotyping risk ($1-R_m$) was not correlated with absolute differences in population

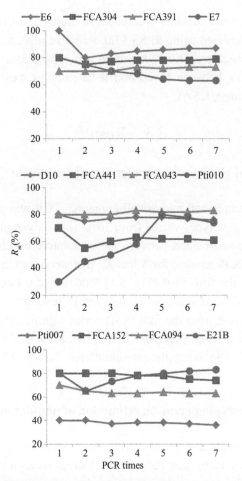

Fig. 8.1 Association between accumulative matching rate of genotypes (R_m) and repeated PCR for each locus

parameters between blood and fecal samples (He: $a = 0.431$, $P = 0.161$; Ne: $a = 0.354$, $P = 0.259$; PIC: $a = 0.411$, $P = 0.184$; DP: $a = 0.385$, $P = 0.217$; and EPP: $a = 0.365$, $P = 0.244$).

We computed pairwise relatedness (r_R) among individuals using blood and fecal genotypes. The range of r_R among individuals using blood genotypes was −0.36 to 0.22, and the range of r_R among individuals using fecal genotypes was −0.41 to 0.23. The regression coefficient should be 1.0 if r_R values of a given individual pair generated from blood and feces are equal, demonstrating that genotyping errors for feces do not influence the estimation of pairwise relatedness. Our results showed that R^2 was only 0.491, although the linear trend was significant ($P < 0.001$; Fig. 8.2). This violated the prediction and suggests that genotyping errors in fecal samples influenced the estimation of individual genetic relatedness.

Table 8.2 Characteristics of 12 microsatellite loci and comparison between fecal and blood samples in a captive tiger population ($n = 10$)

Locus		Allelic frequency						Ho	He	A	Ne	PIC	DP	EPP	R_m (%)	$1-R_m$ (%)
E6	Alleles	150	153	159											87.1	12.9
	Blood	0.650	0.250	0.100				0.700	0.532	3	2.020	0.443	0.308	0.602		
	Feces	0.650	0.250	0.100				0.700	0.532	3	2.020	0.443	0.308	0.602		
E21B	Alleles	156	158	166											82.9	17.1
	Blood	0.750	0.150	0.100				0.500	0.426	3	1.681	0.368	0.391	0.655		
	Feces	0.650	0.250	0.100				0.500	0.532	3	2.020	0.443	0.308	0.602		
D10	Alleles	136	138	144	146	148									75.7	24.3
	Blood	0.150	0.050	0.050	0.100	0.650		0.600	0.568	5	2.174	0.508	0.244	0.484		
	Feces	0.200	0.000	0.050	0.050	0.700		0.500	0.490	4	1.869	0.421	0.331	0.600		
E7	Alleles	138	142	148	152										62.9	37.1
	Blood	0.100	0.250	0.550	0.100			0.900	0.647	4	2.597	0.562	0.201	0.458		
	Feces	0.100	0.100	0.700	0.100			0.600	0.505	4	1.923	0.450	0.300	0.548		
Fca304	Alleles	124	126	130	132	136									78.6	21.4
	Blood	0.200	0.200	0.400	0.150	0.050		0.700	0.774	5	3.774	0.694	0.111	0.310		
	Feces	0.150	0.300	0.400	0.100	0.050		0.600	0.753	5	3.509	0.668	0.128	0.342		
Fca043	Alleles	119	121	125	127	129									82.9	17.1
	Blood	0.050	0.100	0.200	0.050	0.600		0.700	0.616	5	2.410	0.544	0.213	0.456		
	Feces	0.050	0.100	0.200	0.100	0.550		0.800	0.668	5	2.740	0.595	0.173	0.401		
FCA391	Alleles	198	202	206	210	218	222								72.9	27.1
	Blood	0.050	0.450	0.100	0.250	0.050	0.100	0.900	0.747	6	3.448	0.671	0.123	0.319		
	Feces	0.000	0.400	0.050	0.350	0.100	0.100	0.800	0.732	5	3.279	0.643	0.145	0.371		

Table 8.2 *(continued)*

Locus		131	137	139	Allelic frequency 141	143		Ho	He	A	Ne	PIC	DP	EPP	R_m (%)	$1-R_m$ (%)
Fca152	Alleles	131	137	139	141	143									74.3	25.7
	Blood	0.100	0.000	0.150	0.100	0.650		0.400	0.563	4	2.151	0.498	0.253	0.505		
	Feces	0.050	0.050	0.200	0.000	0.700		0.500	0.490	4	1.869	0.421	0.331	0.600		
Pti007	Alleles	141	175	177	191	193									35.7	64.3
	Blood	0.150	0.700	0.150	0.000	0.000		0.600	0.490	3	1.869	0.420	0.331	0.610		
	Feces	0.000	0.600	0.100	0.050	0.250		0.800	0.595	4	2.299	0.509	0.245	0.517		
FCA441	Alleles	136	148	152	156	160									61.4	38.6
	Blood	0.000	0.400	0.250	0.250	0.100		0.700	0.742	4	3.390	0.652	0.141	0.379		
	Feces	0.050	0.450	0.200	0.250	0.050		0.700	0.726	5	3.226	0.640	0.146	0.376		
Fca094	Alleles	197	201	203	205	207	211								62.9	37.1
	Blood	0.150	0.150	0.100	0.150	0.400	0.050	0.700	0.800	6	4.167	0.730	0.088	0.250		
	Feces	0.200	0.100	0.000	0.150	0.550	0.000	0.600	0.658	4	2.667	0.578	0.188	0.436		
Pti010	Alleles	124	126	128											74.3	25.7
	Blood	0.150	0.150	0.700				0.400	0.490	3	1.869	0.420	0.331	0.610		
	Feces	0.200	0.100	0.700				0.400	0.484	3	1.852	0.410	0.341	0.624		

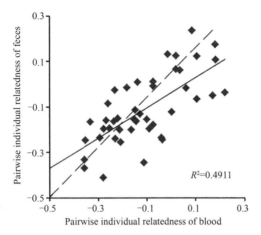

Fig.8.2 Comparison of Lynch and Ritland-estimator (r_R) of pairwise relatedness among individuals between blood and feces genotypes. Dotted line represents the equation $y = x$

8.4 Discussion

The efficiency of genetic analysis using non-invasive samples, particularly feces, is often restricted by microsatellite genotyping errors, leading to inaccurate estimation of population genetic parameters and unreasonable conservation and management strategies (Garshelis et al., 2008). Some claim that improvement of genotyping accuracy largely depends on effective sample preservation (Roon et al., 2003); however, the quality of feces collected from the wild cannot be guaranteed because of exposure to degradation factors prior to collection (Nsubuga et al., 2004; Brinkman et al., 2010). Effective preservation and DNA extraction would have limited positive impact on the analysis of poor quality feces (Huber et al., 2003). Although multi-tube PCR could reduce the effects of genotyping errors (Taberlet et al., 1996), we chose ordinary PCR so that we could effectively control factors (such as reaction temperature) for each locus. Therefore, fecal samples have to be discarded when they are inferior in quality, and laboratory procedures need to be optimized to improve overall genotyping accuracy for samples chosen for analysis. The first step in optimizing laboratory procedures is to assess genotyping risk and any possible impacts on subsequent analyses.

We identified two sources of genotyping error resulting from laboratory procedures, viz. PCR amplification and the nature of the microsatellite fragment. Our index cumulative matching rate of genotypes (R_m) contained information about both sources of error.

For PCR amplification, the R_m of 12 microsatellite loci plateaued after 3–5 PCR. This suggests that the amplification efficiency of alleles fluctuated due to a poor effective template, and each PCR had a certain genotyping risk. The quality of fecal samples used in this study was good overall because they were collected within 12h after defecation and stored at −20℃. In northeast China, cold weather and deep snow can also preserve samples quite well in winter in the field. However, PCR repeats may be required to achieve R_m plateau in fecal samples of poor quality collected in other seasons.

Rm also demonstrated that different microsatellite loci have different genotyping accuracies, even when the same batch of samples is used (Table 8.2, Fig. 8.1). This suggests that each locus might have its own tolerance limit for poor DNA templates. Perhaps uneven degradation patterns of DNA molecules means that some parts are more sensitive to degrading factors (e.g. bacterial nuclease) (Deagle et al., 2006) and therefore for the same DNA sample, the effective DNA template content would be lower for some microsatellite loci. Or perhaps secondary structures at a certain primer site differentially impact annealing efficiency (Nazarenko et al., 2002). Even though annealing can be improved by optimizing annealing temperature or PCR components, such effects cannot be completely removed. Either way, unlike previous reports that assessed overall PCR repeats for a whole panel of microsatellite loci (He et al., 2011), our results suggest that genotyping risk should be assessed for each locus.

Genotyping errors did alter population parameters such as Ne, He, PIC, DP and EPP (Table 8.2), but not in a statistically significant way, and there was no significant association with genotyping risk $(1-R_m)$. However, drawing the conclusion that genotyping errors do not impact the estimation of population genetic parameters may be premature. For example, when looking at R_m values for the 12 microsatellite loci, the lowest R_m was 0.357. We performed PCR 7 times for 10 tigers and according to the R_m equation there should be 24.99 tigers correctly genotyped, almost 3 times the real number of tigers we used ($n = 10$). This roughly demonstrates that a tiger could be correctly genotyped 3 times when PCR is performed 7 times. For other loci with higher R_m values, genotyping would be more reliable; however, the quality of feces varied among tigers and coverage of 3 times might not have covered all tigers. This means that not all tigers would be correctly genotyped and variation in feces quality requires more microsatellite loci with high R_m values, and vice versa.

In conclusion, genotyping risks for microsatellite loci from fecal DNA are derived from the PCR process and the nature of the microsatellite. Our pilot study designed for Amur tigers was part of a current Amur tiger population monitoring program, and the result will be applied to wild Amur tiger fecal DNA analyses in order to conduct secure genotyping.

8.5 Summary

In modern wildlife ecological research, feces is the most common non-invasive source of DNA obtained in the field and polymerase chain reaction (PCR) technology based on microsatellite markers is used to mine genetic information contained within. This is especially the case for endangered species. However, there are risks associated with this genotyping method because of the poor quality of fecal DNA. In this study, we assessed genotyping risk across 12 microsatellite loci commonly used in previous tiger studies using blood and fecal DNA from captive Amur tigers (*Panthera tigris altaica*). To begin, we developed an index termed the accumulated matching rate of genotypes (R_m) between positive DNA (blood samples) and fecal DNA to explore the correct genotyping probability of a certain microsatellite locus. We found that different microsatellite loci had different genotyping risks and required different PCR amplification protocols. The genotyping errors we detected altered population genetic parameters and potentially impact subsequent analyses. Based on these findings, we recommend that: ①four loci (E7, Fca094, Pti007 and Pti010) of 12 loci are not suitable for Amur tiger genetic research because of a low R_m and difficulty reaching a stable status; ②the R_m of the 12 microsatellite loci plateaued differently, and considering limited budgets, amplification times of some loci could be increased when using fecal samples; and ③ future genetic analysis of wild Amur tigers should be corrected by genotyping error rates ($1-R_m$).

Acknowledgements We thank the strong support of staff of Heilongjiang Amur Tiger Park. The authors declare no conflict of interests.

References

Arandjelovic M, Guschanski K, Schubert G, *et al.* 2009. Two-step multiplex polymerase chain reaction improves the speed and accuracy of genotyping using DNA from noninvasive and museum samples. Mol Ecol Resour, 9(1): 28-36.

Bhagavatula J, Singh L .2006. Genotyping faecal samples of Bengal tiger *Panthera tigris tigris* for population estimation: a pilot study. BMC Genetics, 7(1): 48.

Bradley BJ, Boesch C, Vigilant L. 2000. Identification and redesign of human microsatellite markers for genotyping wild chim-panzee (*Pan troglodytes verus*) and gorilla (*Gorilla gorilla gorilla*) DNA from faeces. Conservation Genetics, 1(3): 289-292.

Brinkman TJ, Schwartz MK, Person DK, *et al.* 2010. Effects of time and rainfall on PCR success using DNA extracted from deer fecal pellets. Conservation Genetics, 11(4): 1547-1552.

Deagle BE, Eveson JP, Jarman SN.2006. Quantification of damage in DNA recovered from highly degraded samples–a case study on DNA in faeces. Front Zool, 3(1): 11.

Garshelis DL, Wang H, Wang DJ, *et al.* 2008. Do revised giant panda population estimates aid in their conservation. Ursus, 19(2): 168-176.

Hájková P, Zemanová B, Bryja J, *et al.* 2006. Factors affecting success of PCR amplification of microsatellite loci from otter faeces. Mol Ecol Notes, 6(2): 559-562.

He G, Huang K, Guo ST, *et al.* 2011. Evaluating the reliability of microsatellite genotyping from low-quality DNA templates with a polynomial distribution model. Chin Sci Bull, 56(22): 1763-1770. (in Chinese)

Huber S, Bruns U, Arnold W.2003. Genotyping herbivore feces facilitating their further analyses. Wildl Soc B, 31(1): 692-697

Liu D, Ma Y, Li HY, et al. 2013. Simultaneous polyandry and hetero paternity in tiger (*Panthera tigris altaica*): implications for conservation of genetic diversity in captive populations of felids. Chin Sci Bull, 58(16): 1539-1545. (in Chinese)

Marshall TC, Slate J, Kruuk LEB, et al. 1998. Statistical confidence for likelihood-based paternity inference in natural populations. Mol Ecol, 7(5): 639-655.

Menotti-Raymond M, David VA, Lyons LA, et al. 1999. A genetic linkage map of microsatellites in the domestic cat (*Felis catus*). Genomics, 57: 9-23.

Murphy MA, Kendall A, Robinson A, et al. 2007. The impact of time and field conditions on brown bear fecal DNA amplification. Conserv Genet, 8: 1219-1224.

Nazarenko I, Pires R, Lowe B, et al. 2002. Effect of primary and secondary structure of oligodeoxyribonucleotides on the fluorescent properties of conjugated dyes. Nucleic Acids Res, 30(9): 2089-2195.

Nsubuga AM, Robbins MM, Roeder AD, et al. 2004. Factors affecting the amount of genomic DNA extracted from ape faeces and the identification of an improved sample storage method. Mol Ecol, 13(7): 2089-2094.

Panasci M, Ballard WB, Breck S, et al. 2011. Evaluation of fecal DNA preservation techniques and effects of sample age and diet on genotyping success. J Wildl Manag, 75(7): 1616-1624.

Regnaut S, Lucas FS, Fumagalli L.2006. DNA degradation in avian faecal samples and feasibility of non-invasive genetic studies of threatened capercaillie populations. Conserv Genet, 7: 449-453.

Roon DA, Waits LP, Kendall KC . 2003. A quantitative evaluation of two methods for preserving hair samples. Mol Ecol Notes, 3: 163-166

Rozhnov VV, Sorokin PA, Naidenko SV, et al. 2009. Noninvasive individual identification of the Amur tiger (*Panthera tigris altaica*) by molecular-genetic methods. Dokl Biol Sci, 429(1): 518-522.

Sambrook J, Fritsch EF, Maniatis T.1989. Molecular cloning: a laboratory manual, 2nd edn. New York: Cold Spring Harbor Laboratory Press: 1626.

Sambrook J, Russell DW, Russell DW .2001. Molecular cloning: a laboratory manual (3-volume set). New York: Cold Spring Harbor Laboratory Press: 2100.

Santini A, Lucchini V, Fabbri E, et al. 2007. Ageing and environmental factors affect PCR success in wolf (*Canis lupus*) excremental DNA samples. Mol Ecol Notes, 7: 955-961.

Stenglein JL, Barba MD, Ausband DE, et al. 2010. Impacts of sampling location within a faeces on DNA quality in two carnivore species. Mol Ecol Resour, 10(1): 109-114.

Taberlet P, Griffin S, Goossens B, et al. 1996. Reliable genotyping of samples with very low DNA quantities using PCR. Nucleic Acids Res, 24: 3189-3194.

Taberlet P, Luikart G. 1999. Noninvasive genetic sampling and individual identification. Biol J Linn Soc, 68: 41-55.

Woodruff DS.1993. Non-invasive genotyping of primates. Primates, 34: 333-346.

Xu YC, Li B, Li WS, et al. 2005. Individualization of tiger by using microsatellites. Forensic Sci Int, 151(1): 45-51.

Yeh FC, Yang RC, Timothy BJ, et al. 2001. POPGENE, the user-friendly shareware for population genetic analysis. Edmonton: Molecular Biology and Biotechnology Centre, University of Alberta: 10.

Zhang WP, Zhang ZH, Xu X, et al. 2009. A new method for DNA extraction from feces and hair shafts of the South China tiger (*Panthera tigris amoyensis*). Zoo Biol, 28(1): 49-58

Zhang YG, Li DQ, Rao LQ, et al. 2003. Identification of polymorphic microsatellite DNA loci and paternity testing of Amur tigers. Acta Zool Sin, 49(1): 118-123. (in Chinese)

Chapter 9 Sex determination of Amur tiger from footprints in snow

9.1 Introduction

The Amur or Siberian tiger (*Panthera tigris altaica*) is the larges sub-species of tiger and is primarily found in south-eastern Russia and northern China. In the 1960's it was close to extinction but its numbers recovered and are around 450 today (wwf.panda.org). Poaching and habitat destruction with very low prey base have reduced the population, which formerly ranged broadly over northeastern China, to an estimated 20 animals (Sun, 2011). However, the landscape remains potentially valuable habitat (Kang *et al.*, 2010) and the Chinese government is making concerted efforts to encourage Amur tigers to recolonize from the larger but contiguous population in the Russian Far East.

Monitoring Amur tiger in China using conventional techniques has proved to be challenging (Zhang *et al.*, 2012). At present, there are three main methods being used to monitor tigers: ①mark-recapture population estimates based on photographs of tigers obtained using camera-traps in a few selected tiger reserves (Kawanishi and Sunquist, 2004); ②line transect surveying, indices of snow track encounter rates calibrated to tiger densities used in the Russian Far East (Miquelle *et al.*, 1996; Hayward *et al.*, 2002); ③molecular biology and individual identification using DNA of feces and hair (Russello *et al.*, 2004), sometimes used in combination with the two methods above. The line-transect survey method is inappropriate for the assessment of low-density tiger populations in China. For example, survey areas $1735km^2$ and length of survey route 609km were surveyed and no Amur tiger, or sign, was found (WWF-China report (2010-2011) unpublished data). Also, DNA samples from wild tiger are hard to find in the field in China. For example, Hunchun Nature Reserve only collected 7 fecal samples of Amur tiger during two years fieldwork from 2009-2010, but they found many more snow tracks of Amur tiger (Jianmin Lang, Hunchun Nature Reserve, personal communication). Chinese experts have begun using camera-traps to monitor tigers but these are effective only in small areas where there are known routes that tigers use. Zhang *et al.* (2012) reported that the information network combined with footprints identification may work because footprints and trails in snow are easier to find and collect, which provide a potentially richer source of information.

Various studies have indicated that it is possible to use footprints in sand or mud to identify species, individuals and sex(Riordan, 1998; Jewell *et al.*, 2001; Sharma *et al.*, 2003; Sharma *et al.*, 2005; Alibhai *et al.*, 2008; Russell *et al.*, 2009; Law *et al.*, 2013). The process of discriminating footprints by sex has evolved in 3 stages: ①based on the shape description of the footprints (McDougal, 1977; Sankhala, 1978; Sagar and Singh, 1991); ②based on simple comparison of measurements (Gogate *et al.*, 1989; Sagar and Singh, 1991); ③based on the statistical analysis of one or several measurements (Bhattacharya, 1967; Gore *et al.*, 1993; McDougal, 1999; Sharma *et al.*, 2003; Sharma and Wright, 2005).

For the past decades, censuses of the contiguous Russian Amur tiger population have employed a combination of field signs (scat, tree markings, urine etc) and snow footprints. The latter have been employed particularly for sex determination. The width of the front pad has been the identifying standard, which combined with the other signs has been used to assess tiger individual numbers (Abramov, 1961; Matyushkin *et al.*, 1996; Smirnov and Miquelle, 1999). However, this technique, has depended on the manual measurement of a few features directly from snow footprints, and does not provide sufficient resolution to permit discrimination between adult female and sub-adult male tigers with the front pad width range overlapping (Miquelle *et al.*, 2006).

We report here on a robust algorithm for identifying the sex of Amur tigers from their footprints in snow. Initial trials with free-ranging tigers indicate that this will be a community-friendly, cost-effective and non-invasive tool for field monitoring.

9.2   Materials and methods

Footprints were collected from captive and free-ranging Amur tiger in northeast China (Fig. 9.1). Captive Amur tiger footprints were collected from the Hengdaohezi Amur tiger breeding centre in the Heilongjiang Province of northeast China, where records are kept on date of birth and parentage for more than 300 Amur tigers. Footprints from free-ranging Amur tigers were followed in the Dongfanghong Forestry Bureau of Heilongjiang Province in the Wandashan Mountains of northeast China, where a small but relatively stable Amur tiger population has remained for the past ten years.

9.2.1   Definition of terms

Footprint: One footprint.
Trail: An unbroken series of footprints made by a single animal
FIT: The footprint identification technique (Jewell *et al.*, 2001; Alibhai *et al.*, 2008)

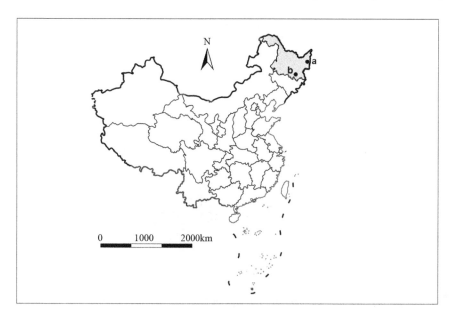

Fig. 9.1 Amur tiger foot prints collecting zones. Heilongjiang Province (blue); location a shows Dongfanghong Forestry Bureau in the Wandashan Mountains; location b shows Hengdaohezi Amur Tiger Breeding Center near Mudanjiang city in China.

9.2.2 Footprint image data collection and analysis methods

We collected footprints using a basic compact digital camera, a carpenter's folding ruler and label paper to record photo information with the image. We analyzed footprint data using the FIT add-in in JMP® (SAS institute) software. We collected footprints from all four feet, for possible future reference, but needed only the left hind footprints for the analysis.

We identified several new challenges in working with footprints in snow. Initially we found that the paper labels, used to record footprint information on the image, became wet in snow. We circumvented this problem by manufacturing a small card slot from soft transparent plastic and glued it to the back of the ruler and used a dry-erase pen to write the detail information directly on the soft plastic card slot. The hand-writing left by the dry-erase pen could be cleaned easily by any cloth. This adapted ruler is a useful addition for the long-term use of the monitoring tool, and like the other equipment is cheap and easy to use (Fig. 9.2 and Fig. 9.3a).

We used only footprints with clear outlines of the four toes and metapodial pad for analysis. We imported them into JMP Software, where a FIT customized script allowed the extraction of 128 measurable variables (distances, angles and areas) of each footprint to provide a comprehensive geometric profile (Alibhai *et al.* 2008 and see below).

Fig. 9.2 Technique for taking the footprint image of an Amur tiger using a compact digital camera.

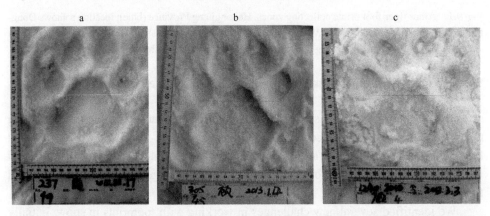

Fig. 9.3 Image(a) is good-quality image (with clear outline) of an Amur tiger footprint, as taken during winter 2011–2012 in northeastern China. Image(b) is poorer quality image but it is still possible to determine the outline. Image (c) could not be used for analysis.

9.2.3 Footprint images collection from captive Amur tigers

Footprints from tigers of known sex were required to develop an effective algorithm to classify by sex. Captive animals, for whom accurate records were held, were ideal for this purpose. We collected sets of left hind footprints in snow from 40 captive Amur tigers whose ages ranged from 3 to 13 years old for both females (277 footprints) and males (246 footprints). The number of left hind footprints collected from known individuals varied from 3 to 21 with a mean value at 13.1.

During the sampling period there was unusually little natural snow in the study area, so we had to develop a method to cover the tiger enclosures with snow which would approximate to natural conditions. We gathered what fresh snow was available to cover the tigers' outdoor enclosures, spread it and then sieved it to imitate natural snow fall. We experimented with different depths of snow and found a depth of 3–5cm (1–2cm firm snow plus 2–3cm uncompacted snow) provided the clearest footprints. The tigers were attracted to walk on it normally with food and calls from their keepers. We collected footprints from trails only if the animals were walking at a normal walking pace and footprint outlines were clear.

We then photographed the footprints according to the FIT protocol (Alibhai et al., 2008). We selected clear footprints and placed the ruler as the scale on bottom and left hand axis in relation to direction of travel, making sure the ruler did not obscure detail at all (Fig. 9.2). A photo label containing the date and name of photographer, animal ID, and footprint number, was included in each image.

We also had to develop an assessment of image quality for photography, based on the clarity of the outline of the metapodial pad and toes. We did this over the research period, gradually refining the technique as we progressed. Figure 9.3 gives information on the clarity of images required.

Another challenge presented by snow conditions was the lack of contrast presented by the high reflectivity of the snow substrate. We were able to work around this limitation by photographing images early morning and late afternoon when there was some contrast, and providing artificial lighting from a flashlight when conditions were overcast and visibility very poor. Images were taken from directly overhead (checked by a second person), ensuring that the frame was filled by the footprint, ruler and label card. We found the best images were obtained taking the image from 30–50cm above the footprint.

9.2.4 Collection of footprints images from free-ranging Amur tigers

To locate and collect footprints from free-ranging Amur tiger, we employed local informants in the winter from the end of 2011 to the beginning of 2012. Information relating to trail sightings was provided by villagers and local patrollers in Amur tiger range areas, and on receipt we went straightly to the reported site to follow the trail as soon as possible. Our objective was to get multiple high-quality left hind footprint images from each trail.

Despite challenging field conditions, it was not difficult to find clear footprints. Experience showed that Amur tiger prefer walking on level terrain such as frozen rivers, roads, ridges or farmland where clear footprints are easily found (Fig. 9.4). Often following these areas with downed dead wood, or neighbouring big rocks led to more favourable substrates. However, occasionally footprints were found in very deep snow, and proved challenging to image. Sometimes, in very deep snow, it was necessary to manually remove some snow around the footprint to keep the ruler and the outline of the footprint in the same plane. There are many causes of snow footprint degradation including fresh snowfall, wind, ice overflow and melt-out, but snow-fall was the most common cause (82%) of footprint degradation in January and February, while footprints created in March degraded faster, lasting on average 2.1 days by warm weather and wind(Hayward et al., 2002). Footprints on snow were relatively stable without new snowfall in January and February which provided good opportunities to collect high-quality footprints during the research period.

Fig. 9.4 Example of the clear trails left by free-ranging Amur tigers in northeastern China during winter 2011–2012.

The method of collecting footprints from free-ranging tigers was the same as that for captive tigers. However, since the identity of the animal was in each case unknown, we recorded trail numbers as follows: the first trail we photographed was trail 1, and the footprints in it were recorded as 1a, 1b, 1c etc. The second trail was trail 2, and footprints in it recorded as 2a, 2b etc. If a trail was obscured or broken at any point, the next set of images was allocated a different number. GPS locations were also recorded for each trail, along with habitat information.

9.2.5 Extracting a geometric profile from digital images

This process has been reported by Alibhai et al. (2008) for white rhino (*Cerato-therium simum*) footprints on sand, and is here reported as a new development and application for Amur tiger snow footprints. Images were processed to extract measurements, the set of which we refer to as the geometric profile. Each image was orientated so that the front toes were at the top. Images were cropped if necessary to remove extraneous space, leaving only the footprint, ruler and photo label in the image.

Brightness and contrast were optimally enhanced for each image, so that toes and metapodial pad could be outlined clearly. Two markers were then placed on the image-one at the lowest point of each side toe. The image was then rotated around a line joining these two points until the line is horizontal, and the image sized, keeping the aspect ratio fixed, until it fitted comfortably into the JMP graphics window. 25 points were marked on each footprint, from which a total of 128 measurable variables were extracted using FIT scripts in JMP software (Fig. 9.5).

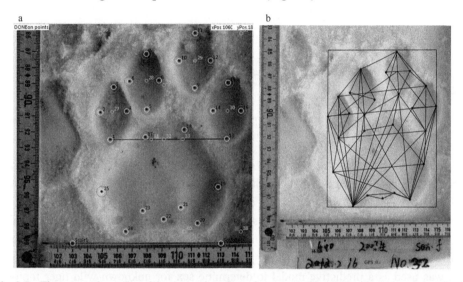

Fig. 9.5 The geometric profile of the Amur tiger footprint extracted by JMP software. a. Amur tiger's left-hind footprint images, as collected during winter 2011–2012 in northeastern China, marked in JMP software with 25 landmark points (marked 1–25) and derived points (26–40). b. The same image showing the derived points and lines, from which 128 morphometric variables are drawn. These constitute the geometric profile of the footprint.

9.2.6 Data analysis

For sex identification, we used a total of 523 footprints from 40 captive tigers, 19 females (277 footprints) and 21 males (246 footprints). Animal ages ranged from 3–13 years for both sexes without significant age differences between groups (female mean age = 8.07±0.18, male mean age = 8.36 ± 0.19, F = 1.18, P> 0.05). We used linear discriminant analysis (LDA) in JMP to discriminate sex. LDA is a technique used to identify a linear combination of variables that characterize or separate different classes of objects, in this case footprint trails. We used the JMP Stepwise Variable Selection function to select the variables which provided best discriminating power based on their F ratios. This function also has the added advantage of excluding highly correlated variables. By plotting the number of variables included in the analysis against the predicted level of accuracy of sex identification, we were able to establish the number of variables which would provide an effective predictive model (Fig. 9.6).

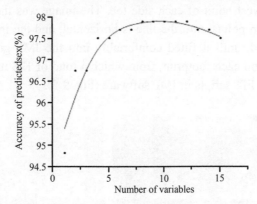

Fig. 9.6 The asymptote of the number of morphometric variables and the footprint accuracy for images of Amur tiger footprints collected during winter 2011–2012 in northeastern China.

To validate the level of accuracy of sex discrimination using linear discriminant analysis, we used 2 holdout methods-Jackknife, and partitioning the data into training and test sets. Jackknife tests sequential subsets of data, excluding every footprint in the dataset in sequence, where partitioning splits the data to enable testing of one set against an algorithm derived from the other. An algorithm derived from all the variables, consisting of the ten best discriminating variables generated by the captive tiger footprint data was used as a predictive model to determine sex for unknown wild tiger footprint data.

9.3 Results

9.3.1 Sex discrimination model for captive Amur tiger

Variables with the highest discriminating power were selected using the stepwise selection feature in JMP. To avoid using too many variables and over-fitting we used linear discriminant analysis to plot the number of variables against the sex-prediction accuracy for all the footprints Figure 9.6 shows the asymptote was reached with 8–10 variables and we thus opted to use 10 variables selected stepwise. Figure 9.7 shows a plot of the first two canonical variables generated using linear discriminant analysis using 10 variables. Out of a total of 523 footprints, 11 were misclassified (7 female footprints classified as male and 4 male as female).

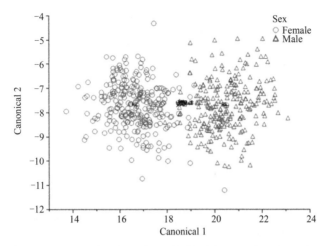

Fig. 9.7 The distribution of male (triangle) and female (circle) Amur tiger footprints, collected during winter 2011-2012 in northeastern China, using 10 morphometric variables. Number of misclassified=11; Percent of misclassified=2.103%; –2 Log Likelihood=63.699.

A Jack-knife validation procedure on the captive tiger dataset using 10 variables selected stepwise resulted in a total of 13 footprint misclassifications, giving an accuracy of sex determination of 97.5%. These few misclassifications were likely have been due to subtly distorted footprints, themselves perhaps due to sudden gait or substrate change. To further validate the level of accuracy, we randomly allocated 50% of the captive dataset to a training set and 50% to a test set using a discriminant validation platform in FIT. This procedure was iterated five times, each time the allocation being carried out randomly. The levels of accuracy for the training sets ranged from 96.9% to 98.4% and for the test sets from 97.4% to 98.5%.

9.3.2 Application in wild Amur tigers

Using the sex discrimination algorithm generated with the captive data set, we attempted to predict the sex of the free-ranging tiger footprints. We selected the 83 qualified wild tiger tracks from 8 trails in 5 forest farms (HYS, QY, XNC, WLD and QS) and performed discriminant analysis using 10 variables. Table 9.1 shows the outcome of sex prediction of footprints from free-ranging tigers using discriminant analysis. The result showed that 5 trails were from female Amur tigers and 3 trails were from males. As Table 1 shows, for all trails, the majority of footprints (92%) were either classified as male or female. Although we couldn't check the individual numbers and real sex of these wild tigers, we still have confidence in this predictive results, because at most only 2 tracks of each trail has different discrimination result with the rest tracks at the same trail (like trail HYS and WLD-A).

Table 9.1 The sex prediction results of the snow footprints of 8 free-ranging Amur tigers' trails, as collected during the winter of 2011-2012 in northeastern China.

Trail name	No. of tracks/trail	No. of tracks classified as female	No. of tracks classified as male	% tracks/trail classified as male or female[a]	Predicted sex
HYS	13	11	02	82	F
QY	12	11	01	91	F
WLD-B	14	13	01	92	F
XNC-A	13	13	0	100	F
WLD-A	07	05	02	71	F
XNC-B	05	0	05	100	M
QS-B	13	0	13	100	M
QS-A	06	01	05	80	M

Note: HYS=Haiyingshan, QY=Qiyuan, WLD=Wulingdong, XNC=Xinancha, and QS=Qingshan. F, Female; M, Male.
[a] The percentage of footprints within the trail (i.e., from one animal) that were assigned to the final predicted sex.

9.4 Discussion

9.4.1 Collecting and measuring footprints

We collected only hind footprints, because during normal walking, the hind footprints often register over those of the front feet, resulting in fewer intact front footprints(Sharma *et al.*, 2005). To facilitate field-collection and minimize the requirement for footprint collection we developed the system to use only left-hind footprints. Riordan *et al.* (1998) found that the misclassification rate for the right hind footprints was slightly higher than that for the left hind. By comparison, Indian experts sampled trails that had at least 5 clear impressions of left and right hind feet(Sharma *et al.*, 2003), and Russian experts just measured width of the front pad, which is a standard measurement used for Amur tigers(Abramov, 1961; Matyushkin *et al.*, 1996; Smirnov and Miquelle, 1999).

We decided to take as many measurements as was practically possible (128) to extract the maximum discriminating power from the footprints, while Sharma *et al.* (2003, 2005) had taken 93 measurements. We included many of the same variables that earlier studies (Gogate *et al.*, 1989; Gore *et al.*, 1993; Das and Sanyal, 1995) had identified as being useful. We always used 10 variables to build and test the sex-discrimination model, but these 10 variables (out of the original 128 variables) might be different every time, depending on the selection of data fed into the model. This experience was also reported by Sharma *et al.* (2003).

9.4.2 Data objectivity of footprint in snow

Field monitors were confident in using the digital cameras to record footprints after a short training period, so there was no element of tracing or copying images (Rishi, 1997). The process of data collection is therefore accessible to local people in northern or snowy landscapes, and can also include a valuable contribution from citizen scientists towards data-collection.

In our experience, when the footprint was in deep snow (> 5cm), snow on the surface obscured the real outline and shape details of footprint below. In addition, the footprint appears smaller in deep snow than the same one in shallow snow, because the distance between the ruler and the outline of footprint increases. Conversely, if the footprint is in very shallow snow (< 3cm) there is insufficient depth to hold detail. Therefore, in captive situation or in the wild, it is advisable to record footprints in a snow depth of around 3-5cm and also keep the ruler and the outline of the footprint at the same plane when taking the footprint image. During this research we also found snow substrates to be relative stable, thus it is possible to find and collect high quality snow footprint data in the field (Hayward *et al.*, 2002).

9.4.3 The effects of sample size and modelling methodology for sex discrimination

This study has developed more rigor in discriminating sex from footprints in several ways. Firstly, with regard to data collection, we have used significantly larger number of footprints for each animal and collected footprints from many more animals than before (Riordan, 1998; Sharma *et al.*, 2003; Sharma *et al.*, 2005). We excluded the influence of age, because we have used a wider age-range of 3–13 years with no significant difference between male and female. Secondly, we have instituted a rigorous quality control photo protocol and filtering system for images to be used. We extracted the data from footprint images, not tracings. Thirdly, using the discriminant validation platform in JMP, which provides a stepwise function, we were able to validate our model effectively using Jackknife and holdout method by randomly apportioning the data into training and test sets.

Lastly, we have developed a single piece of user-friendly software (FIT) that will enable us to complete the image processing and data analysis in one software platform.

9.4.4 Management implications

We consider footprints in the snow the best means of determining the presence of Amur tiger in northeastern China. In decades of monitoring history of Amur tiger in both Russia and China, field workers have identified Amur tiger by sex or individual using simple standard-front pad width measurements, and using assumptions about individual range. The major limitation in this system has been the difficulty in discriminating between adult females and sub-adult males. China faces an additional monitoring challenge due to the very low Amur tiger population density. The new Amur tiger monitoring network (Zhang et al., 2012) requires a more accurate method, and we are confident that FIT can provide this. This technique may also be able to help us assess the number of resident female Amur tigers. Since Female Amur tiger diffusion range is limited and daily movement is shorter than those of males (Goodrich et al., 2010) ; therefore, we believe that female resident tigers are an indicator of local tiger population recovery.

The use of the non-invasive footprint identification technique has positive management implications with respect to monitoring approach. Footprint collection does not disturb or interfere with the natural ecology of the animal, and evidence suggests (Jewell, 2013) that non-invasive approaches lead to better scientific outcomes in conservation monitoring.

In summary, the advantages of FIT are its ease of use for landscape scale monitoring of Amur tiger and a potentially valuable tool for the ongoing assessment of recovery of this species in China. We are currently investigating individual and age identification of Amur tiger and communicating with our Russian colleagues to adopt this non-invasive and cost-effective method as a single unified approach to Sino-Russian transboundary populations, so that this endangered population can be cost-effectively assessed and protected.

9.5 Summary

The Amur tiger(*Panthera tigris altaica*) population in China, once widespread, is now reduced to an estimated 20 individuals widely dispersed over a large area. The Chinese government is making concerted efforts to restore this population from the contiguous Russian population. However, they face a challenge in finding an effective monitoring technique. We report on the development of a robust, non-invasive and cost-effective technique to identify the sex of Amur tiger from snow footprints. Between December 2011 and December 2012, we collected 523 digital images of left hind footprints from 40 known captive Amur tiger (19 females, 21 males), of age range 3–13 years (female mean age = 8.07 ± 0.18, male mean age = 8.36 ± 0.19, $F = 1.18$, $P > 0.05$). Images were captured with compact digital cameras according to a standardized photographic protocol (Alibhai et al., 2008). Using JMP software from the SAS Institute, 128 measurements were taken from each footprint according to the protocol developed by Alibhai et al. (2008), and subjected to a stepwise selection. With just 10 variables, and testing with both jackknife and 50% holdout methods, the resulting algorithm for sex determination gave 98% accuracy for footprints. The algorithm derived from captive tiger footprints of known sex was then used to identify the sex of 83 footprints from 8 trails collected from unknown free-ranging Amur tiger in the winter from the end of 2011 to the beginning of 2012. The algorithm predicted 5 trails from females and 3 from males. This technique is a potentially valuable tool for monitoring the recovery of Amur tiger populations at the landscape scale in northeastern China.

References

Abramov K G. 1961. On the procedure for tiger census. Moscow: Obschestvo Ispytaltely Prirody.

Alibhai S K, Jewell Z C, Law P R. 2008. A footprint technique to identify white rhino *Ceratotherium simum* at individual and species levels. Endangered Species Research, 4: 205-218.

Bhattacharya C G. 1967. A simple method of resolution of a distribution into gaussian components. Biometrics, 23(1): 115.

Das P K, Sanyal P. 1995. Assessment of stable pug measurement variables for identification of tiger. Tiger Paper, (22): 20-26.

Gogate M, Joshi P, Gore A, et al. 1989. Tiger pugmark studies: a statistical perspective. Directorate, Project Tiger Technical Bulletin No.2.

Goodrich J M, Miquelle D G, Smirnov E N, et al. 2010. Spatial structure of Amur (Siberian) tigers (Panthera tigris altaica) on Sikhote-Alin Biosphere Zapovednik, Russia. Journal of Mammalogy, 91(3): 737-748.

Gore A P, Paranjape S A, Rajgopalan G, et al. 1993. Tiger census: role of quantification. Current Science, 64(10): 711-714.

Hayward G D, Miquelle D G, Smirnov E N, et al. 2002. Monitoring Amur tiger populations: characteristics of track surveys in snow. Wildlife Society Bulletin, 30(4): 1150-1159.

Jewell Z C. 2013. Effect of monitoring technique on quality of conservation science. Conservation Biology, 27(3): 501-508.

Jewell Z C, Alibhai S K, Law P R. 2001. Censusing and monitoring black rhino (*Diceros bicornis*) using an objective spoor (footprint) identification technique. Journal of Zoology, 254(1): 1-16.

Kang A, Xie Y, Tang J, et al. 2010. Historic distribution and recent loss of tigers in China. Integrative Zoology, 5(4): 335-341.

Kawanishi K, Sunquist M E. 2004. Conservation status of tigers in a primary rainforest of Peninsular Malaysia. Biological Conservation, 120(3): 329-344.

Law P R, Jewell Z C, Alibhai S K. 2013. Using shape and size to quantify variation in footprints for individual identification: case study with white rhinoceros (*Ceratotherium simum*). Wildlife Society Bulletin, 37(2): 433-438.

Matyushkin E, Pikunov D, Dunishenko Y, et al. 1996. Numbers, distribution, and habitat status of the Amur tiger in the Russian Far East. Moscow: Final report to the USAID Russian Far East Environmental Policy and Technology Project.

Mcdougal C. 1977. The face of the tiger. New York: Rivingon & Deutsch.

Mcdougal C. 1999. You can tell some tigers by their footprints with confidence. Riding the tiger: tiger conservation in human dominated landscapes, 383: 190-191.

Miquelle D, Pikunov D, Dunishenko Y, et al. 2006. A survey of Amur (Siberian) tigers in the Russian Far East, 2004–2005. Wildlife Conservation Society, World Wildlife Fund.

Miquelle D G, Smirnov E N, Quigley H G, et al. 1996. Food habits of Amur tigers in Sikhote-Alin Zapovednik and the Russian Far East, and implications for conservation. Journal of Wildlife Research, 1(2): 138.

Riordan P. 1998. Unsupervised recognition of individual tigers and snow leopards from their footprints. Animal Conservation, 1(4): 253-262.

Rishi V. 1997. Monitoring tiger populations by impression-pad method. The Indian Forester, 123(7): 583-600.

Russell J C, Hasler N, Klette R, et al. 2009. Automatic track recognition of footprints for identifying cryptic species. Ecology, 90(7): 2007-2013.

Russello M A, Gladyshev E, Miquelle D, et al. 2004. Potential genetic consequences of a recent bottleneck in the Amur tiger of the Russian Far East. Conservation Genetics, 5(5): 707-713.

Sagar S R, Singh L A K. 1991. Technique to distinguish sex of tiger (*Panthera tigris*) from pug-marks. The Indian Forester, 117(1): 24-28.

Sankhala K. 1978. Tiger! the story of the Indian Tiger. London: Harpar Collins.

Sharma S, Jhala Y, Sawarkar V B. 2003. Gender discrimination of tigers by using their pugmarks. Wildlife Society Bulletin, 31(1): 258-264.

Sharma S, Jhala Y V, Sawarkar V B. 2005. Identification of individual tigers (*Panthera tigris*) from their pugmarks. Journal of Zoology, 267(1): 9-18.

Sharma S, Wright B. 2005. Monitoring tigers in Ranthambore using the digital pugmark technique. New Delhi: Wildlife Protection Society of India.

Smirnov E, Miquelle D. 1999. Population dynamics of the Amur tiger in Sikhote-Alin State biosphere Reserve // Seidensticker J, Christie S, Jackson P. Riding the tiger: tiger conservation in human-dominated landscapes. Cambridge: Cambridge University Press: 61-70.

Sun H Y. 2011. Amur tiger. Harbin: Northeast Forestry University Publishing House.

Zhang C Z, Zhang M H, Jiang G S. 2012. Assessment of monitoring methods for population abundance of Amur tiger in Northeast China. Acta Ecologica Sinica, 32(19): 5943-5952.

Chapter 10 An adaptive model for determining hierarchical priority conservation areas

10.1 Introduction

Habitat conservation is crucial for saving endangered large mammals and their ecosystems (Ripple *et al.*, 2014). For wide-ranging large mammals like tigers (*Panthera tigris*), habitat conservation is complex and challenging both because of the need to maintain permeable landscapes for effective movement over large spatial scales, and the issues of competing political and socioeconomic land-use decisions (Chapron *et al.*, 2014). An open question is whether existing protected areas can save large free-ranging mammals (McLaughlin, 2016).

By 2015 nature reserves covered 14.8% of China's land surface, which compares favorably to the worldwide average of 8.8% (Zhu *et al.*, 2018). Despite these designations, conservation measures have partially failed to reverse endangered species declines and ecological degradation of China (Liu *et al.*, 2001; Tian *et al.*, 2009). However, Viña *et al.* (Viña *et al.*, 2016) reported the benefits of the natural forest conservation program in China, which led to growth of Amur tiger population sizes from 2000 to 2014, and a large-scale success in restoration of big cats and the forest ecosystems (Jiang *et al.*, 2017). These achievements and experiences led the Chinese government to recognize some shortcomings of the current protected area management systems. Consequently, in December 2015 the central government re-designated some nature reserves as centrally controlled National Parks at large scales (Li *et al.*, 2016).

Although it is encouraging to see the Chinese government's progress towards more rigorous and scientific nature conservation for the national park system (Li *et al.*, 2016), several questions remain: whether these protected areas should be priorities for some species survival, whether they meet their habitat requirements, and whether biological and social conflict goals of conservation actions can be simultaneously met (Sala *et al.*, 2002). There is an urgent need to identify which patch (core areas) and ecological corridors can meet these goals (Fazey *et al.*, 2005). Implementation and determination of priority areas needs to balance resource allocation, social costs or conflicts in human-dominated landscapes (Wikramanayake *et al.*, 2004) and the basic conservation objectives about what should be maximized or minimized (Game *et al.*, 2013). We use the example of Amur tiger conservation in China to ask whether using pre-selected hierarchical priorities in setting conservation goals and planning can enhance conservation.

To address these questions, firstly, we identified the priority habitat patches and corridors by using habitat suitability modelling. Secondly, we established an objective function to maximize habitat suitability probability and the areas of suitable patches while considering conservation cost weights, and minimizing the anthropogenic disturbance intensity. Finally, we select the highest value of the objective function, which represents the best available combination of biological and social factors, resulting in a *P*-value (cut-off probability) that can be used to derive prioritized core areas. Thus, we will establish an adaptive model for determining hierarchical priority conservation areas to ensure enough sparing of land for animal habitat requirements while sharing land (Jiang *et al.*, 2017).

10.2 Research methods

10.2.1 Data collection and habitat patch determination

Firstly, we collected the data on climate, topographical features, vegetation, and anthropogenic factors downloaded from public websites. We conducted line transect surveys in snow for ungulate prey species distribution over four winters (Stephens *et al.*, 2006), collected tiger distribution data from locations of Amur tiger occurrence from 2000 to 2017 (Jiang *et al.*, 2017). Individual identification and movement were confirmed using camera trap surveys from 2012 to 2017 (Wang *et al.*, 2018). We then built the biological habitat model using MaxEnt software to determine the suitable habitat patches by cutoff points of presence probability for species(Phillips *et al.*, 2006). A habitat patch was considered occupied if the presence of Amur tigers was confirmed in it. We modelled the relationship between tiger presence and patch area using logistic regression (Qing *et al.*, 2016). After that, we tested the Amur tiger's sensitivity to area by comparing the observed logistic occupancy curve with a null response curve. The null response curve was defined by the function $1-(1-p_b)^{a/a_b}$, where a_b corresponds to the minimum home range size of an Amur tiger and a ($a > a_b$) is the patch size, p_b is the occupancy rate in a patch with size of a_b(Qing *et al.*, 2016). Then, we estimated Amur tiger area requirements by leveraging dose-response curves from pharmacology in which the key measure is the effective dose (ED) that yields a particular probability. ED_p is the patch area at which it is likely that Amur tigers were present with the probability of *p*%. We used $ED_{90\%}$ as the value of the Minimum Area Requirement (MAR) (Qing *et al.*, 2016).

We collected and used a combination of climate, topographical features, vegetation, and anthropogenic factors (Table 10.1) to model potential distribution of Amur tigers. Snow depth was derived from Canada Meteorological Center (http://nsidc.org/data/NSIDC-0447).

The Geographic Information System (GIS) shapefiles of villages, roads, rivers and railway were obtained from the Environmental and Ecological Science Data Center from West China, National Natural Science Foundation of China (http://westdc.westgis.ac.cn/). The DEM data and Normalized Difference Vegetation Index (NDVI) data were provided by Geospatial Data Cloud site, Computer Network Information Center, Chinese Academy of Sciences (http://www.gscloud.cn/). Slope and aspect were calculated using spatial analyst in ArcMap 10.3. Vegetation data were derived from GlobleL and 30 project, National Geomatics Center of China (NGCC). Vegetation types were classified as forest, shrubland, grassland, cultivated land, and wetlands. All habitat covariate layers were resample as layers with the grid of 3 km to predict prey, and then Amur tiger distribution. Like in Jiang et al. (2017) the following data collection methods were used.

Table 10.1 Habitat variables tested for association with Amur tiger and ungulate distribution.

Habitat facto	Description of the habitat factor	Data type	Unit
Snow depth	Mean snow depth of December to March next year, 2005–2015 (source grid has 24-km resolution)	Continuous	cm
Elevation	Digital Elevation Model grid with 30-m resolution	Continuous	m
Aspect	Aspect gird with 30-m resolution derived from the digital elevation model above	Continuous	
Slope	Slope grid with 30-m resolution derived from the digital elevation model above	Continuous	(°)
Distance to railway	Distance from the central point of each pixel to the nearest railway	Continuous	km
Distance to village	Distance from the central point to the nearest village	Continuous	km
Road density	Road density in a 10-km circular window	Continuous	km/km^2
Distance to river	Distance from the central point to the nearest village	Continuous	km
NDVI	Mean NDVI value of September, 2012–2017 (source grid has 500-m resolution)	Continuous	%
Forest ratio	Area ratio of forest in that pixel	Continuous	%
Cropland ratio	Area ratio of cultivated land in that pixel	Continuous	%

We collected ungulate prey snow track data from the snow transect surveys over four winters (November to March of 2013–2014, 2014–2015, 2015–2016, and 2016–2017). Line transect surveys were conducted with an intensity of 36km per 100km^2 in the survey areas of known Amur tiger range. Each transect was at least 5km long and the minimum distance between two transects was 3km. Scientific personnel and forestry workers were trained in species identification and data collection protocols before going to the field. Animal's tracks made within the last 24h were counted if they crossed transect routes but subsequent recrodings were ignored if investigators could confirm that they were made by the same animal (Stephens et al., 2006; Qi et al., 2015). We conducted 345 transects, and the total length of transects was 1876 km, covering 11 928km^2 (i.e., we sampled 16.4% forest area within the study area) of typical habitat types (Fig. 10.1). Considering the low population density of red deer (*Cervus elaphus*) and sika deer (*Cervus nippon*)in the northeastern China, we only used roe deer (*Capreolus pygargus*) and wild boar(*Sus scrofa*) data as potential prey and used their potential population density distribution as covariates influencing the presence of Amur tigers.

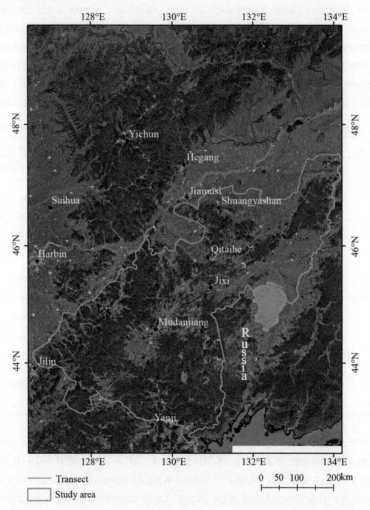

Fig. 10.1 Sampling line transect distribution in study area. Red lines denote line transects. (Base map: Google Satellite Image). Please scan the QR code on the back cover to see the color image.

Locations of Amur tiger occurrence from 2000 to 2017 were obtained from records of compensation for livestock predation, transect survey, kill sites, DNA extraction from fecal or hair samples, and information reported from forestry bureaus and nature reserves confirmed by experienced Amur tiger experts. Camera trap surveys were conducted from 2012 to 2017 in the current range for Amur tiger. The current tiger range was divided into 3km×3km grid cells and camera trap locations were usually chosen on forest road, animal trail or other places which maximize the possibility of capturing tiger images in those grid cells. A total of 1052 camera trap locations were chosen in the tiger range (i.e., 9468km^2 was surveyed). Camera trap locations are shown in Fig.10.2. There were two cameras in one camera trap location, placed on both sides of an animal trail to attempt to capture tiger flank stripe patterns of both sides of the body. Storage cards and batteries of cameras were replaced every four months. By identification of individuals within photos, we assessed the total number of tigers only by confirming the two sides of stripe patterns from each capture event (i.e., only one body side pattern individual was not counted into the total number).

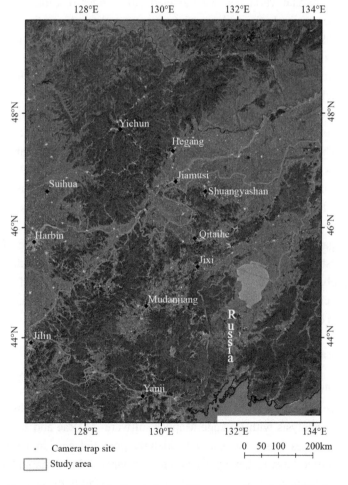

Fig. 10.2 Locations of camera traps in the study area. Base map: Google Satellite Image. Please scan the QR code on the back cover to see the color image.

Additionally, dispersal of new born individuals, and movement patterns of each tiger by tracking their photos were recorded with times and locations from 2012 to 2017. Minimum age of first reproduction of female Amur tiger is 4 years ±0.4 years and mean interval between litters of Amur tiger is 21.4 months±4.4 months (Kerley *et al.*, 2003). We think there should be at least 4 years or maybe more to detect a female tiger to redistribute and reproductive successfully, or determine some individuals' dispersal location, direction or distances to check the potential effectiveness of corridors for dispersal.

Prey distribution is a common and crucial limiting factor for the survival and diffusion of large predators (Karanth and Chellam, 2009). First, we used prey presence data and habitat variables to model prey distribution. Then, with a combination of prey distribution and habitat variables, we modeled Amur tiger distribution. We used the number of presence pixels for each ungulate species and for Amur tigers on the spatial scale of 3km×3km grid cells as independent variables. We tested for multicollinearity among variables using a combination of variance inflation factors (VIF) and Spearman's rank correlation (r) using the *R* package car. Uncorrelated ($r < 0.6$) habitat predictor variables were chosen as initial input to the models (Ramsay *et al.*, 2003). The maximum entropy model was used to model species distribution(Phillips *et al.*, 2006). We used the software MaxEnt(Version 3.4.1) to build habitat models for the prey species and then the Amur tigers. The sampled dataset of tiger occurrence locations was randomly divided into two subsets, training (75% of sampling data set) and test subsets (25% of sampling data set). Training data were used to construct the model and test data were used to validate the model. We used area under the receiver operating characteristic curve (AUC) of test data to evaluate the predictive ability of models. Models with AUC values of 0.7 to 0.9 were regarded as useful and > 0.9 as highly accurate. Importance of each habitat variable was estimated using jackknife tests. Finally, we identified cutoff points (i.e., occurrence probability values) where the sum of model sensitivity and specificity was maximized to classify estimated habitat suitability map into binary (i.e., suitable and unsuitable) grid cells (GUÉNETTE and VILLARD, 2005).

10.2.2 Habitat corridor identification

Based on the value of large patches (>MAR of Amur tiger) as source patch (Walston *et al.*, 2010), we used occurrence probability derived from species distribution models in the connectivity analysis with Linkage Mapper software (McRae and Kavanagh, 2011). Linkage Mapper uses least-cost theory to identify potential corridors for Amur tigers among the identified large suitable patches. The resistance value of habitat to movements between habitat patches was inversely related to the probability of tiger occurrence from the MaxEnt models (Chetkiewicz and Boyce, 2009).

10.2.3 Optimization algorithm model, and identification of prioritized core habitat of suitable patches and corridors

Although larger habitat areas are expected to be better for a species conservation, in reality, the extent of any protected area will be limited by social and economic constraints (Ferson and Burgman, 2006). We aimed to identify the largest suitable habitat patch areas for Amur tiger while minimizing anthropogenic disturbances within them. Then, we developed a model to derive cutoff probabilities that determined core habitat patch by trading off conservation efforts and conflicts(Sala et al., 2002).

Firstly, we used focal statistics in the ArcMap spatial analysis tool with multiple spatial window sizes and species perceptual ranges from 0 to 6 neighboring cells (i.e., perceptual radius from 0 to 18km)(Ribeiro et al., 2017) to process the suitability probability map which resulted from MaxEnt models. Species spatial perceptual range (species perception) is the spatial extent of animal responses to landscape attributes. Experts recommend estimating species spatial perceptual range from animal's movement capacity or home range size(Ribeiro et al., 2017). The mean daily travel distance of female Amur tigers was about 7km (Carroll and Miquelle, 2006), which was covered in the multiple window size range. Under each scenario, we calculated the patch area, average resistance, road density, cropland density, and village density of a patch determined at different cutoff points of probability from 0 to 1 with 0.05 intervals of p-values. The optimization algorithm model, adaptive model, was developed using:

$$Z = \max_i Z_i = \max_i \max_{p \in [0,1]} \lambda_p S_p p \exp\left(-\left(\sum \omega_{pj} \frac{L_{pj}}{\max(L_{pj})}\right)\right) \quad (10.1)$$

where Z is the score of the target function; i refers to scalar scenario range from 0 to 6; p is the Amur tiger occurrence probability predicted by MaxEnt models; λ_p is a penalty factor relating to conservation costs based on average resistance value of each grid of habitat; S_p is the total area of the patches determined when the cutoff point is p; L_{pj} is the area or density of anthropogenic disturbance type j; and ω_{pj} is a selective factor for any human disturbance types ($\omega_{pj} = 0$ when $L_{pj} = 0$, otherwise $\omega_{pj} = 1$). We ran this equation for trade-offs maximizing suitable habitat area and minimizing disturbance intensity considering any combination of human disturbance factors to derive one best p-value to keep the Z-value to be the maximum values for 7 cell window sizes for habitat patches or 10 spatial buffer extents (1–10km) for corridors.

Finally, we used the p-value (cut-off probability) to derive the core habitat patches or corridor locations and area sizes as prioritized conservation areas within current suitable habitats identified by MaxEnt models. We assumed that the 7-km buffer zone can be piloted to provide the example for this adaptive spatial conservation priori extent design for the corridor. All statistics were performed using the SPSS 10.0.

10.3 Results

10.3.1 Amur tigers population and habitat distribution in China

Currently, the Amur tiger range in China is mainly located in three regions: Laoyeling, Zhangguangcailing and Wandashan, covering 137 085km^2, based on 4028 occurrence locations from 2000 to 2017 (Fig. 10.3a). Furthermore, from January 2012

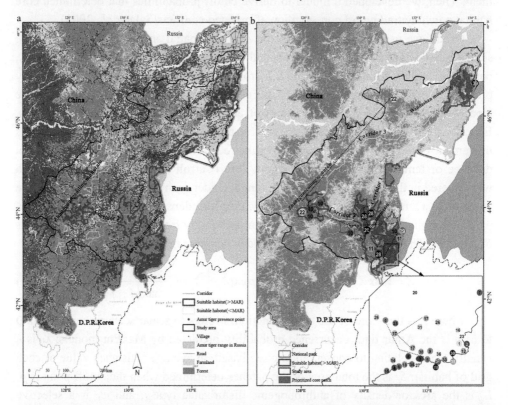

Fig. 10.3 A comparison of biological and adaptive models. a. Traditional biological suitable patch and corridor lines with the locations of Amur tiger occurrence from 2000-2017 (red points). b. Prioritized core patch areas derived by the adaptive model with maximized suitability and minimized human disturbances within biological priority habitat areas larger than MAR using the cut-off point of the maximized Z-value using the 1-cell window size scenario. The new national park area piloted is circled by purple lines; Traditional biological suitable patches were determined by using 0.324 as cut-off probability point, where the sum of model sensitivity and specificity was maximized to classify estimated habitat suitability as suitable or unsuitable The number codes (shown in small circles) of Amur tiger individuals recorded by camera traps from 2012 to 2017 (Each individual is described in Table S1). Individuals are shown in small black squares and some individual diffusion routes are shown connected by dashed black lines.
Base map: World Ocean Reference, website: http://goto.arcgisonline.com/maps/Ocean_Basemap.
Please scan the QR code on the back cover to see the color image.

to December 2017, camera trap surveys recorded that the 27 tigers had the images of both two flanks, while nine had right flank image only and three have left flank image only. Considering the potential duplicates in the tiger individuals with only one side image, there were at least 36 (36–39) different individuals confirmed in this research, and some individuals have had their mother identified (Table 10.2).

Table 10.2 The individual Amur tigers recorded by camera traps in China from January 2012 to December 2017, including flank photographed

ID code	Sex	Flank captured	Capture History						Mother
			2012	2013	2014	2015	2016	2017	
CT1	F	LR	Y	Y	Y	Y	Y	Y	
CT2	M	LR		Y	Y	Y			
CT3	F	LR		Y	Y	Y	Y	Y	
CT4	M	LR	Y	Y	Y	Y			
CT5	F	LR			Y	Y	Y		
CT6	U	R			Y				CT18
CT7	M	LR		Y	Y	Y			
CT8	F	LR		Y	Y	Y			
CT9	M	LR			Y	Y			
CT10	M	LR			Y	Y	Y	Y	
CT11	M	LR			Y	Y	Y		CT1
CT12	F	LR			Y				CT1
CT13	M	LR		Y	Y	Y			
CT14	U	R			Y				CT18
CT15	M	LR			Y	Y			
CT16	F	LR			Y	Y	Y		CT1
CT17	F	LR		Y	Y	Y	Y	Y	
CT18	F	LR		Y	Y	Y	Y		
CT19	U	R			Y				CT3
CT20	M	R	Y						
CT21	U	L			Y				CT5
CT22	M	LR			Y	Y			
CT23	M	LR				Y	Y	Y	
CT24	U	L			Y				CT5
CT25	U	R		Y					
CT26	M	LR			Y	Y	Y	Y	
CT27	M	LR				Y	Y		
CT28	M	LR				Y	Y	Y	
CT29	M	LR		Y	Y	Y	Y	Y	
CT30	U	R	Y						
CT31	F	R					Y	Y	
CT32	F	L						Y	

Continuation Table

ID code	Sex	Flank captured	Capture History						Mother
			2012	2013	2014	2015	2016	2017	
CT33	M	LR					Y		
CT34	M	LR						Y	
CT35	M	LR						Y	
CT36	U	R					Y		CT3
CT37	U	R					Y		CT3
CT38	U	LR			Y			Y	CT8
CT39	M	LR						Y	

Note: F, Female; M, Male; U, Unknown sex; LR, left and right flanks; L, left flank; R, right flank. Mother indicates this tiger belonging to this female's cub.

10.3.2 Habitat modeling and MAR of Amur tiger

Using Amur tiger presence records, environmental variables and predicted prey distribution, we estimated habitat suitability. The area under curve (AUC) value for the test data set was 0.834, indicating that the MaxEnt model of Amur tiger presence had a good discriminatory ability. Results of the jackknife test of variable importance indicate that predicted prey have the most useful information (Fig. 10.4). Using the cut-off value 0.324, we clustered habitat into a binary map of suitable and unsuitable(Fig. 10.5). Habitat suitability modeling identified 4 suitable habitat patches (> MAR) (Fig. 10.3a).

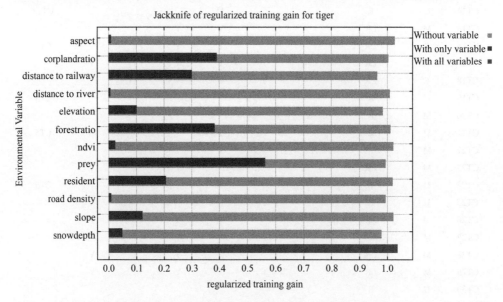

Fig. 10.4 Results of jackknife test of variable importance for habitat modeling. Please scan the QR code on the back cover to see the color image.

Chapter 10 An adaptive model for determining hierarchical priority conservation areas | 323

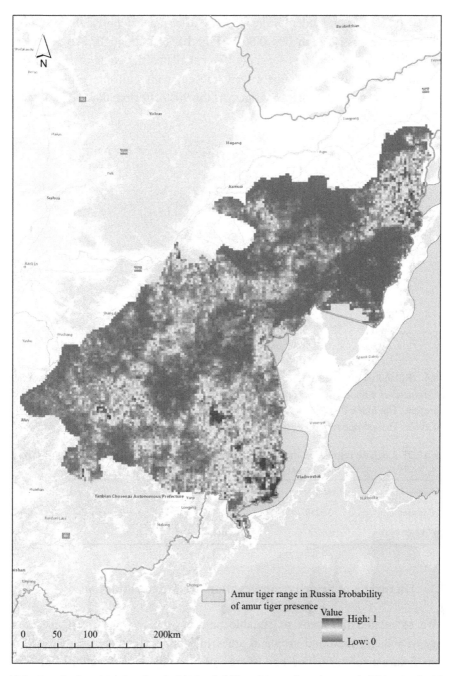

Fig. 10.5 Results from MaxEnt showing the probability of Amur tiger presence in China. Red grid cells indicate high probability of presence while blue indicate low probability. Base map: USA Topo Maps, website: http://goto.arcgisonline.com/maps/USA_Topo_Maps. Please scan the QR code on the back cover to see the color image.

Alogistic regression model for the response of occupancy probability to patch area showed a positive association ($P < 0.001$) (Fig. 10.6, Table 10.3). We conclude that Amur tiger occupancy was area-sensitive. We calculated the value of ED_{90}, the patch area at which it is likely that tigers were present with probability of 90%, as the minimum area requirement for Amur tiger. At the 90% effective dose (ED_{90}), the value of MAR was estimated as 992km^2.

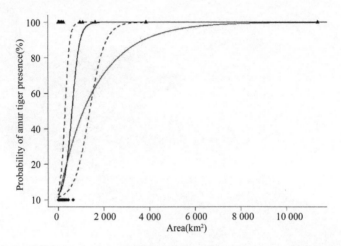

Fig. 10.6 Relationship between the probability of Amur tiger presence and habitat patch size. Filled circles are patches without indicators of Amur tiger presence and triangles are patches where Amur tiger were present. The black line is the logistic regression fit, and blue dashed line shows 95% confidence intervals. The red line represents the null response curve derived from pure passive sampling.

Table 10.3 Logistic regression model of occupancy probability of Amur tiger against habitat patch size. SE, Standard error.

Variable	Coeffcient	SE	t	P> |t|
Constant	−3.5238	0.3211	−10.975	<0.001
Area	0.0058	0.0017	3.402	<0.001

10.3.3 Hierarchical priority conservation areas determination

We calculated the areas of suitable patches, the average resistance, and the density of village, road and cropland areas in identified suitable patches. Multiple human disturbance intensities and scores of the objective function with varying suitability in each scenario are shown in Fig. 10.7 Scores of the objective function were maximal when $P = 0.57$ and at a 1-cell window size (i.e., 3km× 3km)(Fig. 10.8). Habitat suitability modeling identified 4 suitable habitat patches (i.e., conservation units) with patch areas larger than the MAR of Amur tigers, and simulated corridors among them are shown in Fig. 10.3a. Considering social cost in the adaptive habitat modeling, there are three prioritized core patches bigger than MAR (Fig. 10.3b). Distributions of tiger individuals derived from camera trapping are also shown in Fig. 10.3b. The two-level process of determining hierarchical priority areas is shown in Fig. 10.9.

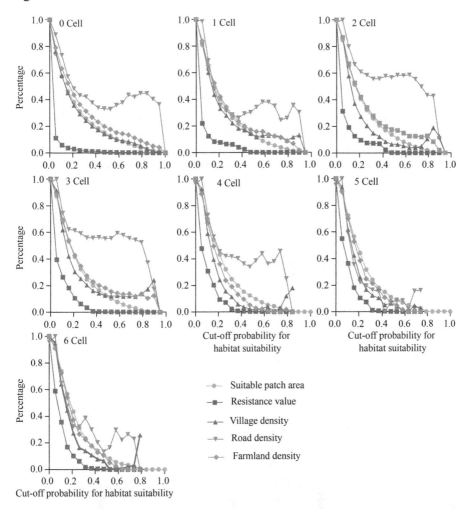

Fig. 10.7 Relationships between occurrence probabilities of tigers from MaxEnt predictions as cut-off points, and the suitable habitat areas, and human disturbance intensification dynamics at multiple spatial window sizes. Village, road and farmland density means, length and areas, respectively within the suitable habitat patches in different cut-off points for habitat suitability with different spatial window sizes. The 0, 1, 2, 3, 4, 5, and 6 cells represent 0km, 3km, 6km, 9km, 12km, 15km and 18km of moving windows as radius, respectively. All the variables were normalized.

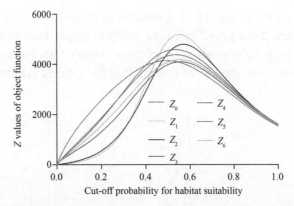

Fig. 10.8 Changes in the Z-values of the objective function (i.e., Equation 1 in Material and methods) with increasing the cut-off probability for habitat suitability to determine the optimal spatial extent of prioritized core habitat patches. The highest value of Z, where the *p*-value will be chosen for deriving prioritized core areas, in each scenario means the best combination of biological and social factors. All the variables were normalized.

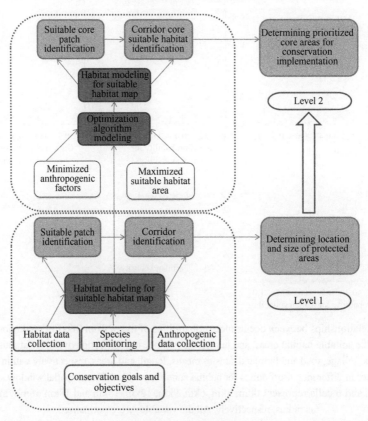

Fig. 10.9 The determining process of hierarchical priority conservation area for selecting suitable patch and corridor areas.

We ran the same algorithm (Equation 10.1) for each corridor scaling scenario to determine the optimal corridor length and total suitable area by maximizing the suitability probability and suitable habitats derived from MaxEnt modeling, and minimizing the anthropogenic disturbance intensity with a 3km window size to derive the core strip habitats within each of 10 buffer-distance corridor areas. While maximizing the Z-value of the objective function in Equation 1, we found that the larger buffer widths resulted in lower cut-off probabilities of habitat suitability (Figs. 10.10, Figs.10.11). This indicated that large corridors provide large areas of low suitability habitat and may need large investment for improvement (Fig. 10.12).

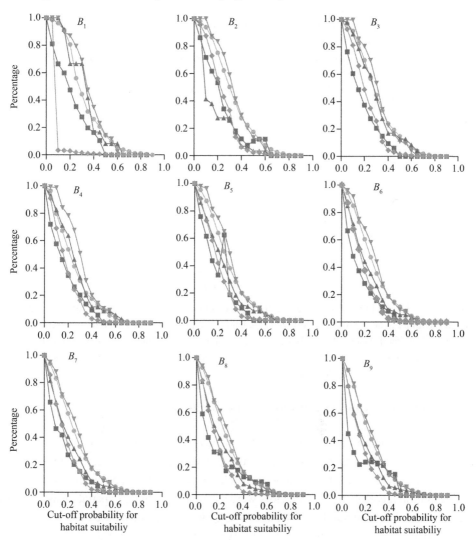

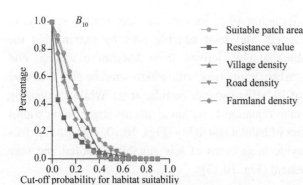

Fig. 10.10 Relationships between different probabilities of MaxEnt prediction as cut-off points, the suitable habitat areas, and human disturbance intensification dynamics at multiple spatial buffered distance corridor scenarios. Village, road and farmland density mean numbers, length and areas, respectively within the suitable habitat patches in different buffered distance corridor scenarios. Note: B_X represents buffer distance, where X is the distance in kilometers.

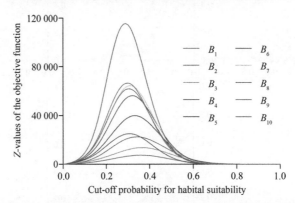

Fig. 10.11 Changes in the Z-values of the objective function with increasing the cut-off probability for habitat suitability to determine the prioritized core habitat within different spatial buffered corridor areas. B_X represents buffer distance, where X is the distance in kilometers. The highest value of Z in each scenario indicates the best combination of biological and social factors, where the P-value will be chosen to derive prioritized core areas. Please scan the QR code on the back cover to see the color image.

10.3.4 Potential conservation function comparison

Based on camera trap database for Amur tiger, we have census the total number of Amur tigers, and the total habitat area occupied was estimated as 2025km^2 by counting the number of cells of habitats with 3-km resolution with tiger occurrences from 2012 to 2017. Anthropogenic disturbance factors were calculated using GIS spatial measurements and censuses. The total number of villages was 8186, total cropland area was 49 205km^2, and the total length of roads was 23 271km. We counted and compared the percentage of

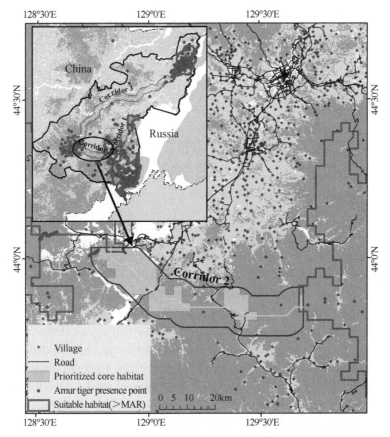

Fig. 10.12 Spatial distribution of stripe habitats and anthropogenic disturbances within the multiple spatial extent scenarios using 1–10km buffer zones of corridors in the Amur tiger range in China. Core habitat of corridor 2 with 7km buffer distance is shown. All the variables were normalized. Base map: USA Topo Maps, website: http://goto.arcgisonline.com/maps/USA_Topo_Maps. Please scan the QR code on the back cover to see the color image.

tigers and their habitat areas, and percentage and identity of quantitative anthropogenic disturbance factors for prioritized core habitat patch areas (Table 10.4).

Habitat suitability modeling indicated that the 4 conservation area patches (> MAR) would protect 7.41% of areas currently occupied by the tiger, all individuals found in these patches, and would allow for the movement of all tiger individuals among the habitat patches. Considering social cost in the adaptive habitat modeling demonstrated that the 3 prioritized core patches areas (> MAR) would protect much more occupied areas (12.06%) by tigers than habitat patch areas occupied (7.4%) in the biological habitat suitability model, and 88.89% of individuals would also be found in these patches (Table 10.4, Fig. 10.3b). Furthermore, in the identified priority core areas (> MAR), human disturbance intensities, that is, number of villages (80 villages), cropland area

Table 10.4 The potential conservation effects of suitable and prioritized core habitats from two levels of habitat models from biological to adaptive models in Amur tiger range of China.

Element	Amur tiger range in China	Suitable patch (> MAR)	Prioritized core patch (> MAR)	National Park
Number of villages	8 186	246	80	249
Cropland area (km^2)	49 205	1 105	320	949
Road length (km)	23 271	2 078	809	1 281
Percentage of habitat areas occupied by tigers in different kinds of habitats with 3km resolution from 2012 to 2017	1.48%($2025km^2$ occupied in the area of $137\,085km^2$ of total range area)	7.41%($1566km^2$ occupied in the $21\,133km^2$ of total suitable patch areas with each patch area more than MAR)	12.06%($954km^2$ occupied in the $7910km^2$ of prioritized core habitat areas identified)	7.31%($1071km^2$ occupied in the $14\,644km^2$ of national park)
Percentage of Amur tiger population (2012 to 2017) within area	100%(36/36)	100% (36/36)	88.89% (32/36)	94.44% (34/36)

($320km^2$) and road length (809km), have obviously decreased compared with those (246 villages, $1105km^2$ of cropland area and 2078km of road length) in habitat patch areas occupied (> MAR) in the biological habitat suitability models, even including the national park area (249 villages, $949km^2$ of cropland area and 1281km of road length) (Table 10.4). Within the national park, although 94.44% of tiger individuals are known to occur there, it only covers 7.31% of areas occupied by tigers, considerably less area than the 12.06% of the prioritized core areas (Table 10.4). The high density of tigers in the national park likely indicates that more conservation work is needed in the large area outside of the national park, for corridor establishment or potential core habitat protection in the Zhangguangcailing (Fig. 10.3a).

10.4 Discussion

Biological models revealed that large-scale corridor conservation is necessary for tiger movement in permeable landscapes (Theobald *et al.*, 2012). In fact, we found that one male tiger (coded 10) was born in Russia and then walked more than 300 km into China along corridor 2 from 2014 to 2017 (Fig. 10.3b). Another male tiger (coded CT 22) from March 2013 to January 2015 moved from the Wandashan to the Zhangguangcailing along corridor 3 (Fig. 10.3b). Even though the sample size is small, this indicates that the two corridors would connect tigers in the key habitat patches. However, some barriers should not be ignored, because it is difficult to exclude roads in prioritizing core habitat with increased habitat suitability (Fig. 10.7). Monitoring data have revealed that the highway together with both croplands and villages blocked two Amur tigers' movements from 2017 to 2018 (unpublished data). Habitat restoration of priority patches and corridors should consider not only converting croplands to forest, relocating villages and closing unpaved roads, but also constructing highway tunnels or bridges for animal passage to connect core habitat patches (Fig. 10.3).

This study suggests that Northeast China Tiger and Leopard National Park (NCTLNP) would play a key role as an important population source (Fig. 10.3b, Table 10.4) (Walston et al., 2010), and we have found five breeding female families with 11 sub-adult individuals in this area till now (Table 10.2). However, we should recognize that the urgent need for Amur tigers in China is for effective dispersal out of this area, where tiger density is almost three times higher than in the surrounding areas of China (Wang et al., 2018). The small tiger population, including in the Sino-Russia border area of neighboring Russia, was only 45 adult individuals recorded from 2013 to 2015 across this Sino-Russia border area, possibly facing a risk of population collapse like one of Russia's small populations that was caused by a disease outbreak (Miquelle et al., 2015). Furthermore, this small tiger population in Russia is genetically isolated from the other big tiger population (95% of total population) in Russia (Fig. 10.3) (Henry et al., 2009) and needs to be connected (Miquelle et al., 2015). In this study, we provide the possibility of linking the big and small Amur tiger population of Russia based on two corridors of China connecting the big population via Laoyingling patch, Zhangguangcailing patch, and Wandashan habitat patch to connect with the Sikhote-Alin patch (Fig. 10.1a, Fig. 10.5). However, we found that NCTLNP has not provided any corridors for their urgent effective dispersal to other large habitat patches (Fig. 10.3b). Therefore, the current national park range is unlikely to effectively promote the Amur tiger population to recolonize and expand into China.

Our study suggested that biological habitat-suitability models could identify large-scale conservation prioritized areas by determining the suitable habitat patches and corridors for the government's conservation policy decision. However, our hierarchical approach provides a more rigorous pre-design procedure by considering both biological requirements and socioeconomic factors. Furthermore, adaptive modeling allows accurate assessment of effects of each of the variables despite the fact that these variables are correlated(Castilla, 1999), and also determines whether patch areas meet animal long-term persistence requirements or not(Qing et al., 2016). Consequently, our approach provides a transparent demonstration to decision makers to assess prioritized conservation design effectiveness considering species needs and socioeconomic constraints into endangered species habitat conservation with practical application to landscape planning and patch management.

10.5 Summary

There is an urgent need to determine priorities for habitat conservation of endangered large animals, especially to simultaneously evaluate both biological requirements and socioeconomic factors. We hierarchically prioritized conservation areas for Amur tigers in China by using optimization algorithms to balance maximizing suitable habitat areas and minimizing anthropogenic disturbances. Adaptive habitat models reveal that hierarchically prioritized core patches with habitat areas exceeding Minimum Area Requirements (MAR) would protect substantially more areas occupied by tigers and have less human disturbance than either suitable habitat patches exceeding MAR from biological models or the current Northeast China Tiger and Leopard National Park (NCTLNP). NCTLNP's design and management should focus on connecting tiger habitat areas. Our hierarchical, quantitative approach demonstrates the use of a broadly applicable transparent tool for siting new protected areas at regional scales, and designing core zones of new protected areas for conserving wide-ranging large terrestrial mammals.

References

Carroll C, Miquelle D G. 2006. Spatial viability analysis of Amur tiger Panthera tigris altaica in the Russian Far East: the role of protected areas and landscape matrix in population persistence. Journal of Applied Ecology, 43(6): 1056-1068.

Castilla J C. 1999. Coastal marine communities: trends and perspectives from human-exclusion experiments. Trends in Ecology & Evolution, 14(7): 280-283.

Chapron G, Kaczensky P, Linnell J D, et al. 2014. Recovery of large carnivores in Europe's modern human-dominated landscapes. Science, 346(6216): 1517-1519.

Chetkiewicz C L B, Boyce M S. 2009. Use of resource selection functions to identify conservation corridors. Journal of Applied Ecology, 46(5): 1036-1047.

Fazey I, Fischer J, Lindenmayer D B. 2005. What do conservation biologists publish? Biological Conservation, 124(1): 63-73.

Ferson S, Burgman M. 2006. Quantitative methods for conservation biology. Berlin: Springer Science & Business Media.

Game E T, Kareiva P, Possingham H P. 2013. Six common mistakes in conservation priority setting. Conservation Biology, 27(3): 480-485.

GuÉnette J S, Villard M A. 2005. Thresholds in forest bird response to habitat alteration as quantitative targets for conservation. Conservation Biology, 19(4): 1168-1180.

Henry P, Miquelle D, Sugimoto T, et al. 2009. In situ population structure and ex situ representation of the endangered Amur tiger. Molecular ecology, 18(15): 3173-3184.

Jiang G, Wang G, Holyoak M, et al. 2017. Land sharing and land sparing reveal social and ecological synergy in big cat conservation. Biological Conservation, 211: 142-149.

Karanth K U, Chellam R. 2009. Carnivore conservation at the crossroads. Oryx, 43(1): 1-2.

Kerley L L, Goodrich J M, Miquelle D G, et al. 2003. Reproductive parameters of wild female Amur (Siberian) tigers (Panthera tigris altaica). Journal of Mammalogy, 84(1): 288-298.

Li J, Wang W, Axmacher J C, et al. 2016. Streamlining China's protected areas. Science, 351(6278): 1160.

Liu J, Linderman M, Ouyang Z, et al. 2001. Ecological degradation in protected areas: the case of Wolong Nature Reserve for giant pandas. Science, 292: 98-101.

McLaughlin K. 2016. Can a new park save China's big cats? https://www.sciencemag.org/news/2016/08/can-new-park-save-chinas-big-cats[2017-1-10].

Mcrae B, Kavanagh D. 2011. Linkage mapper connectivity analysis software. Seattle WA: The Nature Conservancy.

Miquelle D G, Smirnov E N, Zaumyslova O Y, et al. 2015. Population dynamics of Amur tigers (*Panthera tigris altaica*) in Sikhote-Alin Biosphere Zapovednik: 1966–2012. Integrative zoology, 10(4): 315-328.

Phillips S J, Anderson R P, Schapire R E. 2006. Maximum entropy modeling of species geographic distributions. Ecological modelling, 190(3-4): 231-259.

Qi J Z, Shi Q H, Wang G M, et al. 2015. Spatial distribution drivers of Amur leopard density in Northeast China. Biological Conservation, 191: 258-265.

Qing J, Yang Z, He K, et al. 2016. The minimum area requirements (MAR) for giant panda: an empirical study. entific Reports, 6: 37715.

Ramsay T, Burnett R, Krewski D. 2003. Exploring bias in a generalized additive model for spatial air pollution data. Environmental Health Perspectives, 111(10): 1283.

Ribeiro J W, Silveira Dos Santos J, Dodonov P, et al. 2017. LandScape Corridors (LSCORRIDORS): a new software package for modelling ecological corridors based on landscape patterns and species requirements. Methods in Ecology and Evolution, 8(11): 1425-1432.

Ripple W J, Estes J A, Beschta R L, et al. 2014. Status and ecological effects of the world's largest carnivores. Science, 343(6167): 151-162.

Sala E, Aburto-Oropeza O, Paredes G, et al. 2002. A general model for designing networks of marine reserves. Science, 298(5600): 1991-1993.

Stephens P A, Zaumyslova O Y, Miquelle D G, et al. 2006. Estimating population density from indirect sign: track counts and the Formozov-Malyshev-Pereleshin formula. Animal Conservation, 9(3): 339-348.

Theobald D M, Reed S E, Fields K, et al. 2012. Connecting natural landscapes using a landscape permeability model to prioritize conservation activities in the United States. Conservation Letters, 5(2): 123-133.

Tian Y, Wu J, Kou X, et al. 2009. Spatiotemporal pattern and major causes of the Amur tiger population dynamics. Biodiversity Science, 17: 211-225.

Viña A, Mcconnell W J, Yang H, et al. 2016. Effects of conservation policy on China's forest recovery. ence Advances, 2(3): e1500965.

Walston J, Robinson J G, Bennett E L, et al. 2010. Bringing the tiger back from the brink—the six percent solution. PLoS Biology, 8(9): e1000485.

Wang T, Royle J A, Smith J L, et al. 2018. Living on the edge: opportunities for Amur tiger recovery in China. Biological Conservation, 217: 269-279.

Wikramanayake E, Mcknight M, Dinerstein E, et al. 2004. Designing a conservation landscape for tigers in human - dominated environments. Conservation Biology, 18(3): 839-844.

Zhu P, Huang L, Xiao T, et al. 2018. Dynamic changes of habitats in China's typical national nature reserves on spatial and temporal scales. Journal of Geographical Sciences, 28(6): 778-790.

Chapter 11 Ecological threshold on precise management of Amur tiger population and habitat

11.1 Introduction

Large carnivores which are top predators play an important role in structuring and regulating ecosystems and influencing the strength of ecosystem functioning (Ripple et al., 2014). Factors such as slow life histories, low population densities and reproductive rates, and the need for large home ranges to obtain enough prey resources all make large carnivores vulnerable to persecution and recovery difficult (Ripple et al., 2014). Moreover, the conflicts between large carnivores and humans or livestock are also detrimental to the survival of large carnivores (Madden, 2008). In 2014, 61% of the world's large carnivore species (average adult body masses ≥15kg) were listed as threatened (vulnerable, endangered, or critically endangered) by the International Union for the Conservation of Nature (IUCN), and 77% of total species were undergoing continuing population declines (Ripple et al., 2014). Large carnivore conservation is a continuing worldwide dilemma.

Amur tiger (*Panthera tigris altaica*) is a large carnivores distributed in northeast China and the Russian Far East. It is very endangered and protected by governments of China and Russia. Beginning in December 2015, the Chinese government started to design a centrally controlled national park that stretches across Heilongjiang and Jilin provinces. The aim is to provide a coherent management system to protect the Amur tigers and leopards (*Panthera pardus orientalis*). Beyond the problems faced by all large carnivores, the conservation of these two species and construction of the national park are characterized by prohibition of natural resource exploitation in the national park. There are also livestock losses and human fatalities caused by wild animals, which have created economic losses and aggravated human-wildlife conflicts in this developing and populous country (Liu et al., 2011).

Wildlife populations and ecosystems both have the ability to compensate for fluctuations in environment factors to maintain population or ecosystem conditions (Vogt et al., 2015). This ability is known as the flexibility, and it only works within certain limits (Vogt et al., 2015). Changes in environmental factors, which exceed the range the ecosystem or wildlife populations could compensate for, would cause a state shift. State shift often occurs nonlinearly with rapid speed and recovery may be hard. Ecological threshold is the concept used to describe the maximum or minimum values of environmental factors the ecosystem or wildlife populations could tolerate, at which once a small change in one or more key factors would result in rapid state change (Bennett and Radford, 2003). Sometimes, as there may be more than one stable state over a range of environment conditions, and a state may change with a slight fluctuation in any point of this range (Scheffer and Carpenter, 2003), all this range also could be treated as ecological threshold. Crossing threshold leads to loss or recovery of ecosystem functions and biodiversity (Bestelmeyer, 2006). The formation of current Saharan is a good example to illustrate this concept (Demenocal *et al.*, 2000). The fluctuations of vegetation cover around a smoothly decreasing trend had been lasted for thousands of years, before it became to desert from verdant landscape nearly completely vegetated with annual grasses and shrubs between 5000 and 6000 years ago. The abrupt collapse just happened within decades to centuries when summer insolation crossed a threshold value of $470W/m^2$ (Demenocal *et al.*, 2000).

Since the concept of ecological thresholds was introduced to ecology by May (1977), it has been repeatedly scrutinized by ecologists (Briske *et al.*, 2006; Li and Yuan, 2017). The practical application in biological conservation and sustainable ecosystem management also has been particularly attractive (e.g. Huggett, 2005). In general, the utilization of knowledge about ecological thresholds in conservation has been classified in four ways: ① helping to define the sensitivity of a species to threatening processes, ② helping manage the tradeoffs between biodiversity conservation efforts and human activities, ③ helping set targets for habitat retention, restoration and recovery, and ④ helping develop landscape designs for biodiversity conservation (Huggett, 2005).

As threshold reflects the maximum or minimum tolerance of the ecosystem or wildlife population to different environment factors or other drivers. It ideally suits to set standard rules or goals for natural resource management to keep or recover the environmental condition into a tolerated range (Choquenot and Parkes, 2001). In this sense, ecological threshold concept correspond well to the "ecological protection red line" emphasized by the Chinese government (Li, 2014). "Ecological protection red line" was defined as spatial boundaries and management limit values need to be implemented strictly, in terms of natural ecological services functions, environmental quality and safety, and the utilization of natural resources (Li, 2014). It represent the quantitative indexes in ecological conservation that must be maintained, and once overstepped may cause irreversible damage. For Amur tiger conservation and the national park construction, knowledge about the possible threshold effect on this large carnivores and the whole ecosystem by any of the various environmental factors could help a lot to nation park design, institution of regulations, and conservation practice. For example, the threshold of habitat fragmentation for Amur tiger survival or dispersal would be important reference for national park location design and ecological corridors construction; and if managers know how frequent of human activities would cause Amur tiger to avoid this area, rules could be set out to limit human activities in reasonable range; Otherwise, we could organize supplementary feeding only once the snow depth exceeds the threshold which would result in food searching difficulties then cause mass mortality of ungulates, rather than every winter.

In this part, we examine how different forms of ecological thresholds affect Amur tiger populations and their ecosystems directly or indirectly, primarily through a perspective of bottom-up effects. The factors which could cause a threshold effect were classified into climate and weather events, habitat and anthropogenic disturbance, vegetation, prey, and competitors. Examination of which of these factors are important to conservation of the species and ecosystems, and why, allows us to provide practical answers to two overarching questions that are relevant to conservation of large carnivores around the world: ① How to make conservation and management measures more explicit and practical for the Amur tiger? ② What general recommendations can be made to applying concepts about ecological thresholds to the conservation of large carnivores? We suggest that ecological thresholds research should be taken as an important part of Amur tiger conservation and national park construction, as well as applied to large carnivore conservation in other areas.

11.2 Possible ecological thresholds in Amur tiger conservation based on bottom-up effects

11.2.1 Thresholds caused by climate and weather events

Climate change has drawn wide societal interest. Both changes in mean climate state or related increases in extreme events could have profound effects on the distribution and behavior of species (Seddon et al., 2016), and on ecosystem structure and function (Doughty et al., 2015). The effects of climate change stresses on plants and their ecosystems are often nonlinear. Examples of changes include that a global increase of temperature of about 3–4 °C is predicted to result in dieback of the Amazon rainforest (Cox et al., 2004). Similarly, the decreased frequency of extreme cold events may cause a poleward expansion of mangroves in response to geographical shifts in the location of their minimum temperature response threshold (Cavanaugh et al., 2014). Climate effect on vegetation, such as the Changes in plant communities and harvesting time, would in turn affect the animal community, like interspecific interaction and distribution. For example, as climate warms, thermophyte would gain an advantage in competition with hardy plant, consumers that are more dependent on hardy plants would be more limited, while species that are highly dependent on thermophilic plants get more opportunities to survival, the changes in consumer community would affect higher species through food web, finally the Amur tiger as the top predator would be affected. Although there are no evidences straightly reflect how this effect works no top predators, the effects on species or community at the bottom of the food web have been studied a lot (Visser, 1998; Kaiser-Bunbury et al., 2017). Seddon et al. (2016) used an autoregressive model to quantitatively analyze the relative response rates of ecosystems to climatic variability (air temperature, water availability and cloud cover). They identified nine sensitive ecological biomes that were close to critical thresholds for ecosystem change, including parts of the boreal forest belt worldwide that encompass Amur tiger habitats. Many of the forests in Amur tiger ranges in China are secondary or planted forests, which are expected to be especially sensitive to the environmental disturbance (Lugo, 2010). Thus, thresholds should be further studied to learn about how much change in climate would result in forest degradation, then cause a domino effect, endangering the Amur tiger.

For animals, rapid temporal shifts in temperature or other environmental conditions may result in changes in daily activity patterns to match the energetic costs of different conditions (Bellard et al., 2012). Ecologists have also begun to explore the effects of climate change on life-history attributes of animals, including the timing of reproduction, reproductive success, the timing of autumn migration (Visser, 1998). Other studies have looked at effects on geographical distributions, species interactions (Both and Visser, 2001), and population dynamics (Pounds et al., 1999). In a meta- analysis, Root et al. (2003) concluded that on average animals and plants showed an average of 5.1 days earlier timing per decade in key phenological events, responding to climate change during the past 50 years. Climate change may also increase asynchrony in predator-prey systems (Parmesan, 2006), and then may lead to local extinction of the predator. The influence of climate change on Amur tiger could be direct, or indirect. On the one hand, long-term evolution helps Amur tiger adapt to the cold climate, effect of climate change on other animals like the changes on activity pattern, reproduction time, reproductive success, geographical distributions, species interactions is likely would happen to Amur tiger. On the other hand, prey condition directly determines the survival and distribution of predators, variation of distributions, activity patterns, life-history attributes of prey would obviously affect the survival and activity. Thus climate-driven threshold effects could be complex for Amur tigers.

Extreme climate events may also play an important role in driving ecosystems or wildlife populations to shift state(s) (Parker et al., 1984; Visser, 1998). A small amount of evidence suggests that ecosystems with large predators may be particularly vulnerable to shifts resulting from extreme events. Extreme weather may destroy germinating seeds, reducing forest regeneration (Bradford, 2005), and potentially reducing the production of plants or creating a mismatch between peak food availability and peak food demands of the animals (Visser, 1998). Extreme weather like snowstorms also may cause direct death of ungulates. It was reported that a snow depth of 40%–60% of the chest height of an ungulate greatly increased the energy cost of activity, and could threaten winter survival (Parker et al., 1984). Ungulate deaths reduced the availability of prey for large carnivores and might eventually cause a shift in the predator's geographical distribution (Serrouya et al., 2015). While the potential effects of these thresholds on Amur tigers remain unknown.

It has been predicted that northern China in winter will experience greater climate warming than other areas, and precipitation will increase in the northern region of northeast China but decrease in the southern part of the region (Qin et al., 2005). Threshold researches on effects of climate change and weather events on vegetation, ungulate populations and Amur tiger themselves could help the managers develop preventive and emergency measures in advance to cope with the impact of climate change and extreme weather on tigers and the whole ecosystem.

11.2.2 Thresholds due to habitat loss and anthropogenic disturbance

Many species have highly specific ecological and environmental requirements. A specific range of conditions is needed for the daily activities of most mammal species, like food gathering and mating. In particular, top predators need a large and relatively stable ranges, with a sufficient prey population for their survival (Powell, 2000). For Amur tigers, GPS collar data showed that the average fixed kernel range of adult female Amur tigers was as large as $401km^2$ ($401km^2 \pm 205km^2$; Hernandez-Blanco et al., 2015). The core area of an adult female Amur leopard was estimated to be $23.3km^2$ (Rozhnov et al., 2015). Males usually had larger ranges both for Amur tigers and leopards. Learning about the minimum habitat requirement for the persistence of these two large carnivore species is the first step to design a national park to maintain the Amur tiger and leopard populations.

In general, the expansion of human activities increased habitat loss and imposed pressure on remaining habitat (Fahrig, 2001), making the identification of relationship between patch (or habitat) size and population persistence in wildlife conservation more complex. On the one hand, human activities may result in habitat fragmentation or loss directly, which may represent principal contemporary threat to species persistence, and are important determinants of the individual spreading in landscape and dispersal (Fahrig, 2001). Landscapes with highly fragmented habitat also show obvious and marked changes in the pattern of the occurrence of a species in the remnants of suitable habitat (Andren, 1994). On the other hand, wild animals often leave areas with human disturbance to engage in normal activities elsewhere. For instance, it was reported that red deer normally shy away from human facilities within a certain distance, and they preferred areas that were more than 8.2km from villages and avoided forest roads and abandoned roads within 1.6km and 2.2km (Jiang et al., 2007). Jiang et al. (2014) also found that the presence of the Amur tiger was seriously limited by railways built within 15 km, and the lowest relative probability of tiger presence occurred in a pixel of area 196km^2 which contained over 25% farmland in Wandashan region in Heilongjiang province (Jiang et al., 2014). The avoidance of human disturbance results in habitat loss and consequently could be more impactful for survival than direct killing off the wild animals (Jaeger et al., 2005). Thus, habitat requirement study on Amur tiger have to take human disturbance into consideration. Estimate a habitat loss or fragmentation threshold is also crucial in designing a national park to maintain the Amur tiger populations.

Besides habitat loss, human activities affect all other aspects of the ecosystem at unprecedented rates (Leemans and Groot, 2003). For Amur tiger conservation, human activities would not only destroy the habitat, but also be lethal factors directly or indirectly. Human hunting was the main reason leading Amur tiger to become endangered in China (Zhou, 2008). The survival of adult female Amur tigers was greatly influenced by the size and type of roads (Kerley et al., 2002). Livestock damage vegetation and compete with ungulates, decreased the ungulates population and then threaten the survive of local Amur tiger (Li et al., 2016). Although hunting in China was forbidden, to manage human activities and reduce human-animal conflict, we must better quantify the threshold values of the frequency, intensity of various human disturbances that the animals, or ecological system itself, can endure, including livestock density, vehicle frequency, and so on.

Many researchers and managers have tried to identify the ecological thresholds which indicated the maximum or minimum tolerance of focal wildlife or ecosystems to human disturbance, in order to better manage human activities (Spooner and Allcock, 2006; Rodrigues et al., 2016). The attempts to use ecological thresholds to guide the frequency and intensity of human disturbances in rangeland and woodland management to keep systems healthy and utilization sustainable provide good examples (Spooner and Allcock, 2006). Species vary in their sensitivity to human activities. Some generalist species were able to tolerate 90% habitat loss, whereas specialists could only survive in landscapes with 60% loss (Rompré et al., 2010). Our current research analyzed 3428 historical records of tigers from A. D. 218 to 2015 revealed the existence of threshold effect of human population density on tiger extinction probability, the result showed that once human population density exceeded 400persons/km², the tigers suffered an local extinction of over 60% within a period of 50 years (unpublished data). Such basic information is very important for Amur tiger conservation in China.

11.2.3 Thresholds of vegetation

The importance of vegetation (or plants) to animals is not only reflected in the food supply, but also affects the availability of thermal (heat energy) and cover (hiding) (Reimoser and Gossow, 1996). Hence, vegetation could act as driver of threshold effects through various mechanisms. Effects of vegetation on herbivores and omnivores often happened when the quality and quantity of vegetation as food were too poor to support their survival. Alternately if vegetation quality or quantity falls below a minimum dietary requirement, the consumers may cease to forage in an area, as described by the idea of a giving-up density (Brown, 1988). The effects of vegetation on herbivores as food supply would decide prey abundance of predator then influence the survival of large carnivores indirectly through bottom-up effects. Additionally, it was suggested that herbivore diversity was often greatly correlated with plant diversity, and this relationship also may be nonlinear (Siemann *et al.*, 1998), and a more diverse consumer group is more conducive to the coexistence of Amur tiger (Richman and Lovvorn, 2009).

Providing shelters for animals is another key function of vegetation, and the number and quality of shelters the vegetation could provide often shows nonlinear correlation with animal occurrence. Smith and Long (1987) uncovered that the vegetation at a summed diameter at breast height above about $3.14 cm/m^2$ or crown diameters greater than $4.75 cm/m^2$ was required as hiding cover by elk (*Cervus elaphus nelsonii*) and the thermal cover requirement was more restrictive. For many other ungulate species, a shelter effect was also generally measurable and the thresholds for hiding cover availability were smaller (Mysterud and Østbye, 1999). These effort drew attention to the importance of thermal and hiding cover availability for animals to resist the harsh weather and predators. Furthermore, consumers have to balance the predation risk and foraging requirements, the trade-off and choice would then changes the prey's activity pattern and the foot availability (Esparza-Carlos *et al.*, 2016). This relationship would in turn influence the outcomes of vegetation management.

Therefore, vegetation is vital for the survival of prey or herbivores, which is the basis for the survival of tiger. Once ungulates disappeared or left when the vegetation could not support their survival, it's impossible to talk about the conservation of Amur tiger. Vegetation restoration in Amur tiger ranges should attach importance to quantification studies and threshold detection on the influence of vegetation to the biodiversity and abundance of ungulates, and overwintering of individuals in this frigid region. At the same time, the structure of vegetation and its availability as food need to be considered.

11.2.4 Thresholds of prey or herbivores

Beyond the indirect effects of vegetation on predator discussed above, predator-prey relationships have always been an important part of wildlife conservation, especially for large predators (Johnson *et al.*, 2007). Tigers are known to adjust their range sizes according to prey abundance, and their density was related to prey density (Karanth *et al.*, 2004). Similarly, lynx (*Lynx canadensis*) in southwestern Yukon, Canada, expanded their range size from 13.2km^2 to 39.2 km^2 when snowshoe hare (*Lepus americanus*) density declined from 14.7 hares/0.01 km^2 to 0.2 hares/0.01km^2 (Ward and Krebs, 1985). Some lynxes abandoned their home ranges to be nomadic after the density of the snowshoe hare fell below 0.5 hares/0.01km^2 (Ward and Krebs, 1985); this accords with ideas of a threshold giving up density (e.g. Nolet *et al.*, 2006).

Increasing prey population density has been seen as a crucial component of large predator conservation. A study focused on the threshold effect of prey density on the interaction of two prey species, the Cape buffalo (*Syncerus caffer*) and the warthog (*Phacochoerus africanus*), and their predator, the lion (*Panthera leo*) suggested that, to attain co-existence, the buffalo population density should be maintained above half of the habitat's carrying capacity (Ddumba *et al.*, 2012). Miller *et al.* (2014) advised that prey management calls for insight into the minimum density required to ensure energy requirements of predators for survival and reproduction. The average successful hunt frequency of Amur tigers was one time each 6.5 days, and each individual needed to consume about 8.9kg of prey biomass per day (Miller *et al.*, 2013). A female individual would stop reproduction if the prey population density was less than 0.5/km^2 and prey abundance of 250–500 individuals of ungulates, although such prey numbers may sustain Amur tiger food requirements (Li *et al.*, 2010). Such information indicates that prey density thresholds are important for the conservation of Amur tigers and leopards, and further studies are both challenging and important.

Different to that prey density decides the survival and reproduction of individual predators, the prey diversity plays an important role in affecting the coexistence of different predators (Tucker and Rogers, 2014). Prey of different sizes are available to predators of different body sizes (Dickman, 1988). As Amur tigers are known to prefer wild pig as their main prey, while the roe deer were more likely predated by Amur leopards (Yang et al., 2018a). The difference in prey preference between Amur tiger is thought to reduce competition between these two large carnivores for food resources (Yang et al., 2018a). Moreover, spatial heterogeneity in the distribution of prey with different body sizes also facilitates preys' coexistence with predators (Richman and Lovvorn, 2009), and then contributes to maintaining species diversity and ecosystem stability.

On the other hand, beside the role in affecting predators based on bottom-up effects, the preys, which are often herbivores, also affect vegetation greatly based on top-down effects. An excessive amount of herbivores, higher than the carrying capacity, can destroy the forest understory shrub layer and cause a decline in the abundance and distribution of preferred food plants (Rooney et al., 2004). To maintain biodiversity in a harvested deciduous forest, De Leo and Levin (1997) concluded a population management objective for white-tailed deer was 4 deer/km^2. For maintenance of balsam fir boreal forests in the early successional stage, Tremblay et al. (2006) estimated a compatible deer density was 7.5 deer/km^2 to 15 deer/km^2. More generally, it is expected that there would be a threshold effect of herbivore density for vegetation change, and that changes in vegetation state would influence the herbivore population density, which would in turn affect predator density (Sasaki et al., 2008). Such indirect effects might explain why Qi et al. (2015) found that leopard density would decline once the total biomass of wild pig and roe deer exceeded 100–150kg/0.25km^2. Therefore, prey or herbivore populations should be managed to avoid too much pressure on vegetation, which could hinder the conservation of predators. A comprehensive consideration of preys diversity and abundance, bottom-up and top down effects, makes prey thresholds study complex.

11.2.5 Thresholds in coexistence of the Amur tiger and Amur leopard

For sympatric large carnivores, rapid increase of one species would be expected to intensify competition. Accordingly, as tiger density increased from 3.31/100km^2 to 5.81/100km^2, the diet of leopards showed an obvious shift and population densities declined sharply in some areas in India (Harihar et al., 2011). The Amur tiger and leopard are two sympatric large predators and competition between them would greatly influence the outcomes of conservation efforts. In general, research on the coexistence of predators has focused on potential coexistence mechanisms involving differences in preferred prey species (Andheria et al., 2007), activity patterns (Fedriani et al., 1999) and utilization of space (Palomares et al., 1996). Coexistence of Amur tiger and leopard are also influence by these factors. For example, Amur leopards tend to use more cliffy spatial habitat components to avoid tigers (Jiang *et al.*, 2015), and prey preferences and temporal activity patterns are also distinct (Yang *et al.*, 2018a, 2018b).

Studies in South Africa illustrate the possible consequences of competitive interactions. The minimum area required for 10 cheetah (*Acinonyx jubatus*) varied greatly given different competition pressures. It was 48–466km^2 without other predators, whereas when with equal numbers of lions (*Panthera leo*) it was 166–2806km^2, and when with equal numbers of leopards (*Panthera pardus*), spotted hyenas (*Crocuta crocuta*), wild dogs (*Lycaon pictus*) and lions was 727–3739km^2 (Lindsey *et al.*, 2011). For Amur tiger and leopard in Russia Far East, Pianka's niche overlap index for prey resources was estimated to be as high as 0.77 (Sugimoto *et al.*, 2016). Based on this and what we know about competitive dominance, we would expect that population increases of tiger and leopard would cause the Amur leopard to leave the area. Consequently, managers could take steps to create more leopard preferred micro-habitats or promote the spread of leopard individuals to other areas once tiger population densities increase. Furthermore, if we knew what value the niche overlap index of food would mean that it might be possible for one of these two predators to go extinct regionally, then we could artificially adjust the composition and density of preferred prey in advance to facilitate coexistence.

11.3 Future challenges of threshold research and applications

As illustrated by the foregoing discussion and published literature, threshold responses are often caused by several interacting factors rather than one particular driver (van der Ree *et al.*, 2004; Huggett, 2005). For example, climate change may influence the food availability and then influence the density and distribution of animals (Both *et al.*, 2006). Species interactions are also directly affected by climate change (Tylianakis *et al.*, 2008). Determining threshold range sizes for predator persistence may be misconstrued due to habitat fragmentation, competitors, climate change, and the availability of prey. None of these factors are independent from human activities. The potentially confounding effect of multiple variables interaction produces complex threshold responses making the threshold identification much more difficult (Bennett and Radford, 2003).

Interaction between variables also make the application of thresholds more context dependence. For instance, it would be floundering or even failing to extrapolate the threshold values of minimum habitat area for Amur tiger survival in some area to other landscapes where the frequency of human activities, degree of habitat fragmentation and prey abundance or other factors that may affect the habitat utilization of Amur tigers may greatly different. Because of temporal and spatial variation in these processes and charactcristics, there is not a threshold value is universally applicable (Huggett, 2005). We therefore argue that ecological threshold studies should focus on a specific limited area and that limiting factors should be studied integrally to find those threshold values relevant to each step in conservation and management.

The construction of a national park for tiger and leopard conservation, aiming to restore these two species and the whole ecosystem. It requires a huge investment of money, and natural and human resources, the ongoing enhancement and expansion of human activities and huge costs raise more challenges. What fortunately is that threshold method could be an appropriate tool to realize efficient conservation with limited resources under some conditions. For instance, knowing the minimum forest habitat area required to ensure the survival and coexistence of Amur tigers and leopards could avoid to allocate too much land for conservation which means more migration, and more resource extraction forbidden? Controlling human disturbances under the threshold value in key habitat patches and corridors areas to maintain an extensively permeable habitat landscape for the dispersal of such large predators would be better than total ban considering local residents' support for conservation (Chapron et al., 2014). Furthermore, threshold knowledge provides quantitative indicators for developing reasonable objectives and measures to manage the habitat, vegetation and prey population. Amur tiger and leopard national park construction represents a good opportunity to practice such information and methods, knowledge of the mechanisms of biome resilience from threshold study would finally benefit to the construction.

For the research and applications of threshold in Amur tiger conservation as well as other large carnivores species. Firstly, we should have a map of bottom-up effects in their habitat landscape and consider all of the relevant drivers for a particular area and situation. These drivers include climate and weather events, habitat loss and other anthropogenic disturbances, vegetation, prey, and intrigued competition between the two predators. Consideration of both direct and indirect effects is important, such as that vegetation will not only provide shelter for predators directly but also provide food for prey survival, which indirectly benefits the predator. Then attempt to identify key factors that could influence, or alter the condition of concerned species. The final step is to detect if these factors, and their interactions, exhibit threshold responses and what the threshold values are (Fig. 11.1). Regardless of the extent to which the ambitious agenda for research and conservation can be conducted, it is clear that applying theory and thinking about thresholds can make conservation policy and practices more complete. Threshold research should be a regular content of the national park construction and be executed throughout the Amur tiger conservation.

Chapter 11　Ecological threshold on precise management of ... | 345

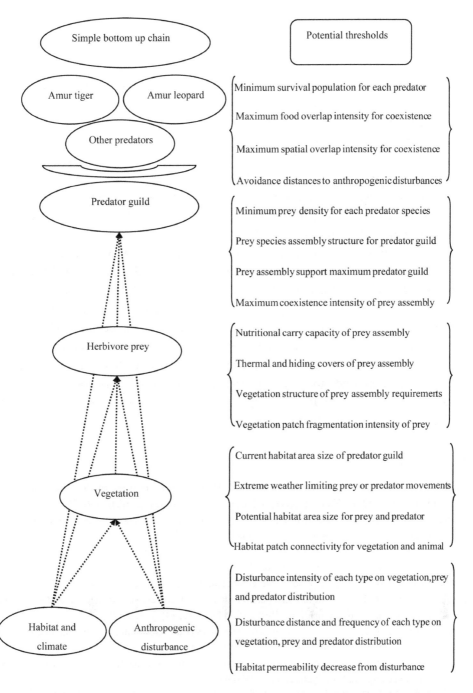

Fig. 11.1　Some potential thresholds were summarized along this simple bottom-up chain for top predators in ecosystem.

11.4 Summary

The ecological threshold describes how small changes in the value of one or more factors can result in a complete transformation of an ecosystem. This concept focuses us on the study of limiting factors that reflect the tolerance of systems or organisms to changes in them. Increasing empirical evidence for the existence of ecological thresholds has created favorable conditions for their practical application in wildlife conservation. Applying the ecological thresholds concept has the potential to enhance conservation of two large carnivores, Amur tiger, through guiding construction of a new national park. In this part, the relevant forms of ecological thresholds resulting from a paradigm of bottom-up control were evaluated for their potential to contribute to the conservation of Amur tiger, to show the great potential of applying ecological threshold into Amur tigers and leopards conservation. It should be noted that, large carnivores as species at the top of the food chain, ecological thresholds often work on them through an upward effect, possible divers include climate and weather events, habitat, anthropogenic disturbances, vegetation, prey, and competitors. What's more, interactions between factors and context dependence are 2 important precondition have to be considered in threshold research and conservation practice. Application of the thresholds concept leads to a more thorough evaluation of conservation needs, and could be used to guide future Amur tiger research and conservation efforts in China. Such application may inform the conservation of other.

References

Andheria A P, Karanth K U, Kumar N S. 2007. Diet and prey profiles of three sympatric large carnivores in Bandipur Tiger Reserve, India. Journal of Zoology, 273(2): 169-175.

Andren H. 1994. Effects of habitat fragmentation on birds and mammals in landscapes with different proportions of suitable habitat: a review. Oikos, 71(3): 355-366.

Bellard C, Bertelsmeier C, Leadley P, et al. 2012. Impacts of climate change on the future of biodiversity. Ecology Letters, 15(4): 365-377.

Bennett A F, Radford J Q. 2003. Know your ecological thresholds. Thinking Bush, 2: 1-3.

Bestelmeyer B T. 2006. Threshold concepts and their use in rangeland management and restoration: the good, the bad, and the insidious. Restoration Ecology, 14(3): 325-329.

Both C, Bouwhuis S, Lessells C M, et al. 2006. Climate change and population declines in a long-distance migratory bird. Nature, 441(7089): 81-83.

Both C, Visser M E. 2001. Adjustment to climate change is constrained by arrival date in a long-distance migrant bird. Nature, 411: 296-298.

Bradford K J. 2005. Threshold models applied to seed germination ecology. New Phytologist, 165(2): 338-341.

Briske D D, Fuhlendorf S D, and Smeins F E. 2006. A unified framework for assessment and application of ecological thresholds. Rangeland Ecology and Management, 59(3): 225-236.

Brown J S. 1988. Patch use as an indicator of habitat preference, predation risk, and competition. Behavioral ecology and sociobiology, 22(1): 37-47.

Cavanaugh K C, Kellner J R, Forde A J, et al. 2014. Poleward expansion of mangroves is a threshold response to decreased frequency of extreme cold events. Proceedings of the National Academy of Sciences of the united states of America, 111(2): 723-727.

Chapron G, Kaczensky P, Linnel J D, et al. 2014. Recovery of large carnivores in Europe's modern human-dominated landscapes. Science, 346: 1517-1519.

Choquenot D, Parkes J. 2001. Setting thresholds for pest control: how does pest density affect resource viability? Biological Conservation, 99(1): 29-46.

Cox P M, Betts R A, Collins M, et al. 2004, Amazonian forest dieback under climate-carbon cycle projections for the 21st century. Theoretical and Applied Climatology, 78(1-3): 137-156.

Ddumba H, Mugisha J Y T, Gonsalves J W, et al. 2012. The role of predator fertility and prey threshold bounds on the global and local dynamics of a predator–prey model with a prey out-flux dilution effect. Applied Mathematics and Computation, 218(18): 9169-9186.

De Leo, G A, Levin S. 1997. The multifaceted aspects of ecosystem integrity. Conservation Ecology, 1(1): 3.

Demenocal P, Ortiz J, Guilderson T, et al. 2000. Abrupt onset and termination of the african humid period: rapid climate responses to gradual insolation forcing. Quaternary Science Reviews, 19(1-5): 347-361.

Dickman C R. 1988. Body size, prey size, and community structure in insectivorous mammals. Ecology, 69(3): 569-580.

Doughty C E, Metcalfe D B, Girardin C A J, et al. 2015. Drought impact on forest carbon dynamics and fluxes in Amazonia. Nature, 519(7541): 78-82.

Esparza-Carlos J P, Laundré J W, Hernández L, et al. 2016. Apprehension affecting foraging patterns and landscape use of mule deer in arid environments. Mammalian Biology-Zeitschrift für Säugetierkunde, 81(6): 543-550.

Fahrig L. 2001. How much habitat is enough? Biological Conservation, 100(1): 65-74.

Fedriani J M, Palomares F, Delibes M. 1999. Niche relations among three sympatric Mediterranean carnivores. Oecologia, 121(1): 138-148.

Harihar A, Pandav B, Goyal S P. 2011. Responses of leopard *Panthera pardus* to the recovery of a tiger *Panthera tigris* population. Journal of Applied Ecology, 48(3): 806-814.

Hernandez-Blanco J A, Naidenko S V, Chistopolova M D, et al. 2015. Social structure and space use of amur tigers (*panthera tigris altaica*) in southern russian far east based on gps telemetry data. Integrative Zoology, 10(4): 365-375.

Huggett A J. 2005. The concept and utility of 'ecological thresholds' in biodiversity conservation. Biological Conservation, 124(3): 301-310.

Jaeger J A G, Bowman J, Brennan J, et al. 2005. Predicting when animal populations are at risk from roads: an interactive model of road avoidance behavior. Ecological Modelling, 185(2): 329-348.

Jiang G, Qi J, Wang G, et al. 2015. New hope for the survival of the amur leopard in China. Scientific Reports, 5: 15475.

Jiang G, Sun H, Lang J, et al. 2014. Effects of environmental and anthropogenic drivers on Amur tiger distribution in northeastern China. Ecological Research, 29(5): 801-813.

Jiang G, Zhang M, Ma J. 2007. Effects of human disturbance on movement, foraging and bed selection in red deer *Cervus elaphus xanthopygus* from the Wandashan Mountains, Northeastern China. Acta Theriologica, 52(4): 435-446.

Johnson C N, Isaac J L, Fisher D O. 2007. Rarity of a top predator triggers continent-wide collapse of mammal prey: dingoes and marsupials in Australia. Proceedings of the Royal Society of London B: Biological Sciences, 274(1608): 341-346.

Kaiser-Bunbury C N, Mougal J, Whittington A E, et al. 2017. Ecosystem restoration strengthens pollination network resilience and function. Nature, 542(7640): 223.

Karanth K U, Nichols J D, Kumar N S, et al. 2004. Tigers and their prey: predicting carnivore densities from prey abundance. Proceedings of the National Academy of Sciences of the United States of America, 101(14): 4854-4858.

Kerley L L, Goodrich J M, Miquelle D G, et al. 2002. Effects of roads and human disturbance on amur tigers. Conservation Biology, 16(1): 97-108.

Leemans H B J, Groot R S D. 2003. Millennium ecosystem assessment: ecosystems and human well-being: a framework for assessment. Physics Teacher, 34(9): 534.

Li C, Yuan Z. 2017. Research development and application in ecological threshold. Journal of Hebei Forestry Science and Technology, 3: 54-57. (in Chinese)

Li G. 2014. "ecological protection red line"—the lifeline to ensure the national ecological security. Qiushi, 2: 44-46. (in Chinese)

Li Z, Kang A, Gu J, et al. 2016. Effects of human disturbance on vegetation, prey and Amur tigers in Hunchun Nature Reserve, China. Ecological Modelling, 353: 28-36.

Li Z, Zimmermann F, Hebblewhite M, et al. 2010. Study on the potential tiger habitat in the Changbaishan area, China. Beijing: China Forestry Publishing House. (in Chinese)

Lindsey P, Tambling C J, Brummer R, et al. 2011. Minimum prey and area requirements of the vulnerable cheetah *Acinonyx jubatus*: implications for reintroduction and management of the species in South Africa. Oryx, 45(4): 587-599.

Liu F, Mcshea W J, Garshelis D L, et al. 2011. Human-wildlife conflicts influence attitudes but not necessarily behaviors: factors driving the poaching of bears in china. Biological Conservation, 144(1): 538-547.

Lugo A E. 2010. Visible and invisible effects of hurricanes on forest ecosystems: an international review. Austral Ecology, 33(4): 368-398.

Madden F M. 2008. The growing conflict between humans and wildlife: law and policy as contributing and mitigating factors. Journal of International Wildlife Law and Policy, 11: 189-206.

May R M. 1977. Thresholds and breakpoints in ecosystems with a multiplicity of stable states. Nature, 269(5628): 471-477.

Miller C S, Hebblewhite M, Petrunenko Y K, et al. 2013. Estimating amur tiger (*panthera tigris altaica*) kill rates and potential consumption rates using global positioning system collars. Journal of Mammalogy, 94(4): 845-855.

Miller C S, Hebblewhite M, Petrunenko Y K, et al. 2014. Amur tiger (*Panthera tigris altaica*) energetic requirements: implications for conserving wild tigers. Biological Conservation, 170: 120-129.

Mysterud A, Østbye E. 1999. Cover as a habitat element for temperate ungulates: effects on habitat selection and demography. Wildlife Society Bulletin, 27(2): 385-394.

Nolet B A, Fuld V N, Rijswijk M E C, et al. 2006. Foraging costs and accessibility as determinants of giving-up densities in a swan-pondweed system. Oikos, 112(2): 353-362.

Palomares F, Ferreras P, Fedriani J M, et al. 1996. Spatial relationships between Iberian lynx and other carnivores in an area of south-western Spain. Journal of Applied Ecology, 33(1): 5-13.

Parker K L, Robbins C T, Hanley T A. 1984. Energy expenditures for locomotion by mule deer and elk. The Journal of Wildlife Management, 48(2): 474-488.

Parmesan C. 2006. Ecological and evolutionary responses to recent climate change. Annual Review of Ecology Evolution and Systematics, 37(1): 637-669.

Pounds J A, Fogden M P L, Campbell J H. 1999. Biological response to climate change on a tropical mountain. Nature, 398(6728): 611-615.

Powell R A. 2000. Animal home ranges and territories and home range estimators // Research Techniques in Animal Ecology: Controversies and Consequences. Cambridge: Colombia University Press: 64-110.

Qi J, Shi Q, Wang G, et al. 2015. Spatial distribution drivers of Amur leopard density in northeast China. Biological Conservation, 191: 258-265.

Qin D, Ding Y, Su J. 2006. Assessment of climate and environment changes in china (i): climate and environment changes in china and their projection. Advances in climate change research,1(1): 4-9. (in Chinese)

Reimoser F and Gossow H. 1996. Impact of ungulates on forest vegetation and its dependence on the silvicultural system. Forest Ecology and Management, 88(1): 107-119.

Richman S E, Lovvorn J R. 2009. Predator size, prey size and threshold food densities of diving ducks: does a common prey base support fewer large animals? Journal of Animal Ecology, 78(5): 1033-1042.

Ripple W J, Estes J A, Beschta R L, et al. 2014. Status and ecological effects of the world's largest carnivores. Science, 343(6167): 1241484.

Rodrigues M E, Roqueb F O, Quintero J M O, et al. 2016. Nonlinear responses in damselfly community along a gradient of habitat loss in a savanna landscape. Biol Conserv, 194: 113-120.

Marciel E R, Fabio de O R, Jose M O Q, et al. 2016. Nonlinear responses in damselfly community along a gradient of habitat loss in a savanna landscape. Biological Conservation, 194: 113-120.

Rompré G, Boucher Y, Bélanger L, et al. 2010. Conserving biodiversity in managed forest landscapes: the use of critical thresholds for habitat. Forestry Chronicle, 86(5): 589-596.

Rooney T P, Wiegmann S M, Rogers D A, et al. 2004. Biotic impoverishment and homogenization in unfragmented forest understory communities. Conservation Biology, 18(3): 787-798.

Root B A, Price J T, Hall K R, et al. 2003. Fingerprints of global warming on wild animals and plants. Nature, 421: 47-60.

Rozhnov V V, Chistopolova M D, Lukarevskii V S, et al. 2015. Home range structure and space use of a female amur leopard, *panthera pardus orientalis*, (carnivora, felidae). Biology Bulletin, 42(9): 821-830.

Sasaki T, Okayasu T, Jamsran U, et al. (2008) Threshold changes in vegetation along a grazing gradient in Mongolian rangelands. Journal of Ecology, 96(1): 145-154.

Scheffer M, Carpenter S R. 2003. Catastrophic regime shifts in ecosystems: linking theory to observation. Trends in Ecology and Evolution, 18(12): 648-656.

Seddon A W R, Macias-Fauria M, Long P R, et al. 2016. Sensitivity of global terrestrial ecosystems to climate variability. Nature, 531(7593): 229-232.

Serrouya R, Wittmann M J, McLellan, B N. et al. 2015. Using predator-prey theory to predict outcomes of broadscale experiments to reduce apparent competition. The American Naturalist, 185(5): 665-679.

Siemann E, Tilman D, Haarstad J, et al. 1998. Experimental tests of the dependence of arthropod diversity on plant diversity. The American Naturalist, 152: 738-750.

Smith F W, Long J N. 1987. Elk hiding and thermal cover guidelines in the context of lodgepole pine stand density. Western Journal of Applied Forestry, 2(1): 6-10.

Spooner P G, Allcock K G. 2006. Using a state-and-transition approach to manage endangered

Eucalyptus albens (White Box) woodlands. Environmental Management, 38(5): 771-783.
Sugimoto T, Aramilev V V, Nagata J, *et al.* 2016. Winter food habits of sympatric carnivores, Amur tigers and Far Eastern leopards, in the Russian Far East. Mammalian Biology-Zeitschrift für Säugetierkunde, 81(2): 214-218.
Tremblay J P, Huot J, Potvin F. 2006. Divergent nonlinear responses of the boreal forest field layer along an experimental gradient of deer densities. Oecologia, 150(1): 78-88.
Tucker M A, Rogers T L. 2014. Examining the prey mass of terrestrial and aquatic carnivorous mammals: minimum, maximum and range. PLoS ONE, 9(8): e106402.
Tylianakis J M, Didham R K, Bascompte J, *et al.* 2008. Global change and species interactions in terrestrial ecosystems. Ecology Letters, 11(12): 1351-1363.
van der Ree R, Bennett A F, Gilmore D C. 2004. Gap-crossing by gliding marsupials: thresholds for use of isolated woodland patches in an agricultural landscape. Biological Conservation, 115(2): 241-249.
Visser M. 1998. Warmer springs lead to mistimed reproduction in great tits (*Parus major*). Proceedings of the Royal Society of London B: Biological Sciences, 265(1408): 1867-1870.
Vogt N D, Pinedo-Vasquez M, BrondãZio E S, *et al.* 2015. Forest transitions in mosaic landscapes: smallholder's flexibility in land-resource use decisions and livelihood strategies from World War II to the present in the amazon estuary. Society and Natural Resources, 28(10): 1043-1058.
Ward R M P, Krebs C J. 1985. Behavioural responses of lynx to declining snowshoe hare abundance. Canadian Journal of Zoology, 63(12): 2817-2824.
Yang H, Dou H, Baniya R K, *et al.* 2018a. Seasonal food habits and prey selection of Amur tigers and Amur leopards in Northeast China. Scientific Reports, 8(1): 6930.
Yang H, Zhao X, Han B, *et al.* 2018b. Spatiotemporal patterns of Amur leopards in northeast china: influence of tigers, prey, and humans. Mammalian Biology, 92: 120-128.
Zhou X. 2008. Protection status of wild Amur tiger (*panthera tigris altaica*) and their conservation strategy. Chinese Journal of Wildlife, 29(1): 40-43. (in Chinese).

Chapter 12 Challenges, opportunities, and measures for precise protection and management of the Amur tiger

12.1 Introduction

China is the birthplace of tigers in the world, and the country with the largest distribution of tiger subspecies and leopard subspecies in the world. The health and stability of the tiger and leopard population reflects the integrity of the food chain and ecological processes, which means that the specific natural ecosystems remain authentic and complete, indicating that the entire natural ecosystem is healthy and stable and the ecological services are functioning normally. In the 1950s, there were still nearly 200 wild Amur tigers in China which were widely distributed in the forest area of northeastern China. At that time nevertheless, there were only 30-40 individuals in Russia. In the late 1990s, the number of individuals distributed in China was reduced to roughly less than 20 (Jiang *et al.*, 2017), and the number in Russian populations increased to nearly 500 through protection and restoration.

Fortunately, due to the implementation of the natural forest protection project, the habitat of the Amur tiger and leopard in China has gradually recovered, and the population has also steadily increased. Despite this, the two species still face the danger of extinction. The main threats are: ① the habitat is fragmented and the population is isolated, especially the only two Amur tiger populations in the world are isolated in Russia. Gene flow between the two populations is difficult; ② Due to the obstacles of artificial landscapes such as highways, railways, villages and farmland, there is no effective connection between Wandashan, Laoyeling and Zhangguangcailing populations. However, the Amur tiger population density in the Hunchun-Wangqing- Suiyang region in the south of Laoyeling near the Sino-Russia border is relatively high, though this population faces the risk of inbreeding, disease outbreaks and even population collapse; ③ Remaining impact of historical forest harvesting, and current forest grazing, collection and development of non-timber forest products and other anthropogenic activities are still intense; ④ Due to the need to revitalize the Northeast economy and national defense needs, border fences, road construction (such as railways and highways, etc.), farmland development and mining are still developing rapidly; ⑤ In the distribution area of the Amur tiger and leopard, the types of prey are incomplete and unevenly distributed, and density of prey that Amur tiger and leopard prefer is relatively low; ⑥ Illegal hunting still exists in some areas. In general, from the perspective of the large, landscape scale, barriers to diffusion are large and the forest landscape is not permeable; these hinder free migration and recovery of Amur tiger and leopard. In regards to human-animal relationships, there are still contradictions between the recovery of these two species and the needs of local economic development(Ma *et al.*, 2015).

As top predators in the forest ecosystem, the Amur tiger and leopard have a wide range of activities, and their protection and recovery involves habitat restoration and connectivity, prey population monitoring and recovery, poaching control, and human-tiger conflict resolution, etc. The management and technical issues of addressing these problems are complex, and the help and advancement of international cooperation is also desperately needed, but there is a long way yet to go. The following three areas of work are essential.

12.2 Top-level design and unified planning

The protection of the Amur tiger and leopard requires the establishment of a network of protected patch sites, the construction of international and domestic corridors, and the formation of a permeable landscape.

Amur tigers in China are currently mainly distributed in the three areas of Laoyeling, Zhangguangcailing and Wandashan (Fig. 12.1). China has established Jilin Hunchun Northeast Tiger National Nature Reserve, Heilongjiang Laoyeling Northeast Tiger National Nature Reserve, Jilin Wangqing National Nature Reserve, Heilongjiang Muling Yew National Nature Reserve, Heilongjiang Niaoqingshan Provincial Nature Reserve, and Jilin Tianqiaoling Tiger Provincial Nature Reserve etc. in the Laoyeling area, Heilongjiang Dongfang Red Wetland National Nature Reserve, Heilongjiang Dajiahe Provincial Nature Reserve, Heilongjiang Qixinglazi Tiger National Nature Reserve etc. in the Wandashan area, and Jilin Huangnihe Nature Reserve and Heilongjiang Xiaobeihu National Nature Reserve in the Zhangguangcailing area. These nature reserves cover the core areas where the Amur tiger and leopard occur most. Within the matrix of these three major regions, the Heilongjiang Dongjingcheng Northeast Tiger Nature Reserve is under construction at the junction of Laoyeling and Zhangguangcailing, further consolidating the habitat of the Chinese tiger population while providing an opportunity for the species to spread northward via Zhangguangailing. The habitat restoration of Heilongjiang Qixinglazi National Nature Reserve and its surrounding forestry bureaus in western Wandashan will also effectively facilitate the spread of the tiger population in eastern Wandashan to the west, and then connect the Laoyeling population through Zhangguangcailing. Therefore, the framework for the network of nature reserves in the Amur tiger and leopard distribution range has basically taken shape.

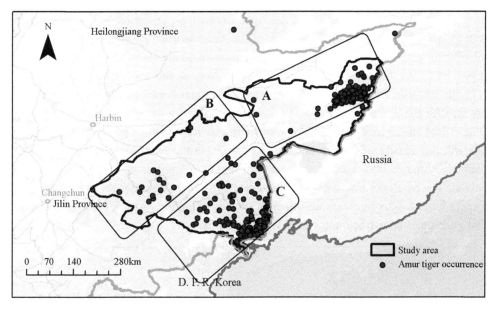

Fig. 12.1 Map of geographical location: A. Wandashan; B. Zhangguangcailing; C. Laoyeling. Please scan the QR code on the back cover to see the color image.

Although the total habitat of Amur tigers in China has reached 137 085km^2, the number of individuals recorded in China has only been near to 40 in the past five years, and the vast majority of individuals are distributed in the Sino-Russian border area of Laoyeling.

The wild Amur tigers in the world mainly survive in two isolated areas. One is the Sikhote-Alin mountain range in Russia and the Wandashan area in China connected to it. The number of Amur tigers in this area accounts for 90% of the entire population (about 500). The other area is the southwestern part of Primorskii Krai in Russia and the adjacent Laoyeling in China. This second, isolated small population of Amur tigers (about 40 individuals) is the main source of the Amur tiger population in China (Fig. 12.2). These two regions have been separated in Russia for a long time due to urban development and wetland areas(Henry et al., 2010). Related studies have shown that this small isolated population in the south has problems such as inbreeding, environmental capacity and susceptibility to disease, thus facing risk of extinction(Miquelle et al., 2015). Since the southeastern part of the southwestern Primorsky Krai is the Sea of Japan, when this small population increases, it can only spread to the hinterland of China, and it can only connect with the northern Russia population (about 500 individuals) through the Wandashan area, Heilongjiang, China (Fig. 12.2). The Northeast Tiger and Leopard National Park (hereinafter NTLNP) covers the core area of the Amur tiger in the Laoyeling distribution area, effectively protecting the small isolated population connected to Russia (Fig. 12.2). Its establishment effectively opened the border corridor and sheltered the Amur tiger's breeding population. The national park is of great significance for the protection of the Amur leopard population, which is only distributed in the southwestern region of Primorskii Kraiin Russia and the Laoyeling region of China. But this does not mean that as long as the tigers and leopards in the NTLNP are protected, China's Amur tigers and leopards can be effectively restored, because for the Amur tiger, the NTLNP only accounts for 11% of its total distribution (Fig. 12.2) and is not connected to the northern Russia population. For the Amur leopard, the density of the Amur tiger population in the National Park also limits its recovery through inter-species competition, and the Amur leopard also needs more habitat. The results of international expert research show that the most effective way to protect large cats is to create an effective connected and permeable habitat landscape that guarantees their free movement(Wikramanayake *et al.*, 2011).

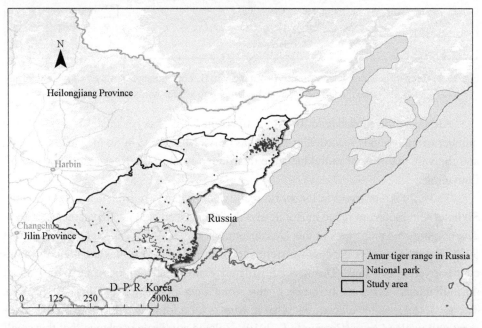

Fig. 12.2 Map of Amur tiger global distribution (note: red dots denote Amur tiger presence since 2000). Please scan the QR code on the back cover to see the color image.

When we put our attention to the entire tiger and leopard distribution area in China, not just the NTLNP, it is natural to realize that the more important protection measures should be to consider how to strengthen the connection of habitat patches in China (Laoyeling, Zhangguangcailing, Wandashan and even Lesser Khingan), and especially how the Amur tiger population in the Laoyeling area is to be connected to the large population of northern Russia. After the large population of northern Russia spreads into China, it is necessary to open the passage from the eastern Wandashan to the western

Wandashan, Wandashan to Zhangguangcailing, and Zhangguangcailing to Laoyeling (Fig. 12.3), thus China's main Amur tiger population can be connected with the large population of northern Russia. However, after years of monitoring, it was found that due to the Jianji highway and large farmland on both sides of this road, individuals from large populations in northern Russia entered the eastern part of Wandashan and

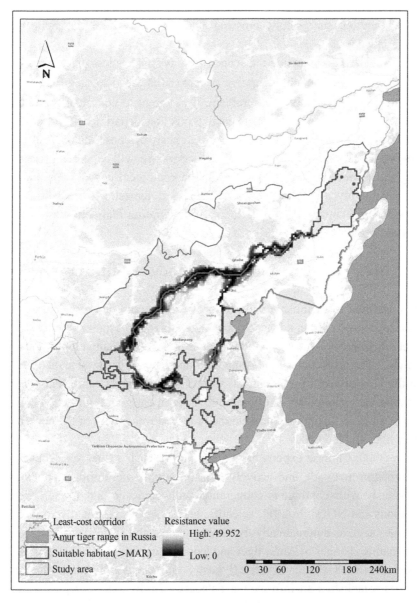

Fig 12.3 Core habitats and corridors of the Amur tiger in China. Please scan the QR code on the back cover to see the color image.

encountered serious obstacles when they intended to spread to western Wandashan. In September 2017, a male adult Amur tiger wandered on the east side of the Jianji highway for half a month before finding a small strip of forested area in which to cross the highway reaching Huanan forest area located in the western part of Wandashan. In November 2017, another adult male Amur tiger from Russia coming to the east of Wandashan tried to spread to the west of Wandashan, but returned back to eastern Wandashan without successfully crossing Jianji highway after traversing along the fence for a few days.

At present, the area outside the scope of the NTLNP is less protected and restricted by factors such as the administrative system—it is easier to ignore the problem of habitat connectivity. To solve this problem, it is urgent to coordinate administrations like transportation, forestry, environmental protection departments and others at the national level to unify top-level design. On the basis of consolidating and improving the network of national parks and nature reserves, we will vigorously carry out the planning and project of the construction of the ecological corridor for the tiger and leopard. This involves opening the international and domestic corridors and establishhing a permeable landscape to ensure the free migration and activities of tigers and leopards in large-scale habitats.

12.3 Expert consultation and scientific norms

It is essential to establish a solid team of experts, and conservation and management needs to be scientific, advanced, and standardized to achieve internationalization of conservation technologies. "Who does not seek the overall situation is not enough to seek a field." The protection of the Amur tiger and leopard is a systematic project. Only unified thinking of central government, scientific planning of competent departments, coordination and participation of multiple departments, and full utilization of the strength of an expert team can promote the orderly development of the conservation of the Amur tiger and leopard.

Since 2000, Chinese experts have begun to establish a tiger and leopard monitoring information network, and actively train grassroots monitoring technology teams simultaneously with continuous cooperation with domestic and foreign government organizations and NGOs. In 2011, scientific teams began to gradually apply advanced technologies such as automatically-triggered cameras and molecular genetics to monitor the dynamic changes of Amur tiger and leopard populations and habitats. At present, the Feline Research Center of State Forestry and Grassland Administration (FRC-SFGA) has established an information database of information- reporting, an image database, a DNA database, a digital footprint database and a disease and parasitism database for Amur tiger sand leopards, and also a database for prey populations and distribution. From 2000 to 2016, FRC-SFGA recorded 4028 tiger presence sites of predation, footprints, images, faeces and other information. The spatial distribution information of species and individuals indicated that the Amur tiger distribution area is expanding steadily, while the Amur tiger has shown a similar trend. These data and achievements are of great significance for the planning and design of the NTLNP, and are an important basis for the design of the national tiger and leopard protection project.

Conservation of the Amur tiger and leopard involves a combination of ecological roads, eco-friendly forests, ecological compensation, ecological corridors, etc. It also requires participation of experts in sociology, economics, architecture, transportation, forestry, geographic information and others (Jiang *et al.*, 2017). The staged achievements of protection are inseparable from the expert teams in various fields, and interdisciplinary research is needed in the future. Therefore, different industries and departments must break constrains of the system, unify thinking, plan scientifically, work together, forge ahead steadily, and actively mobilize society's attention to participate in this protection career. Then people can share ecological rewards brought by the protection of the tiger and leopard.

In addition, due to the distribution of the Amur tiger and leopard across China and Russia, protection and management needs to be internationalized and standardized; there is a special need for deep cooperation with experts of tiger and leopard distribution countries like Russia. For example, with the approval of the former State Forestry Administration and the support of the East Asia Project of the United Nations Asia-Pacific Economic Cooperation Committee, China and Russia jointly launched the Amur tiger and leopard's cross-border research project, which established a good relationship for the joint protection research between the two countries. The international expert team platform also provides technical support for upgrading the protection and management of the NTLNP to an international level (Shevtsova *et al.*, 2018).

In short, the research and protection of tigers and leopards needs a stable scientific research team. While accumulating ecological data, it is necessary to support construction of national key disciplines for wildlife conservation in colleges and universities. By doing this, it can continuously cultivate and deploy professionals for wildlife conservation and management departments. This can ensure that management departments manage wildlife according to scientific management methods and ecological law.

12.4 Innovative measures and human-tiger harmony

We present the need for carrying out refined assessments to explore the survival needs of tigers and leopards, and promoting the harmonious coexistence of people and tigers and leopards to achieve ecological security. At present, the population status of Amur tiger and leopard in China has been basically clarified, and the management work has initially been rationalized. We urgently need to conduct a scientific and systematic assessment of the population structure and habitat quality of the Amur tiger and leopard to promote a healthy and sustained population growth. In view of the main threats facing the current tiger and leopard population, we will focus on the following aspects:

First, it is necessary to quantify the true ecological needs of the Amur tiger and leopard to ensure the protective measures are scientific and effective. For example, to understand how to control the intensity of human disturbance in tiger and leopard habitat, we must first know how strong the human disturbance intensity is and its relationships with negative impacts on tigers, leopards and their prey. As another example, studies have shown that the minimum prey density requirement for an Amur tiger to breed and raise young is not less than 0.5 ungulate prey individuals per square kilometer(Li et al., 2010); so how many areas in the tiger and leopard distribution area are not up to this standard, and where are these regions? Protected areas or national parks must know how to control the intensity of human disturbance and the extent of the affected area to ensure that tigers, leopards and prey are not affected. How large is the minimum area requirements of tigers and leopards? Where are the domestic and international corridors, and what is the width of corridor which enables the spread of tigers and leopards from the Sino-Russia boarder to China'sinterior? How do we build these corridors? To solve these problems needs a true understanding of the ecological needs of the Amur tiger and leopard.

Secondly, on the basis of solving the above management problems and technical standards problems, we must also consider how to effectively protect tigers and leopards while taking into account the needs of economic development. It is necessary to carry out refined assessment inside and outside the protection area within the tiger and leopard distribution, then adopt different habitat and prey restoration measures for different threats. We must find a sustainable way to develop the economy and ensure local residents benefit from the tiger and leopard protection project, thus mobilising their enthusiasm for protection and appropriate ecological economic activities.

Thirdly, there is a need for an innovative monitoring system for wildlife, habitat change, and habitat restoration based on management and assessment needs. Although the design of an integrated monitoring system is combined with a variety of modern advanced technologies, the construction of information monitoring facilities in local areas is also necessary. But it is also necessary to pay attention to the negative impact of monitoring measures and monitoring activities on vegetation and wildlife. Monitoring is a means, not an objective. Based on monitoring objectives and needs, in order to ensure safety of tigers, leopards and other wildlife, and even forest ecological security, it should be encouraged to apply or develop monitoring technologies that have no or little damage to the environment and wildlife, and lower noise or radiation caused by human disturbance and artificial facilities (Jewell, 2013).

Finally, we call for studying the establishment of a protected area system with national parks as the main body. National parks are the most important types of protected area, but they are not unique; there needs to be other types of protected areas as supplementary units. Protection and management measures such as nature reserves, forest parks, and natural forest protection can only be strengthened and must not be weakened.

12.5 Summary

In summary, in order to protect the population of Amur tigers and leopards and the natural ecosystems in which they occur, the conservation ecology law of sustainable survival and the development of species population should be followed, and a unified platform for a management system and technological innovation should be established. Using the good opportunities for natural forest bans, shantytown renovation, national park establishment and other projects in Northeast China to innovate protection measures, we propose to establish a large-scale nature protection network and to promote habitat quality for Amur tiger and leopard. Central to this, we must connect the population of Laoyeling, Zhangguangcailing and Wandashan gradually, and realize the harmonious coexistence between humans and tigers and leopards, making the Amur tiger and leopard an important ecological, economic and cultural wealth.

References

Henry P, Miquelle D, Sugimoto T, *et al.* 2010. *In situ* population structure and *ex situ* representation of the endangered Amur tiger. Molecular Ecology, 18(15): 3173-3184.
Jewell Z. 2013. Effect of monitoring technique on quality of conservation science. Conservation Biology the Journal of the Society for Conservation Biology, 27(3): 501-508.
Jiang G, Wang G, Holyoak M, *et al.* 2017. Land sharing and land sparing reveal social and ecological synergy in big cat conservation. Biological Conservation, 211: 142-149.
Li Z, Zimmermann F, Hebblewhite M. 2010 Study on the potential tiger habitat in the Changbaishan area, China. Beijing: China Forestry Publishing House.
Ma J, Zhang M, Jiang G, *et al.* 2015. Chanllenges and Strategies for Conservation of Tigers and Their Habitat in China. Chinese Journal of Wildlife, 36(2): 129-133. (in Chinese)
Miquelle D, Smirnov E N, Yu Z O, *et al.* 2015. Population dynamics of amur tigers (*P. t. altaica*, Temminck 1884) in Sikhote-Alin Zapovednik: 1966-2012. Integrative Zoology, 10(4): 315-328.
Shevtsova E, Jiang G, Vitkalova A, *et al.* 2018. Saving the Amur tiger and Amur leopard. NEASPEC.
Wikramanayake E, Dinerstein E, Seidensticker J, *et al.* 2011. A landscape-based conservation strategy to double the wild tiger population. Conservation Letters, 4(3): 219-227.